Lecture Notes in Artificial Intelligence 12688

Subseries of Lecture Notes in Computer Science

More information about this subseries at http://www.springer.com/series/1244

Davide Calvaresi · Amro Najjar ·
Michael Winikoff · Kary Främling (Eds.)

Explainable and Transparent AI and Multi-Agent Systems

Third International Workshop, EXTRAAMAS 2021
Virtual Event, May 3–7, 2021
Revised Selected Papers

Editors
Davide Calvaresi
University of Applied Sciences and Arts
Western Switzerland
Sierre, Switzerland

Michael Winikoff
Victoria University of Wellington
Wellington, New Zealand

Amro Najjar
University of Luxembourg
Esch-sur-Alzette, Luxembourg

Kary Främling
Umeå University
Umeå, Sweden

ISSN 0302-9743 ISSN 1611-3349 (electronic)
Lecture Notes in Artificial Intelligence
ISBN 978-3-030-82016-9 ISBN 978-3-030-82017-6 (eBook)
https://doi.org/10.1007/978-3-030-82017-6

LNCS Sublibrary: SL7 – Artificial Intelligence

This Springer imprint is published by the registered company Springer Nature Switzerland AG
The registered company address is: Gewerbestrasse 11, 6330 Cham, Switzerland

Preface

Recent advancements in eXplainable Artificial Intelligence (XAI) are generating new understanding and opportunities. The increasingly complex intelligent agents/robots rely on often opaque machine learning-based algorithms. Explaining such mechanisms is a chief priority to enhance their acceptability, avoid failures, foster trust, and comply with relevant (inter)national regulations.

The 2021 edition of the EXplainable and TRAnsparent AI and Multi-Agent Systems (EXTRAAMAS) workshop continues the successful track initiated in 2019 at Montreal and followed by the 2020 edition in New Zealand (which was virtual due to the COVID-19 pandemic circumstances). In particular, EXTRAAMAS 2021 set the following aims: (i) to strengthen the common ground for the study and development of explainable and understandable autonomous agents, robots, and Multi-Agent Systems (MAS), (ii) to investigate the potential of agent-based systems in the development of personalized user-aware explainable AI, (iii) to assess the impact of transparent and explained solutions on the user/agent behaviors, (iv) to discuss motivating examples and concrete applications in which the lack of explainability leads to problems, which would be resolved by explainability, and (v) to assess and discuss the first demonstrators and proof of concepts paving the way for the next generation of systems. EXTRAAMAS 2021 received 32 submissions. Each submission underwent a rigorous single-blind peer review process (three to five reviews per paper). Eventually, 20 papers were accepted (19 full papers and 1 short paper), which are contained in this volume. Due to COVID-19 restrictions, the workshop and AAMAS (the hosting conference) were held online rather than in London, UK. For each paper, the authors performed live video presentations that, with their consent, are available on the EXTRAAMAS website[1]. Moreover, EXTRAAMAS 2021 included two keynotes: "Social and Ethical Responsibilities of Computing and the Role of Explainability and Transparency" given by Prof. Julie Shah and "Explainable Reasoning in the Face of Contradictions: From Humans to Machines" given by Prof. Emeritus Dov Gabbay, and two panels: the industrial panel "Explainable Agents - Escape from the Ivory Tower" and the panel dedicated to the European Project EXPECTATION which considers XAI in distributed and heterogeneous agent-based recommender systems.

We would like to thank the industrial chair, publicity chairs, and Program Committee for their valuable work. We also thank the authors, presenters, and participants. Particular emphasis goes to Julie Shah and Dov Gabbay for their fantastic keynotes, and to Johanna Björklund, Tathagata Chakraborti, Kristijonas Cyras, Elizabeth Sklar, Andrea Omicini,

[1] https://extraamas.ehealth.hevs.ch/archive.html.

Reyhan Aydogan, and Leon Van der Torre for their participation in very enthusiastic discussion panels.

May 2021

Davide Calvaresi
Amro Najjar
Michael Winikoff
Kary Främling

Organization

General Chairs

Davide Calvaresi	University of Applied Sciences Western Switzerland, Switzerland
Amro Najjar	University of Luxembourg, Luxembourg
Michael Winikoff	Victoria University of Wellington, New Zealand
Kary Främling	Umeå University, Sweden

Industrial Chair

Timotheus Kampik	Umeå University, Sweden

Publicity Chairs

Sviatlana Höhn	University of Luxembourg, Luxembourg
Giovanni Ciatto	University of Bologna, Italy

Advisory Board

Tim Miller	University of Melbourne, Australia
Michael Schumacher	University of Applied Sciences Western Switzerland, Switzerland
Virginia Dignum	Umeå University, Sweden
Leon van der Torre	University of Luxembourg, Luxembourg

Program Committee

Natasha Alechina	Utrecht University, The Netherlands
Amr Alyafi	Grenoble Electrical Engineering, France
Cleber Jorge Amaral	Federal University of Santa Catarina, Brazil
Kim Baraka	Carnegie Mellon University, USA
Suna Bensch	Umeå University, Sweden
Grégory Bonnet	University of Caen, France
Joost Broekens	Leiden University, The Netherlands
Jean-Paul Calbimonte	HES-SO Valais-Wallis, Switzerland
Tathagata Chakraborti	IBM Research AI, USA

Nicolas Cointe	Delft University, The Netherlands
Kristijonas Cyras	Ericsson Research, Sweden
Jérémie Dauphin	University of Luxembourg, Luxembourg
Alaa Daoud	École Nationale Supérieure des Mines de Saint-Étienne, France
Dustin Dannenhauer	Parallax Advanced Research, USA
Lora Fanda	University of Applied Sciences Western Switzerland, Switzerland
Michael W. Floyd	Knexus Research, USA
Stephane Galland	UTBM, France
Önder Gürcan	CEA, France
Maike Harbers	Rotterdam University of Applied Sciences, The Netherlands
Brent Harrison	Georgia Institute of Technology, USA
Salima Hassas	Claude Bernard Lyon 1 University, France
Helen Hastie	Heriot-Watt University, UK
Thomas Hellström	Umeå University, Sweden
Koen Hindriks	Virje Universiteit Amsterdam, The Netherlands
Sviatlana Höhn	University of Luxembourg, Luxembourg
Isaac Lage	Harvard University, USA
Beishui Liao	Zhejiang University, China
Tomer Libal	University of Luxembourg, Luxembourg
Brian Lim	National University of Singapore, Singapore
Avleen Malhi	Thapar Institute of Engineering & Technology, India
Niccolò Maltoni	University of Bologna, Italy
Réka Markovich	University of Luxembourg, Luxembourg
Viviana Mascardi	University of Genova, Italy
Laëtitia Matignon	Claude Bernard Lyon 1 University, France
Yazan Mualla	UTBM, France
Juan Carlos Nieves Sanchez	Umea University, Sweden
Sana Nouzri	University of Luxembourg, Luxembourg
Andrea Omicini	University of Bologna, Italy
Marrina Paolanti	Polytecnic University of Marche, France
Gauthier Picard	École Nationale Supérieure des Mines de Saint-Étienne, France
Patrick Reignier	LIG, France
Francisco Javier Rodríguez Lera	University of León, Spain
Stefan Sarkadi	King's College London, UK
Giovanni Sartor	Università di Bologna and European University Institute, Italy

Sarath Sreedharan	Arizona State University, USA
Cesar A. Tacla	Federal Technological University of Paraná, Brazil
Silvia Tulli	Instituto Superior Tecnico and INESC-ID, Portugal
Rob Wortham	University of Bath, UK
Deshraj Yadav	Georgia Institute of Technology, USA
Jessie Yang	University of Michigan, USA

Contents

Decentralized and Heterogeneous XAI

XAI and Machine Learning

To Pay or Not to Pay Attention: Classifying and Interpreting Visual Selective Attention Frequency Features

Lora Fanda[1(✉)], Yashin Dicente Cid[2], Pawel J. Matusz[1], and Davide Calvaresi[1]

[1] University of Applied Sciences Western Switzerland (HES-SO), Sierre, Switzerland
{lora.fanda,pawel.matusz,davide.calvaresi}@hevs.ch
[2] University of Warwick, Coventry, UK
yashin.dicente@warwick.ac.uk

Abstract. Selective attention is the ability to promote the processing of objects important for the accomplishment of our behavioral goals (target objects) over the objects not important to those goals (distractor objects). Previous investigations have shown that the mechanisms of selective attention contribute to enhancing perception in both simple daily tasks and more complex activities requiring learning new information.

Recently, it has been verified that selective attention to target objects and distractor objects is separable in the frequency domain, using Logistic Regression (LR) and Support Vector Machines (SVMs) classification. However, discerning dynamics of target and distractor objects in the context of selective attention has not been accomplished yet.

This paper extends the investigations on the possible classification and interpretation of distraction and intention solely relying on neural activity (frequency features). In particular, this paper *(i)* classifies *distractor objects* vs. *target object* replicating the LR classification of prior studies, extending the analysis by *(ii)* interpreting the coefficient weights relating to all features with a focus on N2PC features, and *(iii)* retrains an LR classifier with the features deemed important by the interpretation analysis.

As a result of the interpretation methods, we have successfully decreased the feature size to 7.3% of total features – i.e., from 19,072 to 1,386 features – while recording only a 0.04 loss in performance accuracy score—i.e., from 0.65 to 0.61. Additionally, the interpretation of the classifiers' coefficient weights unveiled new evidence regarding frequency which has been discussed along with the paper.

Keywords: EEG · Selective attention · Machine learning

P. J. Matusz and D. Calvaresi—Equal contribution.

D. Calvaresi et al. (Eds.): EXTRAAMAS 2021, LNAI 12688, pp. 3–17, 2021.
https://doi.org/10.1007/978-3-030-82017-6_1

1 Introduction

The comprehension of spatiotemporal brain dynamics can help identify selective attention in both healthy and attention-impaired individuals.

Humans' mental capacity to attend to and process in-depth the incoming information is limited [1]. According to Posner and Peterson [2], attention comprises three subsystems: *(i)* alerting, *(ii)* orienting, and *(iii)* selective attention. The latter is a naturally multisensory set of processes that flexibly engage the limited computational resources according to the task demands. Processing resources are scarce, and the stimuli received by the senses compete for them. Some theories (i.e., *biased-competition* [3]) detailed how the competition for resources is resolved and integrated across different brain processing stages, producing a coherent behavior.

Understanding such concepts requires both Neuroscience and Artificial Intelligence (AI) domain knowledge, with particular emphasis on Machine Learning (ML). The cognitive orchestration of selective attention, its role in enhancing perception and learning skills, and neurocognitive processes engaged by distractor *or* target objects have been widely investigated [4–6]. However, brain mechanisms and classification methods to distinguish selective attention to *target* vs. *distractor* objects have not been understood yet.

Achieving such an understanding can provide essential insights into functional differences in cortical cognitive mechanisms governing attention to object task-relevant and task-irrelevant. Therefore, understanding when and how well individuals pay attention to objects and events can lead to practical tools to better measure attention in the classroom or the workplace (if the ethical issues related to performance tracking vs. privacy are sufficiently addressed).

Current approaches adopt a cortical correlate of the attentional selection, known as the N2PC, of both targets and distractors possessing target features. N2PC is defined as a negative polarity at 200 ms latency appearing over posterior electrodes contralateral to the direction of attention. In other words, the N2PC is reflected by enhanced negativity emerging approximately (200 ms) after stimulus onset over posterior electrodes contralateral to the stimulus location. This measure uses neural signals derived from the acquired electroencephalogram (EEG) to identify selective attention to visual objects. The N2PC is obtained from EEG data via the event-related potential (ERP) technique involving averaging brain responses elicited by one type of stimulus over multiple repetitions of it over time. The averaging amplifies the faint neural signal reflecting neurocognitive processing of that stimulus. The N2PC is known to be particularly modulated by goal-based ("top-down") and visual selective attention processes. Therefore, selective attention is measured in the window between approximately 150 ms and 300 ms after stimulus onset and lasts until the difference between the negative potential between hemispheres is no longer measured.

As of today, the analysis of selective attention via the traditional N2PC analytical approach is a human-intensive task, and as such, it is time-consuming, requires in-depth experience, and is currently not semi-automated [7]. Moreover, due to the high temporal resolution of the EEG signal, human-related errors (i.e., variability in the identification of the start and end of the N2PC time window) can jeopardize the results' precision and accuracy. Prior work has shown that attention to targets and distractors *is* separable in EEG via a classification methodology using linear and non-linear classifiers [8]. However, the classification mentioned above has not been interpreted, leaving incomprehensible *which EEG frequency features* provide the most separable information.

This paper analyzes and interprets the frequency features using the same dataset employed in [8]. In particular, the classifiers' coefficient weights are interpreted to understand which features convey the most relevant information for classification. Finally, we discuss the relevance of our findings within the Neuroscience domain.

The rest of the paper is organized as follows. Section 2 presents the current state of the art about the many mechanisms and abstractions supporting our notion of attention, which is then introduced, defined, and discussed in section Sect. 3. Section 4 elicits the opportunities and presents the challenges related to our definition of attention. Finally, Sect. 5 concludes the paper.

2 State of the Art

The contribution of this paper relies on concepts intersecting neuroscience and machine learning. Therefore, this section provides the necessary background concepts and the related state of the art to facilitate the reader's comprehension of the topic.

EEG: Selective attention has been studied using various modes of data collection, ranging from invasive techniques like electrocorticography (ECoG) [9], to non-invasive techniques like EEG [1,10]. Processes related to attention have distinct markers in the frequency domain [11,12]. Therefore, EEG and ECoG are the preferred data acquisition techniques due to their high temporal resolutions and ability to detect these forms of attention-relevant oscillatory activity. Overall, EEG is preferred to ECoG as it is noninvasive and more convenient for collecting large amounts of electrophysiological data. Thus, EEG neural recordings can be used to classify selective attention to distractor objects, target objects, or non-object stimuli.

N2PC: In traditional methods, selective attention to potentially task-relevant objects is measured through N2PC. N2PC, an event-related potential (ERP) correlate, is a cortical measure of attention to candidate target objects in selective attention task contexts. On the one hand, for target objects, Nobre et al. [13] confirm the presence of changes in the ERP strength over the N2PC period triggered by visual target objects where attention has been captured by visual targets. On the other hand, the target object's properties can be the driving force that determines selective attention to *distractors*. This has been initially shown in behavioral responses in a study by Folk et al. and then confirmed in an EEG study by Eimer et al. [14] and further supported by multiple studies since [14,15]. Therefore, like for the target objects, N2PC is a well-used measure of selective attention for distractor objects. Hence, the N2PC is well suited as a marker of attending towards visual stimuli of distracting and task-relevant (target) type.

Frequency Components of Attention: Attention has a distinct imprint in the frequency domain. Thus, each frequency domain is associated with a class of attentional processing. Changes in the δ band power, 0.5–4 Hz, allow for separation of low and high cognitive load while the θ band power reflects the encoding of new information. The α band power, 8–12 Hz, is higher for target object perception during the attention task. The β band power, 13–30 Hz, increases preceding the correct response. The γ band power, 30–70 Hz, increases by a visual search task when the subject attends to a stimulus [12,16]. Labeling each frequency band with one functionality can be misleading. Thus the range of these frequency bands will be considered as attention-relevant frequency bands.

Discrete Cosine Transform (DCT): A method for frequency feature extraction is the Discrete Cosine Transform (DCT). DCT extraction has previously been used for EEG and MEG datasets [17,18] to extract frequency components of a signal to use in classification. Table 1 summarized ranges of EEG frequency bands and related DCT ranges for convenience.

Table 1. Table of Frequency bands in EEG datasets and their translation to DCT for our EEG dataset.

Frequency bands						
	δ	θ	α	β	γ	High γ
Frequency range (Hz)	(0.5–4)	(4–8)	(8–13)	(13–30)	(30–50)	(50–80)
DCT range	[1:2]	[2:3]	[3:5]	[5:10]	[10:17]	[15:25]

Classifying Neural Data: In the fields of Brain-Computer Interfaces (BCIs) and Epileptic seizure detection, the interest in the classification of neural data is growing. The most common features used in the classification of BCI and Epilepsy data are raw EEG, frequency component extraction, and AutoRegressive features [19,20]. In comparison, the most common features to classify selective attention are raw EEG, frequency component, and N2PC electrode features [21].

Logistic regression (LR) recorded promising performance accuracy for classification of biological brain signals [8,19,22–24]. LR is generally used as a categorical problem-solving method, thus can be applied to multivariate classification [25]. It is deemed a simpler classification technique but can provide unique results if feature vectors are adequately selected and if the data is linearly separable. It is important to note that, unlike BCI/Epilepsy classification applications, selective attention classification had only recently been applied, and it is a steadily growing field.

Classification of Attention Data: Fanda verified that selective attention is separable in the frequency domain using LR and SVM classification and, from manual feature selection, showed that N2PC region electrodes hold the most discriminative information compared to non-N2PC regions [8]. However, in such a study, comparisons across low to high-frequency features are lacking, and the coefficient weights relating to the features have not been interpreted, thus undermining the understanding of the classifier decision-making.

Contribution: In light of the findings mentioned above and due to the lack of interpretability in DCT feature classification of selective attention data, this paper *(i)* replicates the LR classification performed in [8], *(ii)* interprets the coefficient weights relating to all features with a focus on N2PC features, and *(iii)* retrains an LR classifier with the features deemed important by the interpretation analysis.

3 Approach or Method

Overall, to interpret the selective attention frequency features, we will replicate the classification using LR, interpret the model's weights, and retrain the classifier with sub-selected features as extracted from the interpretation data. To facilitate the reader's comprehension, Fig. 1 summarizes the overall pipeline spanning from the EEG data acquisition to the performance analysis. In particular, Fig. 1(a) explicates the step undertaken in prior work, such as EEG dataset and N2PC analysis division [15], and DCT feature extraction and initial LR classification parameters [8]. Figure 1(b) organizes the tasks and results obtained in

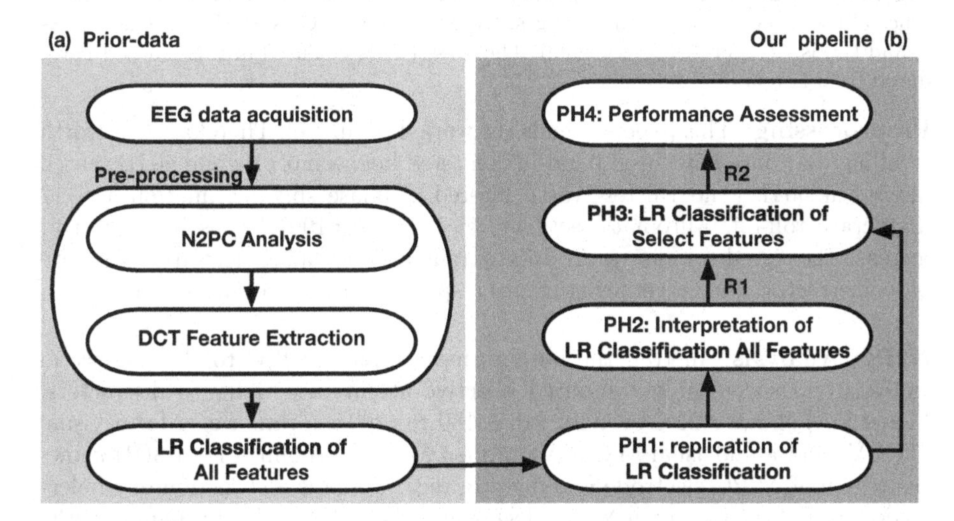

Fig. 1. Overall pipeline from data acquisition to performance analysis: (a) prior work's tasks, (b) methodology pipeline of this contribution.

the specific phases (PHx) of this study, such as (PH1) replication of the initial LR classification, (PH2) interpretation of the features contributing the most information to the replicated LR classifiers' coefficient weights for all features, (PH3) re-learning LR classification with sub-selected featured as extracted by R1, and (PH4) performance assessment and comparison of all LR classifiers.

EEG Data Acquisition: The EEG dataset has been collected using a 129-channel HydroCel Geodesic Sensor Net connected to a NetStation amplifier (Net Amps 400; Electrical Geodesics Inc., Eugene, OR, USA) where 128 electrodes have been used at a 1 kHz sampling rate. During data acquisition, electrode impedances have been kept below 50 kΩ, and electrodes have been referenced online to Cz, a common reference for EEG cap data collection. Participants have been recorded for three hours in a task described in Turoman et al. [15]. This dataset has been collected by Dr. Turoman during her Ph.D. work [26]. To complete the task, the participants have been instructed to search for a predefined color target (target object) in a search array and report the target's orientation (i.e., if the target is horizontal, press right, otherwise, press left). The participants have been instructed about other objects that could appear (distractor objects) and focus solely on reporting the target's orientation. Figure 2 shows examples of the task.

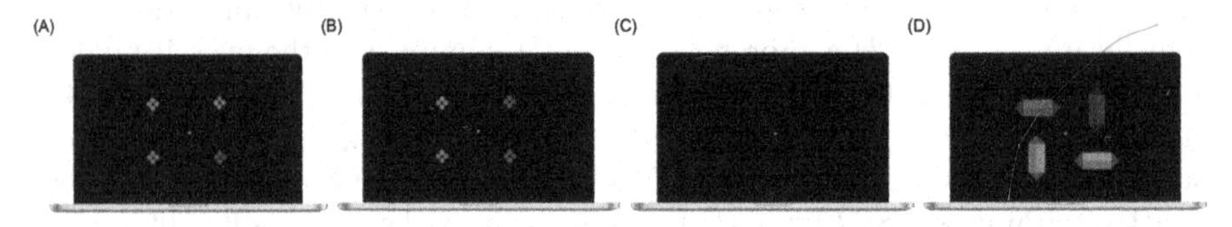

Fig. 2. This figure shows all four stimuli of the paradigm (A–D) and the time importance of the three interested time-ranges. (A) is "Baseline" Class 0, (B) is "Cue" Class 1, and (D) is "Target" Class 2 stimuli. The cross (C) is not used in this study. Reproduced from Fanda [8].

Preprocessing: This paper extends the work done in [26]. Therefore, it is worth recalling that data have been band-pass filtered between 0.1 Hz and 40 Hz, notch filtered at 50 Hz, and Butter-Worth filtered of phase shift elimination at −12 dB/octave roll-off. Automatic artifact rejection of ±100 μV has been used to increase the signal-to-noise ratio. Next, trails have been segmented to separate base, distractor, and target array neural responses for feature extraction.

N2PC Analysis: In [15], the authors have applied N2PC to the dataset in analysis and successfully computed selective attention activity to distractors. Figure 3(A) shows N2PC activity from 180 to 300 ms time range for a visual (TCCV) and audiovisual (TCCAV) property of the stimuli. Figure 3(B) shows the region of N2PC electrode coverage (in red), which is a collection of 14 electrodes around the two main N2PC electrodes (e65 and e90). In this paper, N2PC electrodes refer to the 14 electrodes in the N2PC region from the TCCV condition only, as shown in Fig. 3(B).

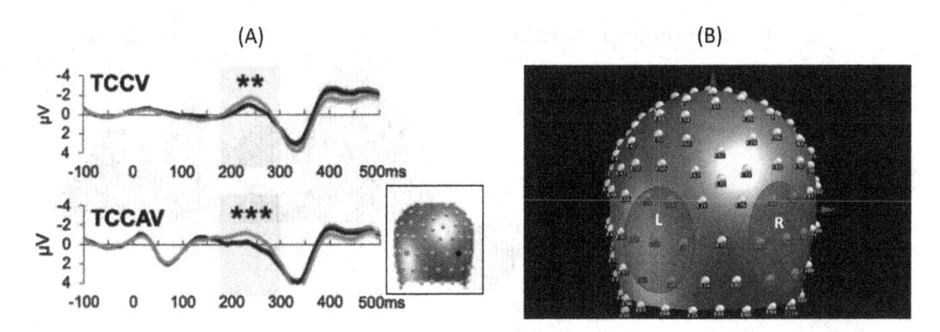

Fig. 3. N2PC analysis: (A) N2PC analysis, indicating presence of N2PC from ~200 to ~300 ms (taken from Turoman et al. [15]). (B) EEG Cap images referencing N2PC region electrodes, taken from Fanda [8].

DCT of N2PC Time-Frame: DCT decomposes and compresses a signal to (*signal-length - 1*) frequency bins. The benefit of using DCT features relies on containing the full frequency identity of the signal while removing biases that can come from time-series and amplitude measures. Equation 1 is one-dimensional DCT, and it is used for feature extraction.

$$y_k = 2 \sum_{n=0}^{N-1} x_n \cos\left(\frac{\pi k(2n+1)}{2N}\right) \tag{1}$$

For example, if a signal x_n is sampled at 1024 Hz for a length of $N = 150$, the frequency components extend up-to 512 Hz. When applying DCT, the 512 Hz frequency components of the signal are cosine-transformed into $(N - 1) = 149$ frequency bins, keeping only the real values of the signal. This results in the y_k vector where the first value $k = 0$ contains prevalence via summation of frequencies ranging from 0 to 3.33 Hz, the second bin from 3.34 to 6.66 Hz, and so on.

Prior knowledge of attentional frequency oscillation dictates that selective attention primarily ranges from 8 Hz to 30 Hz (attention relevant frequency band) [10]. For the DCT extraction of the dataset, this range is contained in DCT bins one to 27 (see Fig. 4). Thus, the lower DCT bins likely contain more relevant information to selective attention than the higher bins. Such a hypothesis is tested by looking at the coefficients of the features in our learned classifier's decision function.

Classification: In this paper, we aim to classify and interpret selective attention EEG data. Specifically, we used LR to classify *distractor objects* vs. *target object*, and interpret the classifier feature coefficient weights. With LR, we used a one-vs-all multi-class structure. The features used for the LR classifier are normalized DCT features, split by participant into train, validation, and test sets. To evaluate the model, performance accuracy scores have been used as a performance evaluation technique.

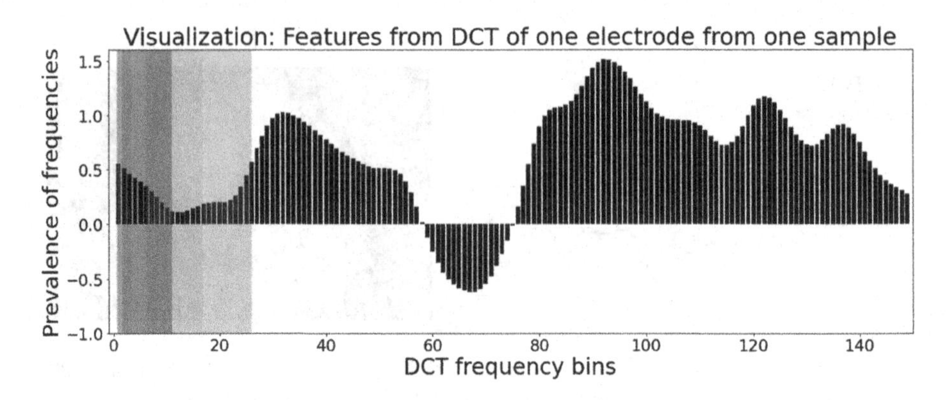

Fig. 4. DCT features visualized for one electrode of one training sample. The shaded regions correspond to frequencies, as translated from Table 1.

Interpretability of LR: In this contribution, interpretability approaches are applied to classifiers to examine the contribution of individual predictors. One method consists of the examination of the regression coefficients of each of the three LR classifiers resulting from our multi-class problem using a one-vs-all setup. LR coefficients are slightly more difficult to interpret as the line of best fit is a `logit` function, the inverse of the sigmoid curve. Thus, the resulting coefficients of LR are *odds ratios* and require exponentiation to convert to regular odds. The odds ratio then corresponds to the β_k coefficients where $k \in [1, n]$ with $n = \textit{total predictors}$ in the LR odds Eq. 2, where x's are values of predictors.

$$\frac{\text{odds}(x_1 + 1)}{\text{odds}(x_1)} = \frac{e^{\beta_0+\beta_1(x_1+1)+\beta_2 x_2+\cdots+\beta_n x_n}}{e^{\beta_0+\beta_1 x_1+\beta_2 x_2+\cdots+\beta_n x_n}} = \frac{e^{\beta_1(x_1+1)}}{e^{\beta_1 x_1}} = e^{\beta_1} \tag{2}$$

Post conversion, the values of the coefficients are positive, and they are interpreted following the rule below:

$$Odds = \begin{cases} e^{\beta_k} \times \text{as likely}, & \text{if } e^{\beta_k} >= 1 \\ \frac{1}{e^{\beta_k}} \times \text{as unlikely}, & \text{if } e^{\beta_k} < 1 \end{cases} \tag{3}$$

As an example, taking k = 1 coefficient β_1 from class A, this rule roughly translates to:

- for $e^{\beta_1} >= 1$: "Each unit increase in x_1, the odds that the observation is in class A are e^{β_1} times as likely as the odds that the observation is not in A."
- for $e^{\beta_1} < 1$: "Each unit increase in x_1, the odds that the observation is NOT in class A are $\frac{-1}{e^{\beta_1}}$ times as *unlikely* as the odds that the observation is in A."

4 Results and Discussions

This section describes the two main results: (R1) the interpreted LR classifiers' coefficient weights as odds ratios for *all* features and N2PC features and (R2) the LR classification accuracy score comparison, using all, N2PC only, and N2PC & selected DCT features. Additionally, the discussions are included in the end.

R1 - Coefficient Weights as Odds Ratios: All Features and N2PC: It is worth recalling that due to the one vs. rest multi-class choice of LR classification, each class has a set of odds ratios extracted from the model's coefficients weights. Thus, the analyses are shown for each classifier (Baseline, Distractor, or Target) individually. To inspect if N2PC electrodes have information *more* valuable than other electrodes to the decision function of the classifier, we have plotted in Fig. 6 the odds ratios of classifiers for Baseline (brown), Distractor (blue), and Target (purple), where the N2PC region electrodes are plotted using a darker color for contrast.

From Fig. 5, N2PC electrodes overall have higher odds ratios compared to other electrodes. To better see the pattern of the distinct shape of the N2PC electrodes, we stacked the coefficient weights vector with respect to DCT frequency bin features, resulting in a 128 by 149 coefficient matrix as plotted in Fig. 6 for each classifier. In addition to identifying the distinct N2PC electrode patterns, we can understand in *which* DCT bins the odds ratios across the three classifiers diverse/stay similar. Then, the odds ratios are analyzed using Eq. 2.

In particular, each unit increase in DCT 1 to 3, the odds that the observation is in-class Target is $\beta_{1-3} \in (\sim 1.004, \sim 1.006)$ times as likely as the odds that the observation is not in class Target. Conversely, each unit increase in DCT,

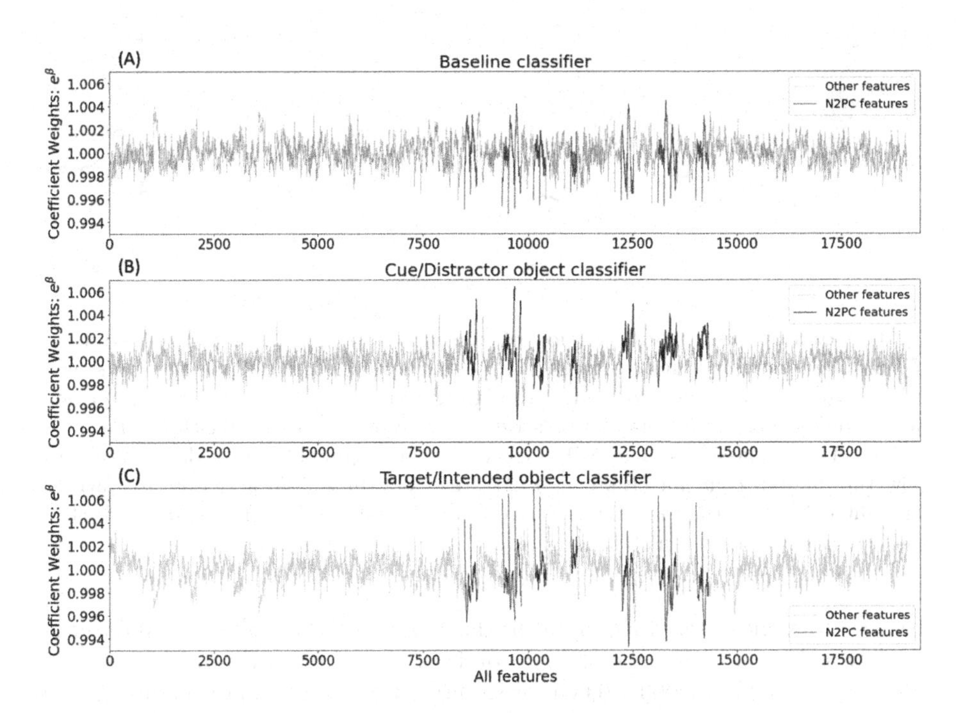

Fig. 5. Coefficients as odds ratios of the features in the decision function of LR classification. In all three classifiers above, for each class, the N2PC electrodes visibly have higher peaks. Figure 6 stacks the feature vector by electrode to better visualize the patterns with respect to DCT frequency bins.

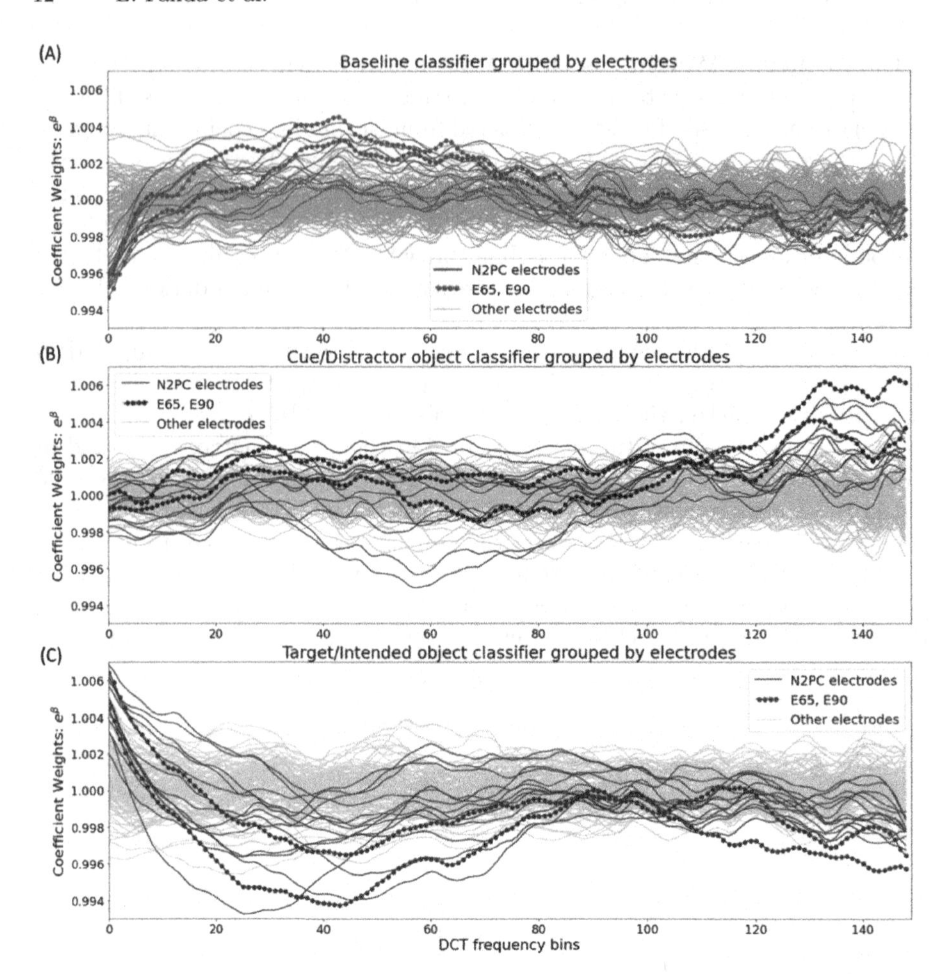

Fig. 6. Coefficients as odds ratios of the features in the decision function of LR, organized by electrode: This figure reorganizes the coefficient weights seen in Fig. 5 by stacking them electrodes (to highlight the patterns w.r.t. DCT frequency bins). The coefficient weights, as odds ratios, are plotted for (A) Baseline, (B) Cue, and (C) Target classifiers. The N2PC electrodes are highlighted in a darker color to show the difference in patterns over DCT frequency bins. E65 and E90 are further highlighted because they are the selective electrodes studied by Turoman et al. [15], from whom the dataset was taken.

the odds that the observation is *not* in class Target is $\frac{1}{\beta} = \frac{1}{0.995} = 1.005$ times as likely as the odds that the observation is in class Target. Following this rule, we set a threshold of 1.000 ± 0.004 for selecting the DCT components for future analysis. A visualization of this threshold is seen in Fig. 7. All DCT features with odds ratio values falling below 0.996 and above 1.004 have been selected for the next iteration of LR classification. This resulted in the selection of the following DCT regions:

Baseline:	DCTs [49:71 and 125:149]
Distractor:	DCTs [0:5 and 32:51]
Target:	DCTs [0:72 and 140–149]

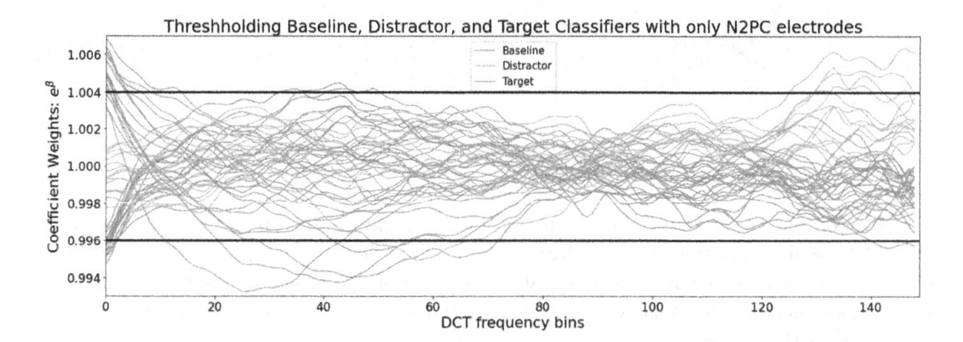

Fig. 7. DCT threshold visualization for DCT feature selection for interpreting which DCT features among N2PC region electrodes are more important. Here, the common denominator of DCT ranges across all three classifiers is DCT 0 to 73 and 124 to 149.

As a result, in the next iteration of LR classification, only the N2PC region features, and DCT features 0 to 73 and 124 to 149 will be used. The rest of the features will be discarded.

Coefficient Weight Analysis for N2PC Electrode Features: To analyze the weight of the coefficients for N2PC electrode features, we used the conversion in Table 1. Frequency bands relating to selective attention activity will not be defined in purpose. Nevertheless, it is known that these frequency ranges' power can correlate with selective attention activity. As such, we expect the odds ratios to have a visibly higher or lower value than 1 for DCT frequency bins from 1 to 27. Taking threshold 1.004 and 0.996, we retrained the LR and compared the performance accuracy scores between all features, only N2PC but all DCTs, and only N2PC and selected DCT.

Confusion Matrices: To compare the performance of the three iterations of learning an LR classifier, accuracy scores and confusion matrices are reported in Fig. 8. Confusion matrices have been calculated to understand better the intra-class classification accuracy and errors for (A) All features, (B) N2PC region features with all DCT components, and (C) N2PC region features with selected DCT components.

Discussions: In this work, we have analyzed the most valuable features of the LR classifier's decision function. In line with what identified in [8], from

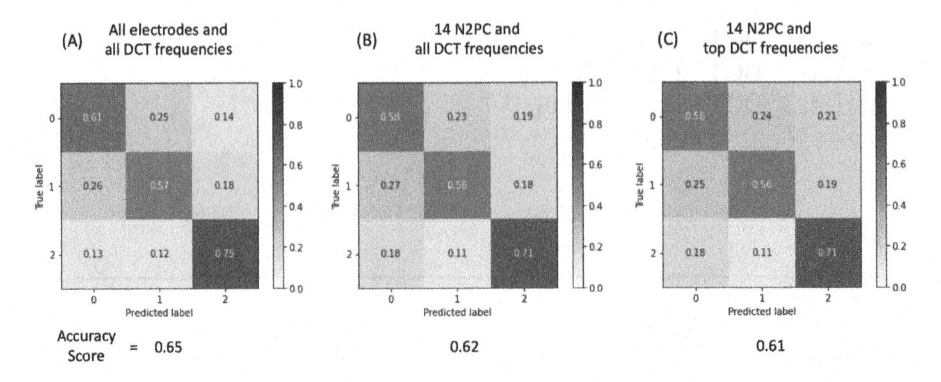

Fig. 8. Confusion Matrices of LR classification accuracy scores using (A) all features, (B) N2PC region features, and (C) N2PC region and select DCT features. The accuracy scores are shown for each class true labels (y axis) and predicted labels (x axis), for Baseline (Class 0), Cue (Class 1), and Target (Class 2). The overall classifier accuracy is written in the bottom.

manual feature selection, N2PC region electrodes hold the most discriminative information compared to non-N2PC regions. Lead by the lack of interpretability in DCT feature classification in EEG datasets, this paper *(i)* replicates the LR classification performed in [8], *(ii)* interprets the coefficient weights relating to all features with a focus on N2PC features, and *(iii)* retrains an LR classifier with the features deemed discriminant by the interpretation analysis.

We verified that N2PC region electrode features hold more discriminative information than non-N2PC region electrodes, as seen by Fig. 6. Additionally, we have identified that DCT frequencies of zero to 73 and 123 to 149 held the most discriminative information than other DCT frequency bins, as seen by Fig. 7. We retrained an LR classifier using only the sub-selection of features (i.e., $\frac{14*99}{19072} = \frac{1386}{19072} = 7.3\%$ of the original features vector size), with only a 0.04 loss in performance accuracy score. Some of the expected outcomes from the interpretations are:

E1. N2PC region electrodes overall have higher odds ratios compared to other electrodes (Fig. 5(A–C)). This verifies that N2PC region electrodes have higher activity associated with selective attention.

E2. In the Baseline classifier, each unit increase in the DCT frequency bins 0 to 7 suggested that the odds that the features are **not** in the Baseline class are larger than the odds that the feature **is** in the Baseline Class (Fig. 5(A)). As DCT frequency bins zero to 7 correspond to selective attention frequencies, this confirms previous work. In other words, given values of <1 in these regions, the odds ratio suggests that these selective attention frequency values are discriminating that they do not belong in the Baseline classifier.

E3. In the Target classifier, each unit increase in the DCT frequency bins 0 to 7 suggested that the odds that the features are NOT in the Target class are larger than the odds that the feature *is* in the Target Class (Fig. 5(A)). As

DCT frequency bins zero to 7 correspond to selective attention frequencies, this confirms previous work.

Additionally, some novel evidence can be extracted from the interpretation of the odds ratios in PH1 and PH2 (see Fig. 1):

E4. The Distractor classifier odds ratios suggest that DCT features from 120 to 149 have high discriminability information. However, such high DCT frequency bins correspond to frequencies 500 Hz. Thus, this requires more investigation as they could be related to artifacts (Fig. 5(B)).

E5. DCT frequency bins 20 to 80 have unexpectedly high odds ratios, which correspond to frequencies of 60 to 360 Hz. To the best of our knowledge, it suggests the presence of discriminating information in those frequency bands that have not been explained yet.

5 Conclusions

Prior work in classifying selective attention identifies the relevance on N2PC electrodes. However, they neglect feature interpretation. This study tackled the interpretation of LR classifiers to better discern the dynamics of target and distractor objects in the context of selective attention.

In particular, this paper has *(i)* classified *distractor objects* vs. *target object* replicating the LR classification of prior studies, *(ii)* interpreted the coefficient weights relating to all features with focus on N2PC features, and *(iii)* retrained an LR classifier with the features deemed important by the interpretation analysis.

The two main results of the interpretation methods are *(i)* successful feature size reduction (decreasing feature size to 7.3% of total features – i.e., from 19072 to 1,386 features – while recording only a 0.04 loss in performance accuracy score), *(ii)* the interpretation of the classifiers' coefficient weights unveiled new evidence.

In particular, such evidence are [E1.] N2PC region electrodes overall have higher odds ratios compared to non-N2PC electrodes; [E2.] In the Baseline classifier, each unit increase in the DCT frequency bins 0 to 7 suggested that the odds that the features are **not** in the Baseline class are larger than the odds that the feature **is** odds ratios of the Baseline classifier suggests that DCTs relating to Selective Attention frequency ranges are more *unlikely* to belong to Baseline class; [E3.] each unit increase in the DCT frequency bins 0 to 7 suggested that the odds that the features are the Target class are larger than the odds that they are not in the Target class.

Future works include the investigation of [E4] high DCT frequency bins having the most discriminant information for Distractor classifiers and [E5] DCT bins from 20 to 80 (approximately 60–360 Hz) having high odds ratios for all classifiers. It is important to understand the nature of the DCT components in relation to selective attention frequencies.

References

1. Carrasco, M.: Visual attention: the past 25 years. Vis. Res. **51**(13), 1484–1525 (2011)
2. Posner, M.I., Petersen, S.E.: The attention system of the human brain. Ann. Rev. Neurosci. **13**(1), 25–42 (1990)
3. Desimone, R., Duncan, J.: Neural mechanisms of selective visual attention. Ann. Rev. Neurosci. **18**(1), 193–222 (1995)
4. Kiss, M., Van Velzen, J., Eimer, M.: The n2pc component and its links to attention shifts and spatially selective visual processing. Psychophysiology **45**(2), 240–249 (2008)
5. Murray, M.M., Thelen, A., Thut, G., Romei, V., Martuzzi, R., Matusz, P.J.: The multisensory function of the human primary visual cortex. Neuropsychologia **83**, 161–169 (2016)
6. Matusz, P.J., Dikker, S., Huth, A.G., Perrodin, C.: Are we ready for real-world neuroscience? (2019)
7. Tivadar, R.I., Murray, M.M.: A primer on electroencephalography and event-related potentials for organizational neuroscience. Organizational Res. Methods **22**(1), 69–94 (2019)
8. Fanda, L.: Classifying attentional dynamics from EEG signals: Feature based perceptual attentional control, February 2021
9. Shum, J., Fanda, L., Dugan, P., Doyle, W.K., Devinsky, O., Flinker, A.: Neural correlates of sign language production revealed by electrocorticography. Neurology **95**(21), e2880–e2889 (2020)
10. Kiss, M., Grubert, A., Petersen, A., Eimer, M.: Attentional capture by salient distractors during visual search is determined by temporal task demands. J. Cogn. Neurosci. **24**(3), 749–759 (2012)
11. Klimesch, W.: Alpha-band oscillations, attention, and controlled access to stored information. Trends Cogn. Sci. **16**(12), 606–617 (2012)
12. Marturano, F., Brigadoi, S., Doro, M., Dell'Acqua, R., Sparacino, G.: A time-frequency analysis for the online detection of the n2pc event-related potential (ERP) component in individual EEG datasets. In: 2020 42nd Annual International Conference of the IEEE Engineering in Medicine and Biology Society (EMBC), pp. 1019–1022. IEEE (2020)
13. Coull, J.T., Nobre, A.C.: Where and when to pay attention: the neural systems for directing attention to spatial locations and to time intervals as revealed by both pet and FMRI. J. Neurosci. **18**(18), 7426–7435 (1998)
14. Folk, C.L., Remington, R.W., Johnston, J.C.: Involuntary covert orienting is contingent on attentional control settings. J. Experimental Psychol. Hum. Percept. Perform. **18**(4), 1030 (1992)
15. Turoman, N., Tivadar, R.I., Retsa, C., Maillard, A.M., Scerif, G., Matusz, P.J.: The development of attentional control mechanisms in multisensory environments. Developmental Cognitive Neuroscience, p. 100930 (2021)
16. Gruber, T., Müller, M.M., Keil, A., Elbert, T.: Selective visual-spatial attention alters induced gamma band responses in the human EEG. Clin. Neurophysiol. **110**(12), 2074–2085 (1999)
17. Kia, S.M., Olivetti, E., Avesani, P.: Discrete cosine transform for MEG signal decoding. In: 2013 International Workshop on Pattern Recognition in Neuroimaging, pp. 132–135. IEEE (2013)

18. Parvez, M.Z., Paul, M.: Features extraction and classification for ictal and interictal EEG signals using EMD and DCT. In: 2012 15th International Conference on Computer and Information Technology (ICCIT), pp. 132–137. IEEE (2012)
19. Lotte, F., Congedo, M., Lécuyer, A., Lamarche, F., Arnaldi, B.: A review of classification algorithms for EEG-based brain-computer interfaces. J. Neural Eng. **4**(2), R1 (2007)
20. Bashar, M.K., Reza, F., Idris, Z., Yoshida, H.: Epileptic seizure classification from intracranial EEG signals: a comparative study EEG-based seizure classification. In: 2016 IEEE EMBS Conference on Biomedical Engineering and Sciences (IECBES), pp. 96–101. IEEE (2016)
21. Fahrenfort, J.J., Grubert, A., Olivers, C.N.L., Eimer, M.: Multivariate EEG analyses support high-resolution tracking of feature-based attentional selection. Sci. Rep. **7**(1), 1–15 (2017)
22. Lotte, F.: A review of classification algorithms for EEG-based brain-computer interfaces: a 10 year update. J. Neural Eng. **15**(3), 031005 (2018)
23. Schlögl, A., Lee, F., Bischof, H., Pfurtscheller, G.: Characterization of four-class motor imagery EEG data for the BCI-competition 2005. J. Neural Eng. **2**(4), L14 (2005)
24. Garrett, D., Peterson, D.A., Anderson, C.W., Thaut, M.H.: Comparison of linear, nonlinear, and feature selection methods for EEG signal classification. IEEE Trans. Neural Syst. Rehabil. Eng. **11**(2), 141–144 (2003)
25. Logistic regression: sklearn. https://scikit-learn.org/stable/modules/generated/sklearn.linear_model.LogisticRegression.html#sklearn.linear_model.LogisticRegression. Accessed 30 Jun 2020
26. Turoman, N.: Early Multisensory Attention as a Foundation for Learning in Multicultural Switzerland. Ph.D. thesis, éditeur non identifié (2020)

GridEx: An Algorithm for Knowledge Extraction from Black-Box Regressors

Federico Sabbatini(✉), Giovanni Ciatto, and Andrea Omicini

Dipartimento di Informatica – Scienza e Ingegneria (DISI),
Alma Mater Studiorum—Università di Bologna, Cesena, Italy
{f.sabbatini,giovanni.ciatto,andrea.omicini}@unibo.it

Abstract. Knowledge-extraction methods are applied to ML-based predictors to attain explainable representations of their operation when the lack of interpretable results constitutes a problem. Several algorithms have been proposed for knowledge extraction, mostly focusing on the extraction of either lists or trees of *rules*. Yet, most of them only support supervised learning – and, in particular, *classification* – tasks. ITER is among the few rule-extraction methods capable of extracting symbolic rules out of sub-symbolic regressors. However, its performance – here intended as the *interpretability* of the rules it extracts – easily degrades as the complexity of the regression task at hand increases.

In this paper we propose GridEx, an extension of the ITER algorithm, aimed at extracting symbolic knowledge – in the form of lists of if-then-else rules – from *any* sort of sub-symbolic regressor—there including neural networks of arbitrary depth. With respect to ITER, GridEx produces *shorter* rule lists retaining higher fidelity w.r.t. the original regressor. We report several experiments assessing GridEx performance against ITER and CART (i.e., decision-tree regressors) used as benchmarks.

Keywords: Explainable AI · Knowledge extraction · Interpretable prediction · Regression · ITER · GridEx

1 Introduction

Nowadays, black-box data-driven predictors such as neural networks or support-vector machines are among the most used tools to solve a wide range of different tasks [35]. Such predictors are *opaque* systems that operate in a sub-symbolic fashion, making it very hard for humans to understand how they manipulate data to compute their outputs. Nevertheless, they are being increasingly adopted in many application fields – including, but not limited to, healthcare, finance, and law – to support forecasting and decision making.

Thus, to enable the exploitation of black-box predictors within critical applications where interpretability is not an option, several methods have been developed to extract *intelligible* knowledge out of black-box predictors [4], aimed at *explaining* the operation and the outcomes of black boxes to humans.

D. Calvaresi et al. (Eds.): EXTRAAMAS 2021, LNAI 12688, pp. 18–38, 2021.
https://doi.org/10.1007/978-3-030-82017-6_2

Virtually all knowledge-extraction methods proposed so far focus on the extraction of either lists or trees of *rules*, and are exploited in many application areas. For instance, knowledge extraction is applied to credit-risk evaluation [6,7,41]. In healthcare, they are used to make early breast cancer prognosis predictions [21], to help the diagnosis of hepatobiliary disorders [25], coronary artery disease or thyroid dysfunctions [10], to determine the type of dermatological diseases, and to discriminate among liver diseases or diabetes [9]. Rule-extraction algorithms are also applied to, e.g., predictive models for credit card screening [39], intrusion detection in computer networks [26], and keyword extraction [5].

However, while most of the algorithms from the literature focus on *classification* tasks – e.g. TREPAN [19], Rule-extraction-as-learning [18], and others [8,31] –, a few are explicitly designed for *regression* tasks—such as ITER [27] and REFANN [40]. To the best of our knowledge, no algorithm has been proposed so far to tackle other branches of machine learning, such as unsupervised learning. REFANN is a *decompositional* extraction procedure which can be only applied to neural networks with just *one* hidden layer, and also requires a reduction of the network aimed at minimising the number of hidden neurons, to simplify the extraction process. It is then poorly suited for modern *deep* neural networks. Conversely, ITER is a *pedagogical* approach [3] which can be applied to regressors of any sort, as it does not make any assumption on the type, structure, and operation of the regressors it is applied to. However, its predictive performance degrades when applied to *high-dimensional* data sets.

Accordingly, in this work we propose GridEx, a new knowledge-extraction procedure extending the ITER algorithm to overcome its limitations and to reduce its computational-time complexity. As an extension of ITER, GridEx inherits a number of relevant features. For instance, they both extract rule lists out of regressors of *any* sort. However, GridEx outperforms ITER in terms of *fidelity* of the extracted rules w.r.t. the underlying regressor, especially when applied to high-dimensional data sets. In other words, GridEx extracts rule lists whose predictive capabilities are generally closer to the original black box.

To demonstrate the effectiveness of GridEx, we present a number of experimental evaluations aimed at comparing GridEx and ITER. The predictions of both extraction algorithms are compared among each other and w.r.t. a decision tree regressor (CART) trained on the same data. This evaluation is repeated on six data sets – having incremental amounts of dimensions and instances –, in order to analyse GridEx scalability, other than predictive performance and its ability to mimic – and therefore explain – the underlying black-box predictor.

2 State of the Art

As the adoption of machine-learning (ML) predictors pervades human activities, critical aspects become more evident and challenging, and require more care. In particular, as widely recognised within the explainable AI (XAI) community, the exploitation of ML comes at the price of relying on *sub-symbolic* algorithms that leverage on poorly-intelligible mechanisms for their operation, since they do not represent knowledge explicitly. Lacking *interpretability*, those algorithms – such as artificial neural networks (ANNs) and support vector machines (SVMs) – are often described as "black boxes" [30]. While it may be negligible or harmless in some application scenarios, interpretability is a critical issue in a growing number of areas. Several solutions have been proposed in the XAI field: the exploitation of (more) interpretable predictors – such as linear models and decision trees – rather than (more) opaque ones – e.g., ANNs and SVMs – in the particular case of supervised learning [36]; or, the exploitation of inspection techniques focusing on either input/output or the black-box internal structure [24].

As discussed in [16], computational systems can be considered as *interpretable* if their operation and outcomes can be easily understood by a human being: unfortunately, most predictors exploited in modern AI tend to sacrifice interpretability, by becoming increasingly complex while seeking for predictive performance. Instead, this paper focuses on those techniques aimed at explaining a sub-symbolic predictor *ex-post*. We restrict our scope to *knowledge-extraction* algorithms which attempt to explain black-box predictors by reverse-engineering their machinery, with the purpose of making their knowledge explicit.

2.1 Knowledge Extraction

Within the scope of supervised learning, knowledge extraction refers to the task of extracting some explicit intelligible representation for the sub-symbolic knowledge acquired by some predictor (either classifier or regressor) via learning from data. Assuming that a procedure for knowledge extraction exists for a particular predictor, any extracted knowledge can then be used as a basis to construct *explanations* – and sometimes as a replacement – for that predictor, provided that such knowledge retains a high *fidelity* w.r.t. the original predictor and the data it has been trained upon [15]. Extracted knowledge, in turn, may then enable further manipulations for the user's benefit—such as merging the knowledge of two or more black-boxes [14]. Unfortunately, no one-size-fit-all solution exists for this task, and several algorithms have been proposed for this purpose [24].

According to [13], virtually all knowledge-extraction methods proposed so far into the literature can be categorised along three orthogonal dimensions, namely: *(i)* the supported sort of learning tasks, *(ii)* the form of the knowledge extracted, and *(iii)* the *translucency* requirements of the black box.

Item *(i)* refers to which supervised learning tasks must be supported by a black box to enable extraction. While most methods support *classification* tasks – e.g. TREPAN [19], Rule-extraction-as-learning [18] and others [8,31] –, only a few are explicitly designed to tackle *regression* tasks—such as ITER [27], REFANN [40], ANN-DT [38] and RN2 [37]. Methods extracting knowledge from black boxes independently of their task are, e.g., G-REX [29] and CART [11].

Conversely, item *(ii)* refers to the form of the extracted knowledge. As decision rules [22,28,32] and decision trees [33,34] are the most widespread human-understandable predictors, most methods produce either decision rules or trees.

Finally, the translucency notion [3] from item *(iii)* refers to the relationship between the extracted rules and the *internal* structure of the underlying black box—and how much of it the extraction procedure can take into account. In particular, there exist two sorts of knowledge extractors w.r.t. translucency. *Decompositional* extractors take into account the black-box internal structure during the extraction process, whereas *pedagogical* ones do not. Therefore, pedagogical approaches are usually more general, despite potentially less precise.

To evaluate the quality of knowledge-extraction methods, different indicators can be exploited, including fidelity and predictive performance measurements [42]. In particular, the former is a meta-measure of how good the extracted knowledge mimics the underlying black-box predictions. The latter measures the predictive power of the explanator with respect to the data. In both cases measurements are taken via the same scoring function used for assessing the performance of the black box—which in turn depends on the black-box performed task. For instance, in the particular case of black-box regressors, the mean absolute error (MAE) and the R^2 scores could be exploited.

2.2 The Iter Algorithm

Iter [27] is a pedagogical knowledge-extraction algorithm explicitly designed for black-box regressors. It extracts knowledge in the form of rule lists, while imposing no constraint on the nature, structure, or training of the regressors.

To extract rules, the Iter algorithm steps through the creation and iterative expansion of several disjoint *hypercubes*, covering the whole input space the regressor has been trained upon. In other words, Iter accepts as input a regressor and the data set used for its training, then iteratively partitions the *surrounding hypercube* containing the whole data set following a bottom-up strategy.

At the end of the process, each partition is converted into a rule of the form

$$\begin{aligned}&\text{if } Var_1 \in [Value_1{}^{Low}, Value_1{}^{High}]\\&\text{and } Var_2 \in [Value_2{}^{Low}, Value_2{}^{High}]\\&\text{and } \ldots \text{ and } Var_k \in [Value_k{}^{Low}, Value_k{}^{High}]\\&\text{then predict some } Constant\end{aligned}$$

where k is the dimension of the input space, i.e. the number of input variables. The predicted output value – *Constant* – is attained by averaging the output values of all samples belonging to the originating hypercube. To compute *Constant* for each hypercube, samples can be both picked from the data set or randomly generated. In the latter case, the underlying regressor is used as an *oracle*.

Pros and Cons. As a pedagogical approach, Iter supports any sort of black-box regressor. For instance, it can be applied to ANNs with any number of hidden layers and neurons, unlike decompositional algorithms as RefAnn.

The if-then-else rules produced by Iter are human-readable and *globally* approximate the underlying black box with high fidelity. Therefore, when the total amount of hypercubes – and rules – found by ITER is relatively small, the resulting rule list is a valuable form of explanation for the underlying black box.

As for the predictive performance, the authors of ITER report very good results with respect to both the data set samples and the underlying black-box outputs. However, the performance of ITER easily degrades when the algorithm is applied to complex data sets—where the complexity is represented by the number of input dimensions. Furthermore, several major ITER drawbacks are also reported, such as *(i)* non-exhaustivity – i.e. the extracted rules do not cover the entire input space –, described with more details in Sect. 2.2 and shown in Fig. 1, *(ii)* the impossibility to handle categorical features, and *(iii)* the impossibility to associate anything than a constant value to each rule—which introduces an undesired discretisation in the predicted values.

In our experience, another limitation of ITER concerns the hypercube expansion mechanism. In fact, as further discussed in Sect. 4, the algorithm may waste a lot of computational efforts processing *irrelevant* regions of the input space – i.e. regions containing no samples from the data set –, and therefore ending up producing several useless rules—which, ultimately, hinder interpretability.

Non-exhaustivity Issue. ITER's hypercube expansion is an iterative procedure strongly affected by the initial conditions, such as the number, position and dimension of the starting cubes. Even by tuning the algorithm parameters, there is always a chance that hypercupe expansion converges to a situation where some portions of the input space are left uncovered. This is undesirable, since the rule list resulting by ITER would then be poorly predictive for data laying in those portions of the input space.

To better clarify the issue, we report in Fig. 1 a trivial example with only two input features taken from [27]. There, the authors show how, after several iterations, the further expansion of the cubes is impossible and a little region of the input space – namely, the central one – is not covered by any of them.

To circumvent this issue, the same authors suggest the creation of additional, smaller cubes to fill the uncovered area. However, despite this solution provides good results for simple data sets, the opposite is true in more complex contexts. In these cases, the uncovered regions require an high number of small hypercubes to be generated. They are smaller and smaller, resulting in an explosion of the amount of rules—which would in turn hinder the interpretability of the final rule list.

The authors suggest another fix for the non-exhaustivity issue: adopting a smaller hypercube update parameter. However, this solution implies more iterations and longer execution time. Thus, this may increase the chance of the algorithm terminating without converging to a valuable partitioning within the maximum iterations limit.

3 GridEx

The design of GridEx aims at overcoming the non-exhaustivity of ITER, other than its inability to discriminate among *interesting* and *negligible* regions of

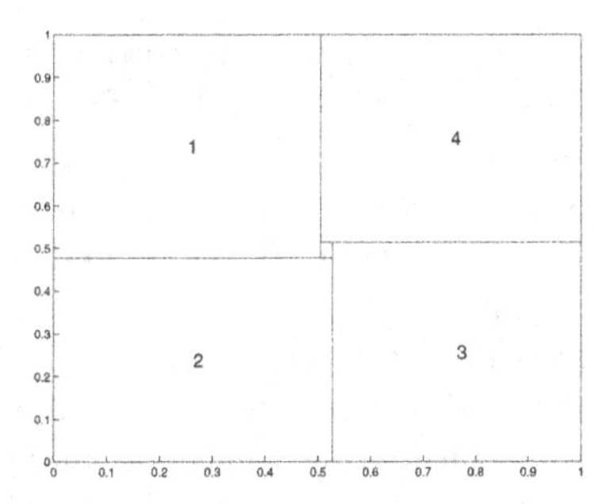

Fig. 1. Example of the ITER non-exhaustivity taken from [27].

the input space. We consider as interesting the regions that contain at least one training-set sample, with the others considered as negligible. Furthermore, GridEx is designed to tackle *complex* data sets—i.e. high-dimensional data sets whose data distribution is non-trivial. In particular, the goal of GridEx is to find bigger and more interesting regions than ITER, while retaining the idea that samples belonging to the same region should have a similar output value. In doing so, GridEx tries to keep the computational and human efforts minimal—where by "human effort" we mean manual parameter tuning, whereas by "computational effort" we mean time and memory requirements for the algorithm execution.

3.1 The Algorithm

GridEx is a pedagogical knowledge-extraction algorithm that – similarly to ITER – produces rule lists out of sub-symbolic predictors, by using them as oracles. In other words, GridEx only takes into account the inputs and outputs of the underlying predictor. For this reason its performance is not tied to the kind or the structure of the model and it can thus be applied to neural networks regardless of their depth, as well as to other sorts of regressors.

Similarly to ITER, GridEx assumes that a black-box regressor R is available, as well as the input data D it has been trained upon. Under that hypothesis, both algorithms strictly operate inside the *surrounding hypercube* containing all data in D, by trying to find a partitioning of the surrounding hypercube such that, for each partition, the output value of R is similar for all samples contained into that partition. In that case, both can produce a list of if-then-else rules approximating the behaviour of R, one for each partition selected by the algorithm.

Of course, finding *fewer* relevant partitions implies producing more concise rule lists, which are more easily grasped by humans. Accordingly, GridEx differs from ITER in the way partitions are computed, and relevant hypercubes are selected. In fact, while ITER relies on a bottom-up strategy – starting from infinitely small hypercubes containing just one input space point and expanding them as much as possible –, GridEx adopts a top-down strategy—starting from a single partition containing the whole input space and recursively splitting it for a user-defined amount of times, into partitions of equal size. Thanks to this strategy, GridEx actually succeeds in finding *fewer* partitions w.r.t. ITER, while producing rules retaining a good fidelity w.r.t. R. After every split, GridEx attempts to merge couples of adjacent partitions—provided that all the samples therein contained yield similar values for R. Split and merge phases are alternated until a stopping criterion is met.

User can choose between two stopping criteria – not necessarily mutually-exclusive –, one based on the similarity (w.r.t. R) among the samples in the current hypercube, the other considering whether a maximum number of iterations has been reached or not. More precisely, if the standard deviation of the R output values of some hypercube exceeds a given threshold, then that hypercube is further partitioned: when all hypercubes are under threshold the algorithm terminates. The threshold value is a trade-off between *sensitivity* – intended as how similar should be samples grouped together – and number of rules extracted so far—i.e., the more increases the sensitivity, the more the output rules will be.

This procedure may eventually bring to the creation of *adjacent* hypercubes with similar averaged values of R. Thus, after each split and before proceeding to the successive iteration, the algorithm tries to pair-wise *merge* similar hypercubes so as to reduce the total amount of hypercubes—and thus to preserve the model interpretability. More precisely, two adjacent hypercubes are merged only if the standard deviation of R for the samples belonging to the merged hypercube does not exceed a given threshold. In this way, GridEx attains larger hypercubes without affecting the predictive performance of the resulting rules.

Overall, GridEx relies upon $n + 3$ user-defined parameters, being $n \in \mathbb{N}_{>0}$ the maximum amount of iterations it performs. Such parameters are: n, θ, m and $p_1, \ldots, p_n$, where $\theta \in \mathbb{R}_{\geq 0}$ is the similarity threshold, m is the minimum amount of samples to be considered in non-empty hypercubes, and p_i is the number of slices the algorithm performs along each dimension of the current hypercube during the i-th iteration. In the remainder of this section, we use P to denote $\langle p_1, \ldots, p_n \rangle$, $k = \dim(D)$ to denote the dimension of the input space D.

Under such hypotheses, a formal definition of GridEx is provided in Algorithm 1. Intuitively, the operation of the algorithm can be described as follows. It firstly computes the surrounding hypercube containing all data in D by finding the minimum and maximum value of each input variable. Then it recursively splits such hypercube into p_i parts along each direction, n times, therefore producing p_i^k adjacent and non-overlapping partitions of equal size, at each step. Only non-empty partitions (w.r.t. the data in D) are taken into account by the algorithm. Similarly, partitions containing samples whose standard deviation (w.r.t. R) is greater than θ are further partitioned in successive steps of the algorithm. It may happen, however, that some partition contains too few samples to provide precise predictions. To prevent this, before computing standard deviation, GridEx generates m new random samples, using R as an oracle.

Algorithm 1. GridEx pseudocode

Require: parameters $n, \theta, m, p_1, \ldots, p_n$ to be provided

1: **function** GRIDEX(R, D)
2: $\quad H_0 \leftarrow$ SURROUNDINGHYPERCUBE(D)
3: $\quad$ **return** SLPIT(1, H_0, R, D)
4: **function** SURROUNDINGHYPERCUBE(D)
5: $\quad$ **return** the minimal hyper-cube that includes all the samples of D
6: **function** SPLIT(i, H, R, D)
7: $\quad$ **if** $i > n$ **then return** $\{H\}$
8: $\quad \Pi \leftarrow \varnothing, \quad \Pi' \leftarrow \varnothing$
9: $\quad$ **for all** $H' \in$ PARTITIONS(H, p_i) s.t. $H' \cap D \neq \varnothing$ **do**
10: $\qquad D \leftarrow D \cup$ GENERATESAMPLESIN(H')
11: $\qquad$ **if** STDDEV(H', R, D) $\leq \theta$ **then**
12: $\qquad\quad \Pi \leftarrow \Pi \cup \{H'\}$
13: $\qquad$ **else**
14: $\qquad\quad \Pi' \leftarrow \Pi' \cup \{H'\}$
15: $\quad \Pi'' \leftarrow$ MERGE(Π, R, D)
16: $\quad$ **for all** $H' \in \Pi'$ **do**
17: $\qquad \Pi'' \leftarrow \Pi'' \cup$ SPLIT($i+1$, H', R, D) $\qquad$ ▷ Recursion!
18: $\quad$ **return** Π''
19: **function** GENERATESAMPLESIN(H)
20: $\quad$ **return** $\{m$ random points in $H\}$
21: **function** PARTITIONS(H, p)
22: $\quad$ **return** {all p^k partitions of H after splitting each edge into p parts}
23: **function** MERGE(Π, R, D)
24: $\quad C \leftarrow$ ADJACENTCOUPLES(Π)
25: $\quad$ **while** ($|C| > 0$) **do**
26: $\qquad (H_1^*, H_2^*) \leftarrow \underset{(H_1,H_2)\in C}{\arg\min} \{$STDDEV$(H_1 \cup H_2, D, R)\}$
27: $\qquad H \leftarrow H_1^* \cup H_2^*$
28: $\qquad$ **if** STDDEV(H, R, D) $\leq \theta$ **then**
29: $\qquad\quad \Pi \leftarrow \Pi \setminus \{H_1^*, H_2^*\} \cup \{H\}$
30: $\qquad\quad C \leftarrow$ ADJACENTCOUPLES(Π)
31: $\qquad$ **else**
32: $\qquad\quad$ **return** Π
33: $\quad$ **return** Π
34: **function** STDDEV(H, R, D)
35: $\quad$ **return** the standard deviation of all $\{R(x) \mid x \in H \cap D\}$
36: **function** ADJACENTCOUPLES(Π)
37: $\quad$ **return** $\{(H_1, H_2) \mid H_1, H_2 \in \Pi \wedge (H_1$ and H_2 are adjacent$)\}$

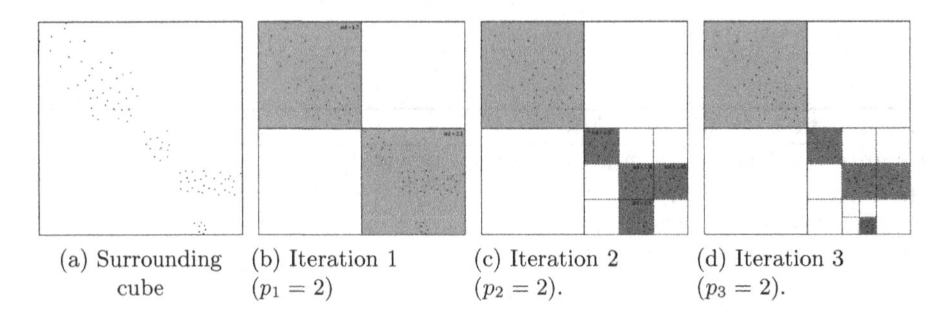

(a) Surrounding cube (b) Iteration 1 ($p_1 = 2$) (c) Iteration 2 ($p_2 = 2$). (d) Iteration 3 ($p_3 = 2$).

Fig. 2. Example of GridEx hyper-cube partitioning (merging step not reported). (Color figure online)

3.2 An Example

An example of hypercube partitioning executed by GridEx is reported in Fig. 2. The merging phase is not represented. The data set has two input variables (i.e. $k = 2$); user-defined parameters are $n = 3$, $P = \langle 2, 3, 2 \rangle$ and $\theta = 2.0$. In particular, Fig. 2a depicts the surrounding cube and the data set samples, represented by red dots. After first iteration (Fig. 2b), the surrounding cube is split into 4 (p_1^k) hypercubes (continuous lines), as $p_1 = 2$. The bottom-left and top-right ones are discarded as they are empty (white background). The top-left hypercube standard deviation (1.5) does not exceed θ (2.0), so it is not partitioned any further. Conversely, the fourth hypercube (standard deviation 2.1) must be further partitioned: thus, in the second iteration, it is split into 9 (p_2^k) partitions (dashed lines, Fig. 2c), as $p_2 = 3$. The same logic is then recursively applied to the 9 new hypercubes, leading to the final stage in Fig. 2d: 5 hypercubes out of 9 are discarded as empty (white background), 3 remain unaffected as their standard deviation is lower than θ (orange background), whereas the remaining one is partitioned into 2^2 smaller partitions (dotted lines), as $p_3 = 2$. Finally, of these 4 partitions, only 1 is non-empty then kept (green background).

3.3 GridEx Adaptive Splitting

Operationally, GridEx partitions a hypercube by relying on the parameters $P = \langle p_1, \ldots, p_n \rangle$, which essentially state how many splits must be performed per dimension, at each step of the algorithm. In practice, the actual values of the many p_i greatly impact on the outcome of GridEx—both in terms of efficiency of the whole procedure, and in terms of predictive performance (and complexity) of the resulting partitioning. However, they still implicitly rely on the assumption that all the input variables present a comparable relevance w.r.t. the output value definition—i.e. the expected output can be more dependent or almost not dependent at all on some variables rather than others.

Accordingly, we argue that a wiser strategy in hypercubes partitioning could rely on the following insight: more *relevant* dimensions of a hypercube should be split into more parts, whereas less relevant ones can be split in fewer parts. Thus, to further optimise the amount of selected hypercubes – without sacrificing predictive performance –, users of GridEx may adopt an *adaptive* splitting. When in adaptive splitting mode, GridEx takes into account the *relevance* of each dimension of D w.r.t. the expected output value. In particular, it uses an importance measure to choose the number of splits for each dimension of a hypercube.

Importance values can be estimated in several ways. Among the many methods available – e.g. [2,44,45] – here we leverage on a simple feature selection method, SciKit-Learn's `feature_selection.f_regression`[1]. It consists of a sequential algorithm aimed at iteratively and greedly selecting the most relevant features of a dataset. It starts by training a temporary regressor on a single feature – namely, the most correlated w.r.t. the output values – and it keeps repeating this operation by adding one feature at a time, always peaking the one that mostly increases the temporary regressor predictive performance. At the end of this process, features are ranked w.r.t. their relevance, and such ranking is used for the adaptive splitting.

Accordingly, users willing to exploit adaptive splitting should specify the number of partitions assigned to each dimension on the basis of the importance calculated w.r.t. D. Importance values are normalised into range $[0, 1]$ so that the value of the most important value is 1. The algorithm should then be provided with an increasingly-monotone function of the form $f : [0, 1] \to \mathbb{N}$ to set the splits to perform. For instance, a reasonable enhancement for standard two-split iterations could be: a single split along dimensions whose importance is <0.1, two splits if the importance is between 0.1 and 0.65, and three splits otherwise.

We remark that there exists no fixed optimal choices for f: yet, a trial-and-error approach can possibly lead to quickly find the most suitable combination.

3.4 Parameter Tuning

As described in Sect. 3.1, GridEx relies on a number of parameters.

The similarity threshold θ depends on the problem under study, i.e. on the problem output value distribution, and on the trade-off between the interpretable prediction performance and the total number of extracted rules. Small values of θ lead to more rules with higher predictive performance. Conversely, large values produce less rules at the expense of the predictive performance.

As for the m parameter – representing the minimum amount of considered samples in each cube –, our experiments showed that it does not notably influence the final results. Values ranging from 10 to 100 are a good choice. We adopted $m = 15$ for the experiments reported in this paper.

For both the standard and adaptive versions of GridEx, one or two iterative partitions are enough to obtain very good results with a limited amount of output rules. Our experiments showed that larger values lead to an explosion of the rule number without any significant enhancement of the predictive performance.

[1] cf. https://scikit-learn.org/stable/modules/generated/sklearn.feature_selection.f_regression.html.

Similarly, we found that values of 2 or 3 are suitable for the various p_i parameters identifying the number of partitions to create on each cube dimension. Larger values should be avoided, adopting instead an additional partitioning step. For instance, $P = \langle 4 \rangle$ gives the same predictive performance than $P = \langle 2, 2 \rangle$, but producing a great excess of hypercubes. This fact occurs because the single-step partitioning is equivalent to a 2-step partitioning where all the hypercubes created during the former iteration are further split during the latter.

When the output is not satisfying, it is possible to take advantage of the adaptive splitting for reducing the number of output rules with negligible deterioration of the predictive performance following the aforementioned suggestions.

4 Assessment of GridEx

In this section we provide a numerical analysis of GridEx under an explainability perspective. More precisely, our analysis is aimed at understanding if and to what extent GridEx: *(i)* is capable to approximate a black-box regressor, *(ii)* performs better than ITER in doing so, and *(iii)* is capable to provide concise and intelligible explanations for regression tasks.

Accordingly, we construct our experiments as follows. We implement both GridEx and ITER in Python and run them on a pool of black-box regressors – trained on many publicly-available data sets of growing size and dimensionality – to compare their rule-extraction capabilities. We also compare the complexity of the partitioning produced by GridEx and ITER with the ones produced by the CART decision tree regressor trained on the same data[2]. Similarly, the data sets for our experiments are summarised in Table 1, providing, for each data set, *(i)* a bibliographic reference, *(ii)* the number of input features, *(iii)* the total number of instances, *(iv)* the percentage of samples taken apart as test set while training a black-box regressor on that data set, and *(v)* the performance of the black-box regressor. In particular, our black-box regressors are ANNs with one or two hidden layers, depending on the data set. The predicting performance of the ANN is reported in terms of MAE and R^2 value averaged on 100 tests.

Generally speaking, our experiments show how GridEx performs better than both ITER – as it produces partitionings containing *fewer* hypercubes, while attaining rule lists with better predictive performances – and CART—as it produces simpler rule lists having choice points.

4.1 ITER Experimental Analysis

As a first step, we run ITER on all the aforementioned black-box regressors and data sets. Experiments concerning each data set are repeated 10 times. The averaged results of such experiments are summarised in Table 2.

In all the experiments, ITER parameters are the same: the number of initial hypercubes is set to 1 for all data sets, while the update parameter is chosen

[2] For the sake of *reproducibility*, the source code of our experiments is publicly available at https://github.com/sabbatinif/GridEx.

Table 1. Overview of the adopted data sets and the performances of the black-box regressors trained upon them.

Data set name	Acron.	Ref.	Features	Instances	Test set (%)	MAE	R^2
ARTI1 ($\alpha = 0$)	ARTI1	[27]	2	1 000	50	0.01	0.99
Combined Cycle Power Plant	CCPP	[17]	4	9 568	20	4.16	0.89
Airfoil Self-Noise	ASN	[1]	5	1 503	20	2.04	0.85
Energy Efficiency	EE	[20]	8	768	20	2.70	0.87
Gas Turbine CO and NOx Emission	GAS	[23]	10	36 733	20	3.16	0.84
Wine Quality	WQ	[43]	11	6 497	20	0.60	0.31

Table 2. Results of ITER applied to the data sets described in Table 1.

Data set	ARTI1	CCPP	ASN	EE	GAS	WQ
# features	2	4	5	8	10	11
Threshold	0.2	7.0	4.0	4.0	15.0	2.0
# iterations	26	600	582	600	600	600
# hyper-cubes	4	329	113	55	44	23
# useful hyper-cubes	4	16	64	25	14	11
Coverage (%)	100.0	91.8	97.9	83.5	76.6	62.3
Left training samples (%)	0.0	8.4	0.0	2.9	6.6	6.1
Missed test samples (%)	0.0	9.1	1.7	3.3	6.7	7.8
MAE (data)	0.04	5.37	4.24	3.52	8.77	0.73
MAE (ANN)	0.04	3.77	3.40	2.37	7.88	0.55
R^2 (data)	0.92	0.83	0.40	0.76	0.10	−0.05
R^2 (ANN)	0.92	0.91	0.55	0.92	0.17	−0.12

as double w.r.t. the predefined one described by the authors of ITER—i.e. 0.1 instead of 0.05. This aims at reducing the amount of iterations required by ITER to converge, especially with the more complex data sets, provided that the algorithm terminates when either 600 iterations are performed, or *all* the training samples have been covered by the created hypercubes.

Accordingly, for each data set we collect *(i)* the overall number of actually useful hypercubes – i.e. those containing at least one training sample – found by ITER, *(ii)* the total number of iterations performed by the algorithm, *(iii)* the selected threshold parameter value, *(iv)* the amount of input space covered by the hypercubes expressed in percentage, *(v)* the percentage of training samples that are not included in any hypercube and, analogously, *(vi)* the percentage of test samples that the output model is not able to predict. To improve readability, the feature number of each data set is reported in Table 2 as well. Finally, the MAE and R^2 scores of all ITER predictions are reported w.r.t. both the original data set and the black-box predictions. These latter measurements are performed using the test-set samples—which in turn are *never* used for training.

Table 2 highlights that in 4 cases out of 6 the algorithm reaches the maximum allowed iterations without covering the whole training set. When this is the case, the resulting rule list produced by ITER is affected by a reduced predictive capability, w.r.t. the black-box regressor it mimics. This effect can be detected by comparing their test-set performance. Such non-exhaustivity issue is particularly evident for the ASN data set as well.

Generally speaking, the overall ITER performance is very good with simple data sets (e.g., the ARTI1 data set). However, we observe a degradation as the complexity of the data set grows. In any case, both performance and computational cost heavily depend on the parameter values and initial conditions, such as the starting cube number and position. Parameter tuning can be performed through a trial-and-error approach [27], but this often implies a trade-off between execution time, number of extracted rules and result accuracy.

To better analyse how ITER attempts to address the non-exhaustivity issue, the left side of Fig. 3 depicts several plots describing executions of the algorithm on different data sets and black boxes. Each plot has two panels. In both panels the horizontal axis refers to the computational time (from left to right), whereas the vertical bars refer to hypercubes. So left-most bars refer to hypercubes which are found *earlier*. Top panels represent the number of samples belonging to each hypercube, expressed as the percentage of the overall training examples. Conversely, bottom panels represent the relative volume of each hypercube, expressed as the percentage of the whole input feature space.

Notably, we observe that later iterations of ITER tend to create smaller hypercubes that include less samples than the ones computed in previous iterations (cf. the GAS and WQ data sets). Exceptions may occur when samples are not uniformly distributed within the input space. This happens for instance in the EE data set, where it is possible to find big hypercubes with few samples and, conversely, very small cubes including up to a fourth of the training set. Finally, the CCPP data set is a perfect example of ITER uncontrolled hypercube expansion towards *irrelevant* input space regions: more than 95% of the hypercubes have no predictive relevance. The algorithm wastes time and resources exploring these regions, reaching the maximum iteration number with almost a 10% of the samples uncovered by the interpretable model rules.

As discussed below, GridEx overcomes those drawbacks by achieving better predictive performance in less time and with a lower computational effort.

4.2 GridEx Experimental Analysis

To fairly compare ITER and GridEx, we evaluate the latter as well against all the aforementioned black-box regressors and data sets. Results are reported in Table 3, where rows and columns retain the same meaning as in Table 2, except for the "Left training samples" row, missing. Indeed, it is not useful to report the number of training samples left out by the algorithm, as GridEx always covers the entire training set, by construction. Furthermore, the user-defined partitions

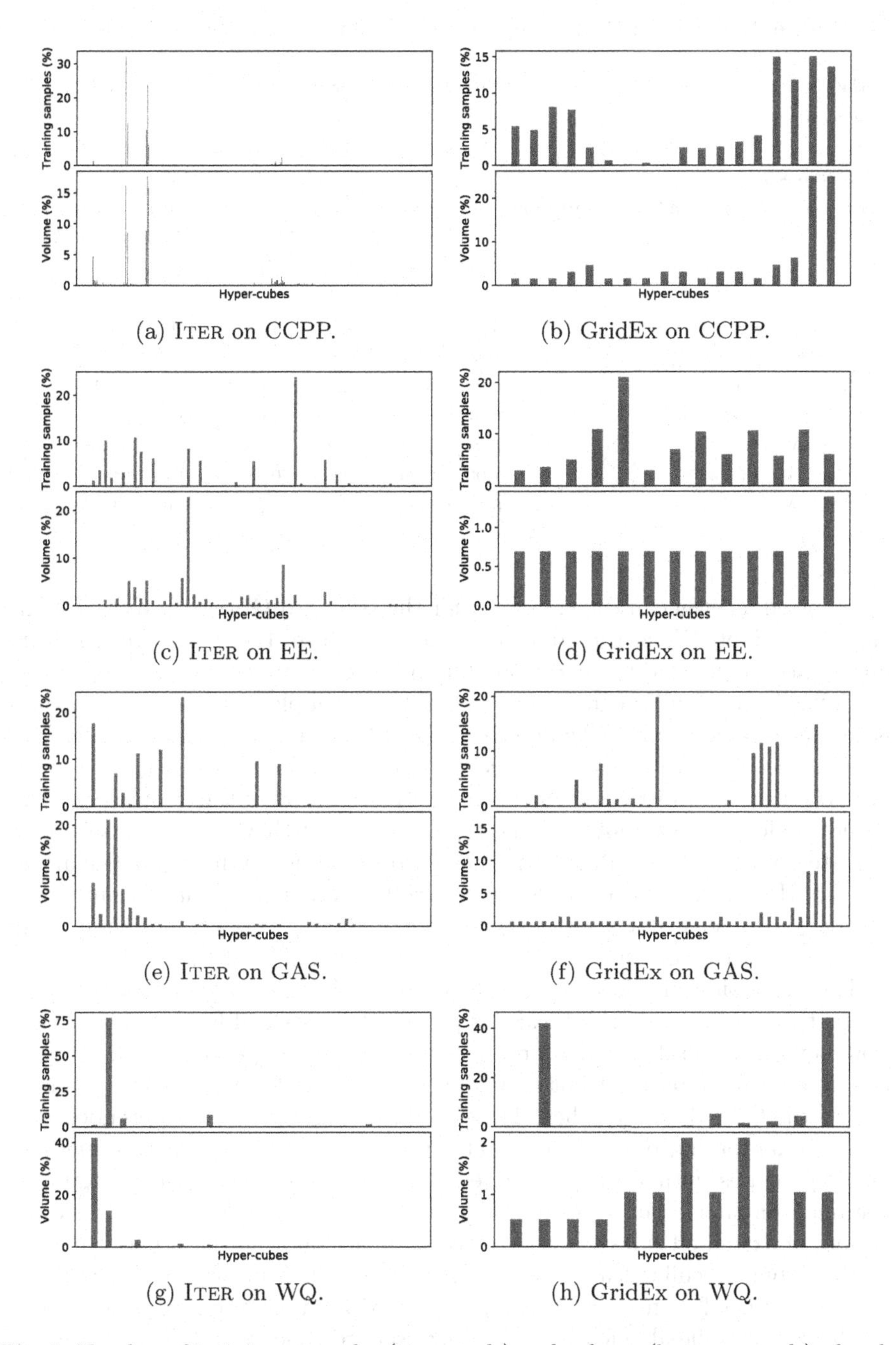

(a) ITER on CCPP.
(b) GridEx on CCPP.
(c) ITER on EE.
(d) GridEx on EE.
(e) ITER on GAS.
(f) GridEx on GAS.
(g) ITER on WQ.
(h) GridEx on WQ.

Fig. 3. Number of training examples (top panels) and volume (bottom panels) of each hyper-cube created by ITER (left plots) and GridEx (right plots). Values are expressed as percentage of the total number of training samples and the surrounding cube volume, respectively. Each row represents a different data set.

Table 3. Results of GridEx applied to the data sets described in Table 1.

Data set	ARTI1	CCPP	ASN	EE	GAS	WQ
# features	2	4	5	8	10	11
Threshold	0.01	4.10	4.00	2.00	7.00	1.00
Partitions	⟨2⟩	⟨a, a⟩	⟨a, a⟩	⟨a⟩	⟨a, a⟩	⟨a⟩
Feature importance, adaptive partitions	–	≤0.04, 1 ≤0.5, 2 ≤1, 4	≤0.2, 1 ≤0.5, 2 ≤0.7, 3 ≤1, 4	≤0.001, 1 ≤0.5, 2 ≤1, 3	≤0.1, 1 ≤0.7, 3 ≤1, 4	≤0.03, 1 ≤0.6, 2 ≤1, 3
# hyper-cubes	4	18	38	13	41	12
Coverage (%)	100.0	93.8	43.2	9.7	84.0	13.0
Missed test samples (%)	0.0	0.0	1.3	0.0	0.0	0.1
MAE (data)	0.01	4.37	3.27	2.65	6.79	0.69
MAE (ANN)	0.01	2.61	2.57	1.23	5.51	0.38
R^2 (data)	1.00	0.89	0.66	0.88	0.48	0.17
R^2 (ANN)	0.99	0.96	0.76	0.98	0.61	0.46

are reported as either integer numbers for the standard GridEx or as 'a' for the adaptive variant. When the adaptive option is chosen, the user-defined feature importance ranges and the corresponding partition numbers are also reported.

Unlike ITER, GridEx can predict almost every sample of the test set. Furthermore, the extracted rules only describe *relevant* regions of the input space. This design choice may lead to a partial coverage of the input space—e.g. for the EE and WQ data sets. Little coverage commonly occurs when the training samples are not uniformly distributed in the input space, but they are, conversely, concentrated only in some sub-regions that GridEx is able to discern and mark as relevant. However, in some corner cases, GridEx may require a larger number of hypercubes to achieve a *significantly* better predictive performance than ITER: then, it produces more rules at the expense of readability of the final rule set.

The right side of Fig. 3 depicts plots describing executions of GridEx on different data sets and black boxes, as previously described for ITER. The plots show how, in general, GridEx produces fewer hypercubes than ITER, proving itself more concise on most data sets. There are, however, notable exceptions—such as the GAS data set, where ITER produces slightly fewer hypercubes than GridEx (considering only the relevant ones). So, also GridEx may sometimes find hypercubes with little predictive relevance, for the shortage of examples contained: in that case, a pruning algorithm could be exploited to reduce the number of extracted rules with no relevant impact on the overall performance.

Also, some peculiar features of GridEx are effective in overcoming ITER, in the general case. For instance, the iterative multi-level partitioning performed by GridEx is a very flexible feature, enabling users to tune the refinement of those inaccurate rules that include samples with larger standard deviations. However, to be effective, it must be carefully tuned, as the number of hypercubes may grow

Table 4. Results of CART applied to the data sets in Table 1.

Data set	ARTI1	CCPP	ASN	EE	GAS	WQ
# leaves	4	15	50	40	50	15
MAE (data)	0.01	4.49	3.06	2.78	4.88	0.65
MAE (ANN)	0.01	2.44	2.08	0.59	3.53	0.24
R^2 (data)	1.00	0.88	0.70	0.86	0.67	0.25
R^2 (ANN)	0.99	0.96	0.80	0.99	0.78	0.71

exponentially when samples are uniformly distributed among the entire input space. Similarly, GridEx merging phase supports the creation of compact rule lists with less, coarser-grained rules and no predictive performance degradation.

4.3 Comparison of ITER and GridEx

Finally, to perform an unbiased comparison between GridEx and ITER, a decision tree regressor [11] is taken as a reference—similarly to [12]. More precisely, we adopt the CART algorithm for each data set in Table 1, with a maximum depth parameter equal to twice the amount of input features. The resulting decision trees are summarised in Table 4, in terms of MAE and R^2 value w.r.t. both the data and the underlying neural network. The number of leaves is reported as well since it determines the number of decision rules exploited by the regressor.

The choice of the maximum depth parameter comes from the following considerations. ITER and GridEx rules are expressed as hypercubes. Every cube has 2 constraints on each input variable, i.e. the lower and upper values. Since each node of the decision tree represents a constraint on a variable, our choice of the maximum depth parameter ensures that the resulting decision rules cannot be subject to more constraints than the rules produced by either ITER or GridEx.

Figure 4 summarises results of the comparison among the three extraction procedures. GridEx (Fig. 4a) produces an equal or smaller number of rules w.r.t. both ITER and CART in almost all the cases—even if we cut off the irrelevant hypercubes produced by ITER. Since interpretability of a rule list decreases with its length, we argue that GridEx produces more interpretable results.

As for the predictive performance, GridEx is always better than ITER in terms of MAE (Figs. 4b and 4c) and R^2 value (Figs. 4d and 4e), with respect to both the data and the underlying black-box predictions. However, CART has generally smaller MAE and larger R^2 than GridEx. This can be explained by the undesired discretisation introduced by the constant output value of GridEx and ITER rules, as well as by the higher amount of rules extracted by CART.

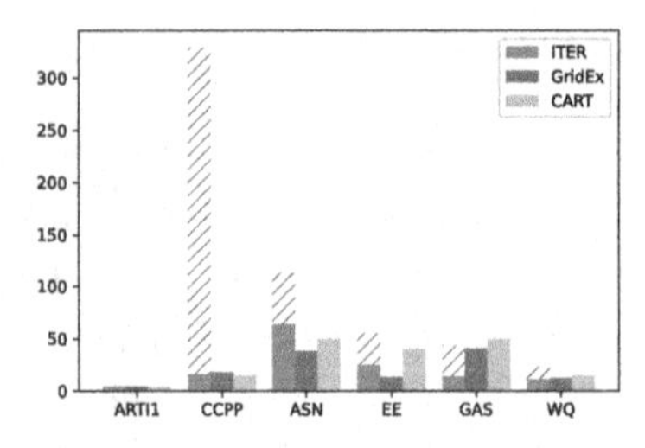

(a) Number of extracted rules.

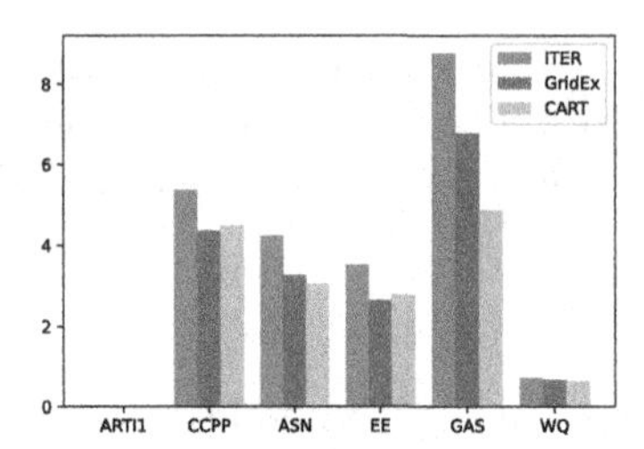

(b) MAE with respect to the data.

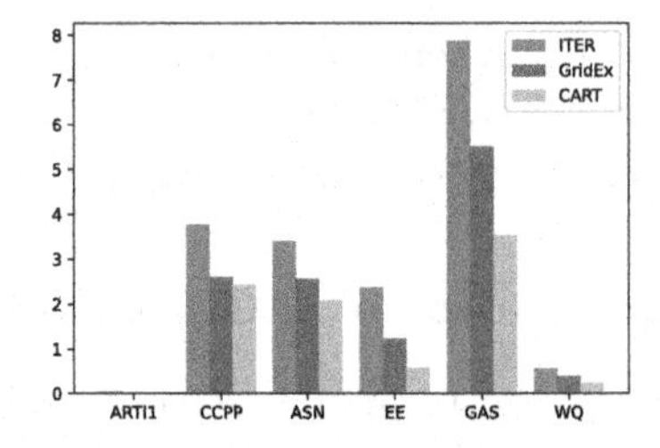

(c) MAE fidelity with respect to the underlying black box.

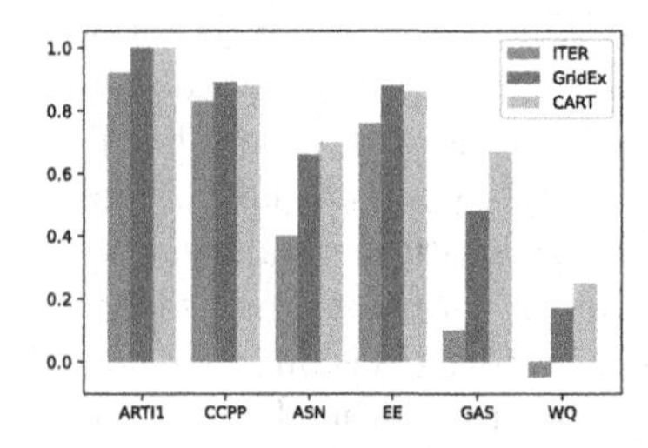

(d) R^2 value with respect to the data.

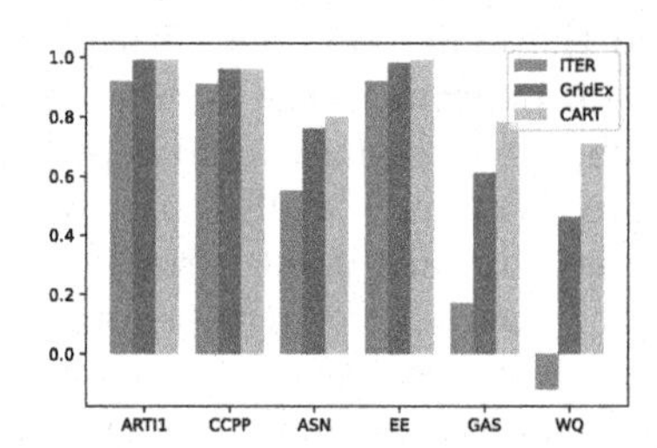

(e) R^2 value with respect to the underlying black box.

Fig. 4. Comparison between ITER, GridEx, and CART using both MAE (the lower the better) and R^2 scores (the higher the better).

5 Conclusions

In this paper we present GridEx, a new pedagogical knowledge-extraction procedure aimed at globally explaining black-box regressors. GridEx extends ITER by overcoming some of its major limitations—namely, its non-exhaustivity and its tendency to focus on non-interesting regions of the input space. GridEx is able to create an interpretable approximation of any black-box regressor in the form of decision rules. W.r.t. ITER, GridEx produces fewer decision rules, while attaining better predictive performance, even when the data set is high-dimensional.

In our next research efforts we plan to extend GridEx so as to address other limitations of ITER, such as the inability to handle categorical input features – with no pre-processing – and the constant output value of its decision rules. Furthermore, we intend to design an automatic procedure regarding the best adaptive splitting parameter selection. We also plan to extend our comparative numerical analysis to the other rule-extraction algorithms for regression mentioned in Sect. 2.1.

Acknowledgements. This paper has been partially supported by *(i)* the H2020 project "StairwAI" (G.A. 101017142), and *(ii)* the CHIST-ERA IV project "EXPECTATION" (G.A. CHIST-ERA-19-XAI-005).

References

1. Airfoil Self-Noise Data Set (2014). https://archive.ics.uci.edu/ml/datasets/Airfoil+Self-Noise. Accessed 19 Jan 2021
2. Altmann, A., Toloşi, L., Sander, O., Lengauer, T.: Permutation importance: a corrected feature importance measure. Bioinformatics **26**(10), 1340–1347 (2010)
3. Andrews, R., Diederich, J., Tickle, A.B.: Survey and critique of techniques for extracting rules from trained artificial neural networks. Knowl. Based Syst. **8**(6), 373–389 (1995)
4. Ayache, S., Eyraud, R., Goudian, N.: Explaining black boxes on sequential data using weighted automata. In: International Conference on Grammatical Inference, pp. 81–103. PMLR (2019)
5. Azcarraga, A., Liu, M.D., Setiono, R.: Keyword extraction using backpropagation neural networks and rule extraction. In: The 2012 International Joint Conference on Neural Networks (IJCNN), pp. 1–7. IEEE (2012)
6. Baesens, B., Setiono, R., De Lille, V., Viaene, S., Vanthienen, J.: Building credit-risk evaluation expert systems using neural network rule extraction and decision tables. In: ICIS 2001 Proceedings, vol. 20 (2001). http://aisel.aisnet.org/icis2001/20
7. Baesens, B., Setiono, R., Mues, C., Vanthienen, J.: Using neural network rule extraction and decision tables for credit-risk evaluation. Manage. Sci. **49**(3), 312–329 (2003)
8. Barakat, N., Diederich, J.: Eclectic rule-extraction from support vector machines. Int. J. Comput. Intell. Appl. **2**(1), 59–62 (2005)
9. Bologna, G.: A study on rule extraction from neural networks applied to medical databases. In: The 4th European Conference on Principles and Practice of Knowledge Discovery (PKDD2000) (2000)
10. Bologna, G., Pellegrini, C.: Three medical examples in neural network rule extraction. Physica Medica **13**, 183–187 (1997)
11. Breiman, L., Friedman, J., Stone, C.J., Olshen, R.A.: Classification and Regression Trees. CRC Press (1984)
12. Calegari, R., Ciatto, G., Dellaluce, J., Omicini, A.: Interpretable narrative explanation for ML predictors with LP: a case study for XAI. In: Bergenti, F., Monica, S. (eds.) WOA 2019–20th Workshop "From Objects to Agents", CEUR Workshop Proceedings, vol. 2404, pp. 105–112. Sun SITE Central Europe, RWTH Aachen University, 26–28 June 2019. http://ceur-ws.org/Vol-2404/paper16.pdf

13. Calegari, R., Ciatto, G., Omicini, A.: On the integration of symbolic and sub-symbolic techniques for XAI: a survey. Intelligenza Artificiale **14**(1), 7–32 (2020). https://doi.org/10.3233/IA-190036
14. Ciatto, G., Calegari, R., Omicini, A., Calvaresi, D.: Towards XMAS: eXplainability through Multi-Agent Systems. In: Savaglio, C., Fortino, G., Ciatto, G., Omicini, A. (eds.) AI&IoT 2019 - Artificial Intelligence and Internet of Things 2019, CEUR Workshop Proceedings, vol. 2502, pp. 40–53. Sun SITE Central Europe, RWTH Aachen University, November 2019
15. Ciatto, G., Calvaresi, D., Schumacher, M.I., Omicini, A.: An abstract framework for agent-based explanations in AI. In: El Fallah Seghrouchni, A., Sukthankar, G., An, B., Yorke-Smith, N. (eds.) 19th International Conference on Autonomous Agents and MultiAgent Systems, pp. 1816–1818. IFAAMAS, May 2020. http://ifaamas.org/Proceedings/aamas2020/pdfs/p1816.pdf
16. Ciatto, G., Schumacher, M.I., Omicini, A., Calvaresi, D.: Agent-based explanations in AI: towards an abstract framework. In: Calvaresi, D., Najjar, A., Winikoff, M., Främling, K. (eds.) EXTRAAMAS 2020. LNCS (LNAI), vol. 12175, pp. 3–20. Springer, Cham (2020). https://doi.org/10.1007/978-3-030-51924-7_1
17. Combined Cycle Power Plant Data Set (2014). https://archive.ics.uci.edu/ml/datasets/Combined+Cycle+Power+Plant. Accessed 19 Jan 2021
18. Craven, M.W., Shavlik, J.W.: Using sampling and queries to extract rules from trained neural networks. In: Machine Learning Proceedings 1994, pp. 37–45. Elsevier (1994). https://doi.org/10.1016/B978-1-55860-335-6.50013-1
19. Craven, M.W., Shavlik, J.W.: Extracting tree-structured representations of trained networks. In: Touretzky, D.S., Mozer, M.C., Hasselmo, M.E. (eds.) Advances in Neural Information Processing Systems 8. Proceedings of the 1995 Conference, pp. 24–30. The MIT Press, June 1996
20. Energy Efficiency Data Set (2012). https://archive.ics.uci.edu/ml/datasets/Energy+efficiency. Accessed 19 Jan 2021
21. Franco, L., Subirats, J.L., Molina, I., Alba, E., Jerez, J.M.: Early breast cancer prognosis prediction and rule extraction using a new constructive neural network algorithm. In: Sandoval, F., Prieto, A., Cabestany, J., Graña, M. (eds.) IWANN 2007. LNCS, vol. 4507, pp. 1004–1011. Springer, Heidelberg (2007). https://doi.org/10.1007/978-3-540-73007-1_121
22. Freitas, A.A.: Comprehensible classification models: a position paper. ACM SIGKDD Explor. Newslett. **15**(1), 1–10 (2014)
23. Gas Turbine CO and NOx Emission Data Set (2019). https://archive.ics.uci.edu/ml/datasets/Gas+Turbine+CO+and+NOx+Emission+Data+Set. Accessed 19 Jan 2021
24. Guidotti, R., Monreale, A., Ruggieri, S., Turini, F., Giannotti, F., Pedreschi, D.: A survey of methods for explaining black box models. ACM Comput. Surv. (CSUR) **51**(5), 1–42 (2018)
25. Hayashi, Y., Setiono, R., Yoshida, K.: A comparison between two neural network rule extraction techniques for the diagnosis of hepatobiliary disorders. Artif. Intell. Med. **20**(3), 205–216 (2000)
26. Hofmann, A., Schmitz, C., Sick, B.: Rule extraction from neural networks for intrusion detection in computer networks. In: SMC 2003 Conference Proceedings. 2003 IEEE International Conference on Systems, Man and Cybernetics. Conference Theme-System Security and Assurance (Cat. No. 03CH37483), vol. 2, pp. 1259–1265. IEEE (2003)

27. Huysmans, J., Baesens, B., Vanthienen, J.: ITER: an algorithm for predictive regression rule extraction. In: Tjoa, A.M., Trujillo, J. (eds.) DaWaK 2006. LNCS, vol. 4081, pp. 270–279. Springer, Heidelberg (2006). https://doi.org/10.1007/11823728_26
28. Huysmans, J., Dejaeger, K., Mues, C., Vanthienen, J., Baesens, B.: An empirical evaluation of the comprehensibility of decision table, tree and rule based predictive models. Decis. Support Syst. **51**(1), 141–154 (2011)
29. Johansson, U., König, R., Niklasson, L.: Rule extraction from trained neural networks using genetic programming. In: 13th International Conference on Artificial Neural Networks, pp. 13–16 (2003)
30. Lipton, Z.C.: The mythos of model interpretability. Queue **16**(3), 31–57 (2018). https://doi.org/10.1145/3236386.3241340
31. Martens, D., Baesens, B., Van Gestel, T., Vanthienen, J.: Comprehensible credit scoring models using rule extraction from support vector machines. Eur. J. Oper. Res. **183**(3), 1466–1476 (2007)
32. Murphy, P.M., Pazzani, M.J.: Id2-of-3: Constructive induction of m-of-n concepts for discriminators in decision trees. In: Machine Learning Proceedings 1991, pp. 183–187. Elsevier (1991)
33. Quinlan, J.R.: Simplifying decision trees. Int. J. Man Mach. Stud. **27**(3), 221–234 (1987). https://doi.org/10.1016/S0020-7373(87)80053-6
34. Quinlan, J.R.: C4.5: Programming for Machine Learning. Morgan Kauffmann 38, 48 (1993)
35. Rocha, A., Papa, J.P., Meira, L.A.: How far do we get using machine learning black-boxes? Int. J. Pattern Recogn. Artif. Intell. **26**(02), 1261001-(1–23) (2012). https://doi.org/10.1142/S0218001412610010
36. Rudin, C.: Stop explaining black box machine learning models for high stakes decisions and use interpretable models instead. Nat. Mach. Intell. **1**(5), 206–215 (2019). https://doi.org/10.1038/s42256-019-0048-x
37. Saito, K., Nakano, R.: Extracting regression rules from neural networks. Neural Netw. **15**(10), 1279–1288 (2002). https://doi.org/10.1016/S0893-6080(02)00089-8
38. Schmitz, G.P., Aldrich, C., Gouws, F.S.: ANN-DT: an algorithm for extraction of decision trees from artificial neural networks. IEEE Trans. Neural Networks **10**(6), 1392–1401 (1999). https://doi.org/10.1109/72.809084
39. Setiono, R., Baesens, B., Mues, C.: Rule extraction from minimal neural networks for credit card screening. Int. J. Neural Syst. **21**(04), 265–276 (2011). https://doi.org/10.1142/S0129065711002821
40. Setiono, R., Leow, W.K., Zurada, J.M.: Extraction of rules from artificial neural networks for nonlinear regression. IEEE Trans. Neural Networks **13**(3), 564–577 (2002). https://doi.org/10.1109/TNN.2002.1000125
41. Steiner, M.T.A., Neto, P.J.S., Soma, N.Y., Shimizu, T., Nievola, J.: Using neural network rule extraction for credit-risk evaluation. Int. J. Comput. Sci. Network Secur. **6**(5), 6–16 (2006)
42. Towell, G.G., Shavlik, J.W.: Extracting refined rules from knowledge-based neural networks. Mach. Learn. **13**(1), 71–101 (1993). https://doi.org/10.1007/BF00993103
43. Wine Quality Data Set (2009). https://archive.ics.uci.edu/ml/datasets/Wine+Quality. Accessed 19 Jan 2021

44. Zhuang, J., Dvornek, N.C., Li, X., Yang, J., Duncan, J.: Decision explanation and feature importance for invertible networks. In: 2019 IEEE/CVF International Conference on Computer Vision Workshop (ICCVW), pp. 4235–4239. IEEE (2019)
45. Zien, A., Krämer, N., Sonnenburg, S., Rätsch, G.: The feature importance ranking measure. In: Buntine, W., Grobelnik, M., Mladenić, D., Shawe-Taylor, J. (eds.) ECML PKDD 2009. LNCS (LNAI), vol. 5782, pp. 694–709. Springer, Heidelberg (2009). https://doi.org/10.1007/978-3-642-04174-7_45

Comparison of Contextual Importance and Utility with LIME and Shapley Values

Kary Främling[1,2(✉)], Marcus Westberg[1], Martin Jullum[3], Manik Madhikermi[1], and Avleen Malhi[2,4]

[1] Department of Computing Science, Umeå University, Umeå, Sweden
{Kary.framling,marcus.westberg,manik.madhikermi}@umu.se,
Kary.framling@cs.umu.se

[2] Department of Computer Science, Aalto University, Espoo, Finland
{kary.framling,avleen.malhi}@aalto.fi

[3] Norwegian Computing Center, Gaustadalleen 23a, 0373 Oslo, Norway
jullum@nr.no

[4] Department of Computing and Informatics, Bournemouth University, Poole, UK
amalhi@bournemouth.ac.uk

Abstract. Different explainable AI (XAI) methods are based on different notions of 'ground truth'. In order to trust explanations of AI systems, the ground truth has to provide fidelity towards the actual behaviour of the AI system. An explanation that has poor fidelity towards the AI system's actual behaviour can not be trusted no matter how convincing the explanations appear to be for the users. The Contextual Importance and Utility (CIU) method differs from currently popular outcome explanation methods such as Local Interpretable Model-agnostic Explanations (LIME) and Shapley values in several ways. Notably, CIU does not build any intermediate interpretable model like LIME, and it does not make any assumption regarding linearity or additivity of the feature importance. CIU also introduces the value utility notion and a definition of feature importance that is different from LIME and Shapley values. We argue that LIME and Shapley values actually estimate 'influence' (rather than 'importance'), which combines importance and utility. The paper compares the three methods in terms of validity of their ground truth assumption and fidelity towards the underlying model through a series of benchmark tasks. The results confirm that LIME results tend not to be coherent nor stable. CIU and Shapley values give rather similar results when limiting explanations to 'influence'. However, by separating 'importance' and 'utility' elements, CIU can provide more expressive and flexible explanations than LIME and Shapley values.

Keywords: Explainable AI · Contextual Importance and Utility · Outcome explanation · Post hoc explanation

The work is partially supported by the Wallenberg AI, Autonomous Systems and Software Program (WASP) funded by the Knut and Alice Wallenberg Foundation.

D. Calvaresi et al. (Eds.): EXTRAAMAS 2021, LNAI 12688, pp. 39–54, 2021.
https://doi.org/10.1007/978-3-030-82017-6_3

1 Introduction

The need for explainability in Artificial Intelligence (AI) has been understood since the very beginnings of AI, as seen for instance in MYCIN [23]. Even though the term Explainable AI (XAI) is quite recent, AI explainability was a very active domain during the 1990's when a list of five general desiderata for any explanation was identified in [24], which were *Fidelity*, *Understandability*, *Sufficiency*, *Low Construction Overhead*, and *Efficiency*. XAI research in the 1990's can be considered to have focused on so-called *intrinsic interpretability* or *interpretable model extraction* [7], i.e. extract rules or other interpretable forms of knowledge from a complex black box model and then use that representation as an explanation. One exception to that trend was the Contextual Importance and Utility (CIU) method for outcome explanation, first presented in [12] and explained in detail in [9]. However, CIU seems to have passed unnoticed by the XAI community because the first paper on CIU since 1996 wasn't published until 2019. Using modern terms of XAI, CIU can be classified as a model-agnostic outcome explanation method. The first objective of this paper is to provide a comparison of CIU with two of the most popular model-agnostic outcome explanation methods available, i.e. Shapley values [16,22] and LIME [21]. The category of use cases and data sets considered in this paper is tabular data only.

CIU's mathematical foundation and underlying philosophy are different from those of Shapley values and LIME. Notably, CIU is not an additive feature attribution method. Furthermore, CIU estimates *Contextual Importance (CI)* and *Contextual Utility (CU)* instead of estimating feature 'influence' like most (or all) comparable methods. However, 'influence' can be calculated directly from CI and CU values, which simplifies the comparison with influence-based methods, such as LIME and Shapley values. The second objective of the paper is to study to what extent the explanations produced by the studied XAI methods provide fidelity towards the true behaviour of the model.

The next section provides a background and definitions used in the paper, as well as an overview of Shapley values and LIME methods. Section 3 describes CIU and its use in this paper. Section 4 shows experimental results and comparisons between the three methods, followed by Conclusion.

2 Background and Definitions

The outcome explanation concept may be divided into two separate settings based on the aim of the explanation task. The first setting seeks explanations of how (each of) the input features influences the outcome solely through the given prediction model. This setting is most relevant when trying to understand the behaviour of the prediction model in itself. The second setting also accounts for the dependence between the input features, and may therefore assign high importance to an input feature that has a minor direct impact on the output through the prediction formula, if the feature is highly correlated with one or more features that *do* have such a high direct impact. This is most relevant when the actual real behaviour of the modelled output is of interest. Leaning on the fidelity criterion described below, we concentrate on the former setting here.

Going forward, it is important that we look at our definitions of the terms 'fidelity' and 'ground truth': When we refer to 'fidelity', what we mean is how accurately the explanation remains faithful/truthful to the underlying black-box model in its representation thereof. This follows similar definitions in [5,19]. Following on that, the 'ground truth' of a model is the actual observed behaviour of that model. Concentrating solely on the model itself, it is generally admitted in the XAI domain that the actual input versus output behaviour of the underlying model is the so called ground truth against which the fidelity of an explanation should be assessed [4,26]. The LIME (Local Interpretable Model-Agnostic Explanations) method [21], for instance, calculates to what extent the generated interpretable linear model gives similar results to the original black box model. That is called the Explanation Fit and is the R^2 error between the linear model and the actual model. Shapley values do not have a proper intermediate model where an Explanation Fit makes sense. However, one may interpret the additive Shapley value explanation as a model which is linear in the set of indicator variables defined as whether each of the input features are observed or not.

In human-to-human communication, an explanation lacking fidelity towards the real underlying model is usually considered to be a lie although it can appear convincing to the explainee. When developing and comparing XAI methods, the fidelity of the provided explanation in regard to the underlying model should be the first and foremost assessment criterion. An explanation lacking in fidelity might be considered easier to understand and accept than a true explanation, as depicted in some human surveys for assessing the goodness of different methods. However, a false explanation or lie that looks or sounds convincing should not lead to consideration that the underlying XAI method is better.

2.1 Core Definitions

The two fundamental concepts of CIU are 'importance' and 'utility' as explained in this section. Their origin is in Multi-Attribute Utility Theory [25], as explained also in [10]. An 'influence' concept can be calculated from 'importance' and 'utility' but it is not a core CIU concept and is here used mainly to simplify comparisons with other methods. In our usage of the terms 'influence', 'importance' and 'utility', the 'importance' of 'something' (such as an input feature of an AI model) denotes the significance of that 'something' but does NOT express adjoining positive or negative judgements. Something like 'good importance', 'bad importance', 'typical importance', etc., are not accurately represented by importance alone. Instead, adjectives such as 'good', 'bad', 'typical', 'favorable', etc., express judgments of the *utility* of feature *values* for the situation or context at hand, as provided by a *utility function* that expresses the *value utility* of both output and input values. LIME and Shapley values do not have a utility concept and typically use the term 'importance' for what we here call 'influence'. Even

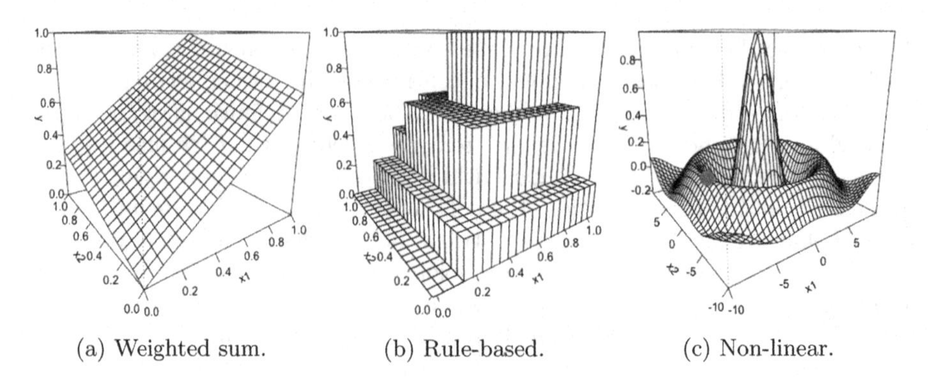

(a) Weighted sum. (b) Rule-based. (c) Non-linear.

Fig. 1. Examples of linear, rule-based and non-linear models.

though [16] also uses the term 'importance', more recent Shapley values literature tends to also use the term feature *influence* [1]. [17] uses the term 'effect' in the same sense as we use 'influence' here.

The influence of a feature will depend on the feature's importance as well as on the utility of the current feature value. A feature with high importance and a good value utility will have great positive influence on the result. A feature with high importance and a bad value utility will have great negative influence on the result. A feature with zero importance for the output will have zero influence on the result, no matter what value utility it has. As such, our definitions for the core concepts of CIU look like this:

- **Importance**: The feature importance *in a particular context* of a factor impacting a particular decision.
- **Utility**: How well the values of the features in the same context match outcome expectations. This follows from the definition of *utility function* in decision theory, where it is a numerical representation of preference/desirability orderings [25].
- **Influence**: A combination value of utility and importance, representing the positive or negative impact of a factor on a particular decision, typically relative to some 'baseline' [6].

In the function $y = b(x) = 0.3x_1 + 0.7x_2$, represented by Fig. 1a, the influence of the x_1 term is $0.3x_1$ and the influence of the x_2 term is $0.7x_2$ if we use a zero baseline. The function could also be expressed in the more generic form $y = w_1 \times x_1 + w_2 \times x_2$. The weights (importances) w_1 and w_2 are in this case 0.3 and 0.7. The utility function in this case is unity because the utility or 'goodness' of an input value is directly the input value itself. For instance, if x_1 and x_2 have different value ranges, then it becomes necessary to apply a utility function to them. If the input value range would be $[0, 5]$ rather than $[0, 1]$, then a utility function $u_i(x_i) = x_i/5$ would be appropriate. For generic black box models, the utility function can be an arbitrarily complex, non-linear function.

Interpretability becomes more challenging when dealing with step-wise functions such as the one illustrated in Fig. 1b, which corresponds to the kind of functions produced by rules and decision trees. Model-agnostic methods like the ones studied in this paper can also deal with such models. However, the significance of the concepts influence, importance and utility becomes more complicated than for the linear case. When dealing with non-linear models such as that in Fig. 1c ($y = \sin(\sqrt{x_1^2 + x_2^2})/\sqrt{x_1^2 + x_2^2}$), the 'influence' concept alone becomes increasingly challenging to use. Thus, we argue for the contextual 'importance' and 'utility' concepts used in CIU.

2.2 LIME

Ribeiro et al. [21] in their research work proposed a method called Local Interpretable Model agnostic Explanations (LIME) for explanation of an individual prediction $b(x)$ made by a black box machine learning model. The approach used to explain the individual predictions can be detailed as: In sampling step (1), a set of normally distributed instances X_{sx} is drawn having same mean and standard deviation as the original feature space of X, which is done independently of the instance x to be explained. For the labels $Y_{sx} = b(X_{sx})$, LIME works with the prediction returned by the model b. In surrogate fitting step (2), the LIME surrogate is trained to locally approximate the decision boundary of the black-box model. The standard version (Linear LIME) uses linear regression with regularization to do this. The local surrogate model centered on x is fitted by having each instance of X_{sx} associated with a weight calculated using an RBF kernel by default, i.e. higher importance will be assigned to instances closer to x during the training [15]. In the last explanation step (3), the explanations for the prediction $b(x)$ are generated by using the trained surrogate s_x's linear regression coefficients. Choosing an adequate and representative sampling strategy for generating the instances to fit the surrogate model has a major impact on the quality of the local approximation of the black-box model and thus on the accuracy of the generated explanation [14]. In particular, the effect of locally important features can be hidden by globally important ones.

LIME's ground truth could be summarized as follows: Find a linear regression function that locally approximates the tangent plane of the underlying model as well as possible for the current instance.

LIME's fidelity towards the LIME ground truth is assessed based on how well the linear regression corresponds to the actual behaviour of the model, which LIME calls the 'explanation fit' and is an R^2 value calculated on the difference between the actual model output and the output given by LIME's linear regression function.

The LIME experiments of this paper have been executed using the R-package `lime`, version 0.5.1 [20].

2.3 Shapley Values

Shapley value is a concept originating from cooperative game theory [22]. The concept was picked up by the XAI community and became popular for producing outcome explanations following [16], and the introduction of SHAP (SHapley

Additive exPlanations). The method distributes the difference between the prediction output and the global mean prediction, additively on the input features according to a formula which is consistent with a set of four theoretical properties. A key ingredient in the Shapley values methodology is the contribution function $v(S)$, measuring expected output $b(x)$ when only a subset S of the input features were available (x_S). Motivated by the fidelity criterion, we have used the so-called interventional conditional expectation, as in [16]. Other choices may be more appropriate in other explanations settings, as explained e.g. in [6]. In the case of interventional conditional expectation, the Shapley value for feature i is a weighted mean over $v(S+i) - v(S)$ for all subsets S, and therefore measures the influence that the act of observing feature i has on the predicted output, with or without each of the other features observed. This allows the Shapley value for a feature to be compared with other features within the individual/instance and also with the same feature for other individuals/instances. A significant drawback with Shapley values is that it is computationally costly when there are many input features. Explanation through Shapley values also requires the availability of the training set, which may not always be easily accessible.

Shapley values' ground truth could be summarized as follows: Distribute the difference between the current and expected (e.g. the global mean prediction) output value to the input features according to a 'fairness estimation' about how much each feature attributed to the output in a positive or negative way.

The fidelity of Shapley values towards the Shapley value ground truth can be guaranteed by a sufficiently great sampling of all value combinations. The main challenge is that the number of such combinations grows exponentially with the number of input features.

The experiments of this paper have been executed using the `iml` (Interpretable Machine Learning) R-package, version 0.10.1 [18].

3 Contextual Importance and Utility (CIU)

A formal presentation of CIU can be found in [11]. In this paper, we will explain the principles of CIU using the so-called 'sombrero' function in Fig. 1c as an example. The studied instance or Context $\vec{C}$ is indicated by the red dot in Fig. 1c and corresponds to the input values $(x_1, x_2) = (-7.5, -1.5)$. Figure 2 shows how the output value y changes as a function of x_1 and x_2 when keeping the other input at the $\vec{C}$ value. The range of possible input values is here $[-10, 10]$ for both x_1 and x_2. In Fig. 2, five values are indicated:

- *absmin*, *absmax*: The minimal and maximal values that the output y can get. In classification tasks these values are typically zero and one for all outputs.
- *Cmin*, *Cmax*: The minimal and maximal values that the output y can take by changing the value of the studied input.
- *out*: The value of output y for the studied instance, i.e. with input values $\vec{C}$.

In Fig. 2, $absmin = -0.217$, $absmax = 1$ and $out = 0.128$. For the x_1 input $Cmin = -0.217$ and $Cmax = 0.664$. The contextual importance CI expresses

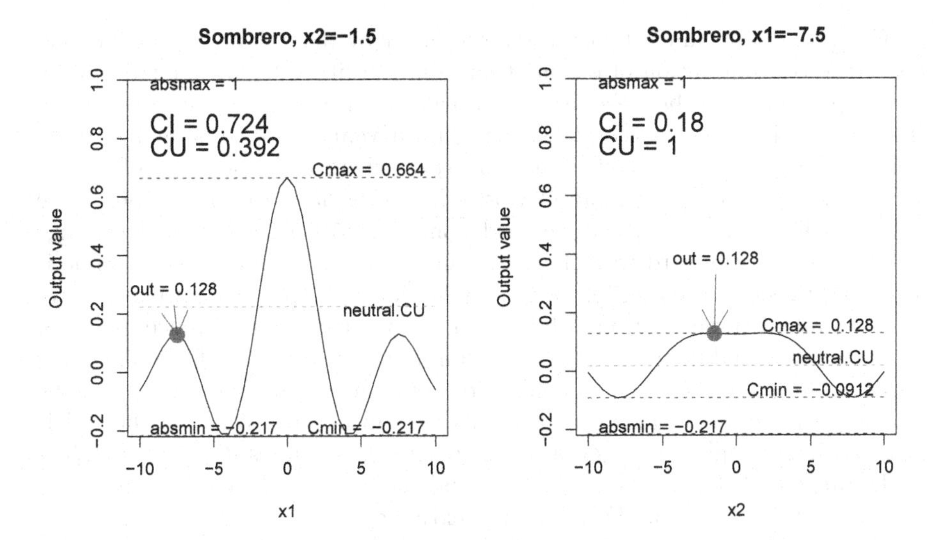

Fig. 2. CIU for sombrero function.

to what extent an input can modify the output value, which leads us to the following definition

$$CI_j(\vec{C}, \{i\}) = \frac{Cmax_j(\vec{C}, \{i\}) - Cmin_j(\vec{C}, \{i\})}{absmax_j - absmin_j}, \tag{1}$$

where the different variables have the same meaning as before but with appropriate indices as follows:

- $\{i\}$ defines the indices of inputs $\vec{x}$ for which CIU is calculated.
- j is the index of the studied output.

For input x_1 in Fig. 2, this gives $CI_{x_1} = \frac{0.664-(-0.217)}{1-(-0.217)} = 0.724$ and $CI_{x_2} = \frac{0.128-(-0.0912)}{1-(-0.217)} = 0.18$ for input x_2. Therefore, x_1 is about four times as important as x_2 in the studied context $\vec{C}$.

CU expresses to what extent the current feature value(s) contribute to a high-utility output value, i.e. what is the utility of the input value for achieving an output value that has a high utility. CU is expressed as

$$CU_j(\vec{C}, \{i\}) = \frac{y_j(\vec{C}) - Cmin_j(\vec{C}, \{i\})}{Cmax_j(\vec{C}, \{i\}) - Cmin_j(\vec{C}, \{i\})}, \tag{2}$$

where $y_j = b(\vec{C})$ corresponds to the *out* value in the example. For input x_1 in Fig. 2, this gives $CU_{x_1} = \frac{0.128-(-0.217)}{0.664-(-0.217)} = 0.392$ and $CU_{x_2} = \frac{0.128-(-0.0912)}{0.128-(-0.0912)} = 1$ for input x_2. Therefore, x_1 has a less than average favorable value, whereas x_2 has the most favorable value possible in the context $\vec{C}$.

CI and CU are limited to the interval $[0, 1]$ by definition. In classification tasks, the transformation of output values into utility values is trivial because the output value can be considered to already be a probability/utility value in the range $[0, 1]$. For regression tasks, the output values need to be mapped into utility values through a utility function $u(y_j)$, where y_j is the value of output j. For instance, in the well-known Boston Housing data set, the output value is the median value of owner-occupied homes in $1000's and is in the range $[5, 50]$. A straightforward way of transforming that value into a utility value is an affine transformation $[5, 50] \mapsto [0, 1]$, assuming that the preference is to have a higher value. However, from a buyer's point of view, the preference might be for lower prices and then the transformation would rather be $[50, 5] \mapsto [0, 1]$. It is important to point out that the definitions of CI and CU in Eqs. 1 and 2 assume that $u(y_j)$ is an affine transformation of the form $u(y_j) = Ay_j + b$ where A is positive. In principle, $u(y_j)$ could have any shape as long as it produces values in the range $[0, 1]$ but that case goes beyond the scope of the current paper.

In the original work by Främling [9], textual explanations were generated by quantifying CI and CU values according to intervals such as $very\_important = [0.9, 1]$ for CI and $very\_good = [0.9, 1]$ for CU. In this paper, CIU explanations are provided by bar plot explanations for simplifying comparisons with LIME and Shapley values. The only subjective parameter in that case is the choice of what CU value is considered 'neutral'. We call that parameter $neutral.CU$ here and it provides a 'baseline' for influence-based explanations using CIU. In Sect. 4, $CU = 0$ corresponds to red, $CU = 0.5$ is 'neutral' and corresponds to yellow, and $CU = 1$ corresponds to dark green, as illustrated by the colours of $Cmin$, $Cmax$ and $neutral.CU$ in Fig. 2.

In order to make the difference between the concepts 'influence', 'importance' and 'utility' more explicit, we here provide a definition of *contextual influence*. Such a 'contextual influence' concept makes it possible to compare directly with the influence value ϕ of LIME and Shapley values, which is the reason why we use the symbol ϕ in Eq. 4. However, using 'influence' makes explanations less expressive and less understandable than when using CI and CU, as illustrated in Sect. 4. We begin by defining contextual influence according to:

$$Cinfluence_j(\vec{C}, \{i\}) = CI_j(\vec{C}, \{i\}) \times CU_j(\vec{C}, \{i\}) \quad (3)$$

Since $Cinfluence$ is relative, it can be freely scaled into any desired range $[rmin, rmax]$. Such a 'scaled contextual influence' can be defined as follows:

$$\phi = (rmax - rmin) \times CI \times (CU - neutral.CU) \quad (4)$$

where '$_j(\vec{C}, \{i\})$' has been omitted from all three terms ϕ, CI, and CU for easier readability. For comparison with Shapley values and LIME, we use $[rmin, rmax] = [-1, 1]$ in Sect. 4. Setting $neutral.CU = 0.5$ also makes it possible to restrict ϕ values to only negative, zero or positive, as for Shapley values and LIME.

A formal study of the relationship between CU, CI and ϕ is out of the scope for the current paper. Other aspects of CIU that are not in the scope of this paper is how *Cmin* and *Cmax* are estimated. The sampling method used in this paper is described in [13]. Främling also introduced so-called *intermediate concepts* in [8,9], which use the fact that CI and CU can be estimated for any joint combination of input features, i.e. the set $\{i\}$ in Eqs. 1 and 2 can contain any number of inputs, from one to all inputs. However, LIME and Shapley values do not have any intermediate concepts so it is not possible to perform a comparison with them, which is the main reason for not including intermediate concepts in this paper.

CIU's ground truth could be summarized as follows: Estimate how much the output can change when modifying the values of one or more input features, on a scale of 0–100% (Contextual Importance). Provide an estimate of how favorable the current value(s) are towards a high-utility output value, as compared to all possible values for the studied input features on a scale 0-100% (Contextual Utility). The fidelity of CIU towards its ground truth depends only on how accurately *Cmin* and *Cmax* values can be estimated.

The CIU experiments in this paper have been executed using the `ciu` R-package, version 0.1.0 [13] and using the latest version at https://github.com/KaryFramling/ciu for 'influence' plots.

4 Experiments

The data sets to be used for assessing the different methods have been selected so that discrete and continuous values are used as inputs and that both classification and regression tasks are taken into consideration. The results of the methods are evaluated mainly using two assessment criteria (AC):

- *AC1* Is the explanation rational and in line with the output value? For a high-utility output value, the total influence of features is expected to be highly positive, and vice versa for a low-utility output value.
- *AC2* Does the explanation correspond to the actual observed behaviour of the model?

4.1 Classification with Continuous Inputs

The Iris data set has been chosen for this category mainly because the limits between the different classes require highly non-linear models for correctly estimating the probability of the three classes for each studied instance. Figure 3 shows CIU, Shapley values and LIME explanations generated for instance number 100 with a random forest model for the Iris data set. Any instance from the data set could be used but for Iris flowers the classes 'versicolor' and 'virginica' are usually the most interesting ones because they are more similar to each other than to 'setosa'. Instance number 100 is a 'versicolor'.

CIU with influence and Shapley values give almost identical results here. The output value is 'one' for the versicolor output and 'zero' for the two other classes, which is well represented also in the explanations so AC1 is fulfilled for CIU and

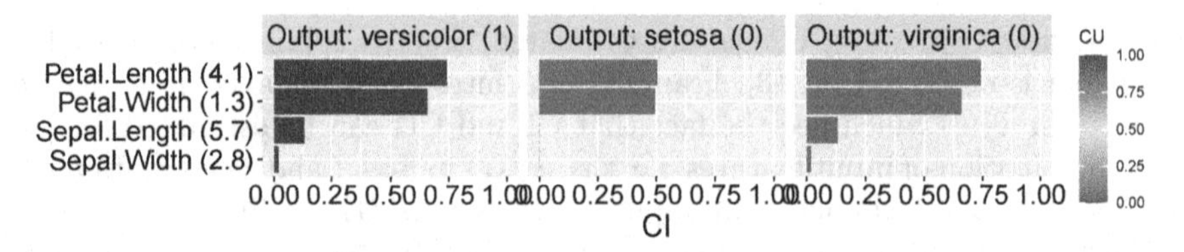

(a) CIU. Bar length shows CI, bar color shows CU according to palette on the right.

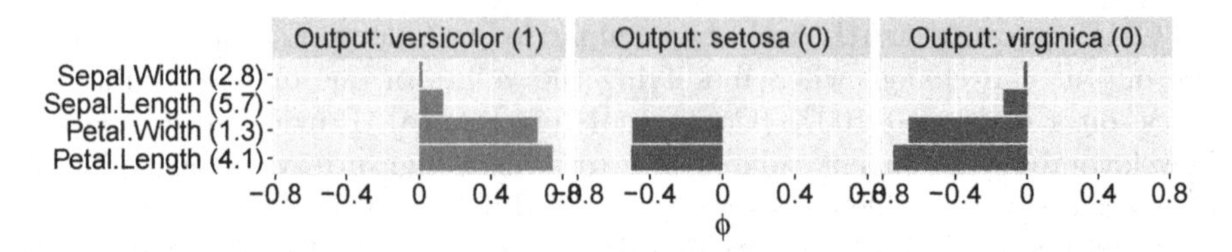

(b) CIU with influence. Positive influence is shown in blue, negative in red.

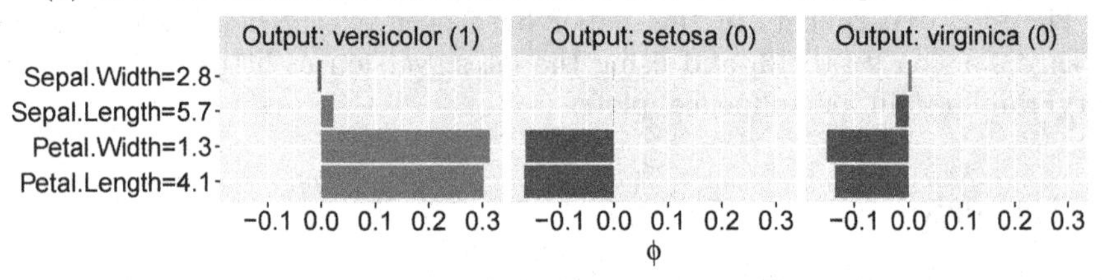

(c) Shapley values. Positive influence is shown in blue, negative in red.

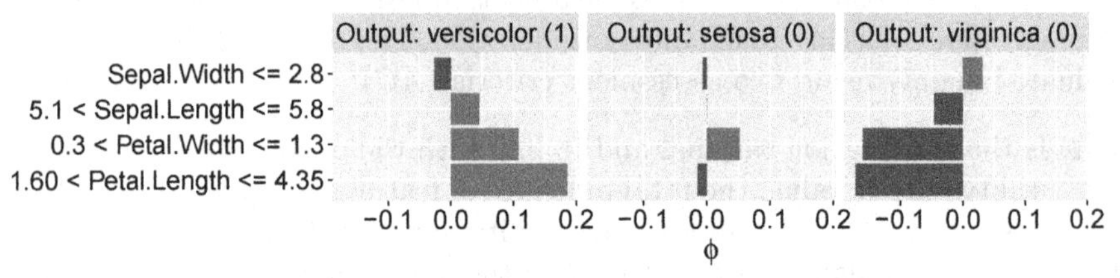

(d) LIME. Positive influence is shown in blue, negative in red.

Fig. 3. Explanations with four methods for instance #100 of Iris data set. Bar length shows CI/ϕ value. CU value determines bar color in CIU plot. In influence plots, negative influence is shown in red and positive influence is shown in blue. (Color figure online)

Shapley values. LIME results differ significantly from CIU and Shapley values for the setosa class, where *Petal Width* has a significant positive influence that is not in line with the output value 'zero'. LIME results also tend to change from one run to the other. Therefore, LIME fails against AC1 for setosa explanation. It is also interesting to note that the 'Explanation fit' indicated by LIME is very low, i.e. < 0.1 for all three classes. Regarding AC2, Fig. 4 shows the 'CIU ground truth' for the *Petal Length* feature. CIU values can be deduced directly

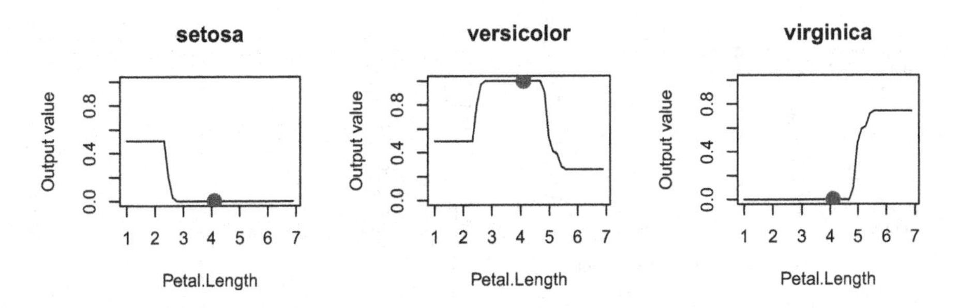

Fig. 4. Output values as a function of 'Petal Length' for the three Iris classes.

from the figure and therefore fulfill AC2. Both Shapley values and LIME can also be considered to fulfill AC2 for *Petal Length*.

4.2 Regression with Continuous Inputs

The Boston Housing data has one continuous-valued output and only continuous-valued inputs. It is a regression task for which a Gradient Boosting Machine model is used here. Figure 5 shows CIU, Shapley values and LIME results for instance #370 of the data set. CIU and Shapley values again obtain quite similar results. Instance #370 has almost the highest possible value (49), which signifies that most input features should have a positive influence (but it could be any other instance too). The influence is here positive for most features with all methods, even though a little bit less so for LIME than for the others. Therefore, all methods satisfy AC1.

Regarding AC2, Fig. 6 shows the 'CIU ground truth' for three input features. Again, CIU values can be deduced directly from these figures and therefore fulfill AC2, which is true also for Shapley values. For LIME, however, the dummy variable 'Charles River' (*chas*) is indicated as the most important one, which is a clear error. For the *lstat* feature, LIME only puts it third. For the *rm* feature, LIME gives a high positive influence (after *chas*), even though it is clear that 6.7 is only an average value for instance #370. Finally, LIME shows strong negative influence for the criminality rate (*crim*), even though the value 5.7 is actually good. Hence, LIME fails to satisfy AC2.

4.3 Classification with Mixed Discrete and Continuous Inputs

The Titanic data set is a frequently used benchmark for machine learning methods. It has two output classes, i.e. survives or not. There are both discrete and continuous-valued input features, which makes it interesting also for this paper. A random forest model was used. The studied instance is an 8-year old boy. The corresponding feature values are shown by the red dots in Fig. 8. With output probabilities of 0.61 for *survives* and 0.39 for *doesn't survive*, it could be expected that there's dominantly positive influence for *survives* and dominantly negative

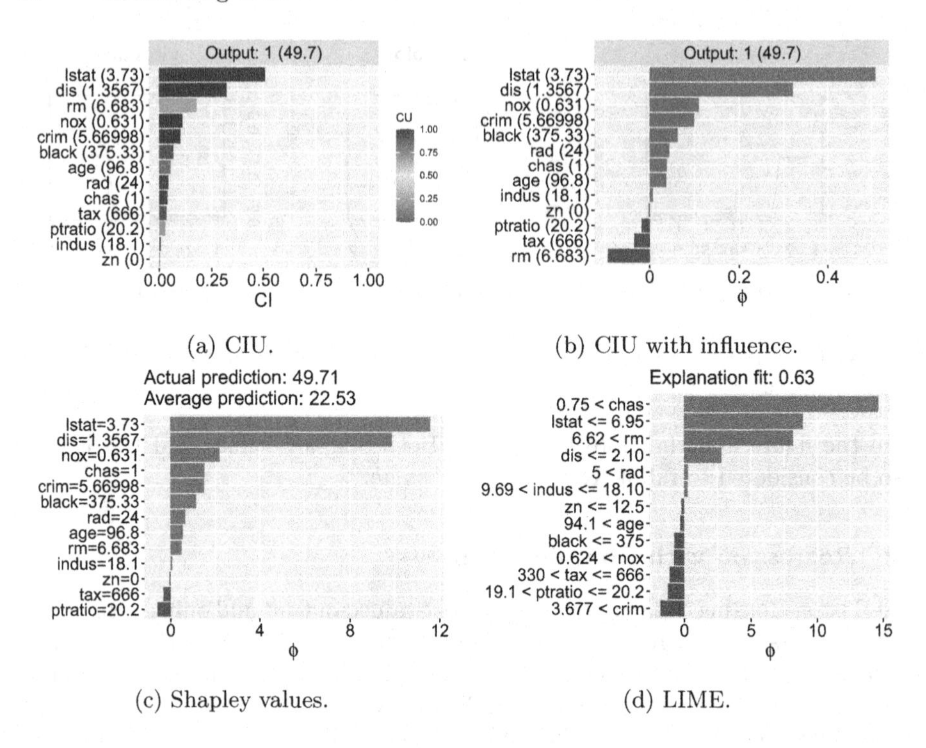

(a) CIU.

(b) CIU with influence.

(c) Shapley values.

(d) LIME.

Fig. 5. Explanations for instance #370 of Boston Housing data set. Bar length shows CI/ϕ value. CU value determines bar color in CIU plot. In influence plots, negative influence is shown in red and positive influence is shown in blue. (Color figure online)

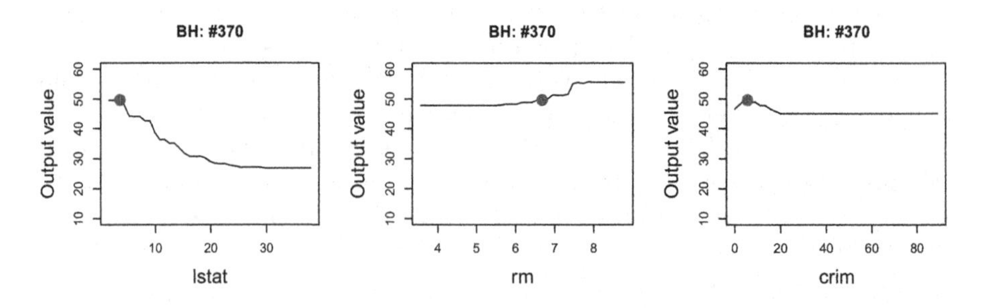

Fig. 6. Boston Housing output value as a function of input value for features 'lstat', 'rm' and 'crim'.

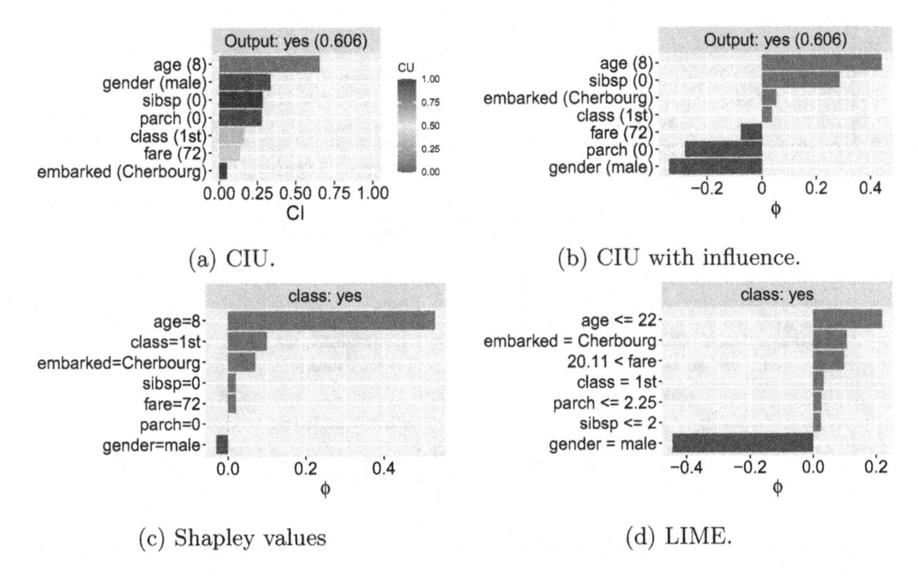

(a) CIU.

(b) CIU with influence.

(c) Shapley values

(d) LIME.

Fig. 7. Bar chart explanations for example person on Titanic. Only explanations for 'survives' have been included. Bar length shows CI/ϕ value. CU value determines bar color in CIU plot. In influence plots, negative influence is shown in red and positive influence is shown in blue. (Color figure online)

influence for *doesn't survive*. This is indeed the case for CIU, whereas Shapley values has almost only positive influence for *survives* and therefore almost only negative influence for *doesn't survive*. It can therefore be questioned whether Shapley values satisfies AC1 here. When studying the effect of input feature values on the probability for *survives*, it seems like the influence of *age* is by far over-estimated by Shapley values in this case, which signifies that the Shapley values explanation does not correspond to the true behaviour of the model. Therefore Shapley values does not satisfy AC2 in this case (Fig. 7).

The LIME explanation again differs from the two others, where 'male' is indicated as the input feature that clearly has the greatest influence. LIME results change slightly at every run and sometimes 'parch' gets a negative influence, which is in line with the results of the other methods.

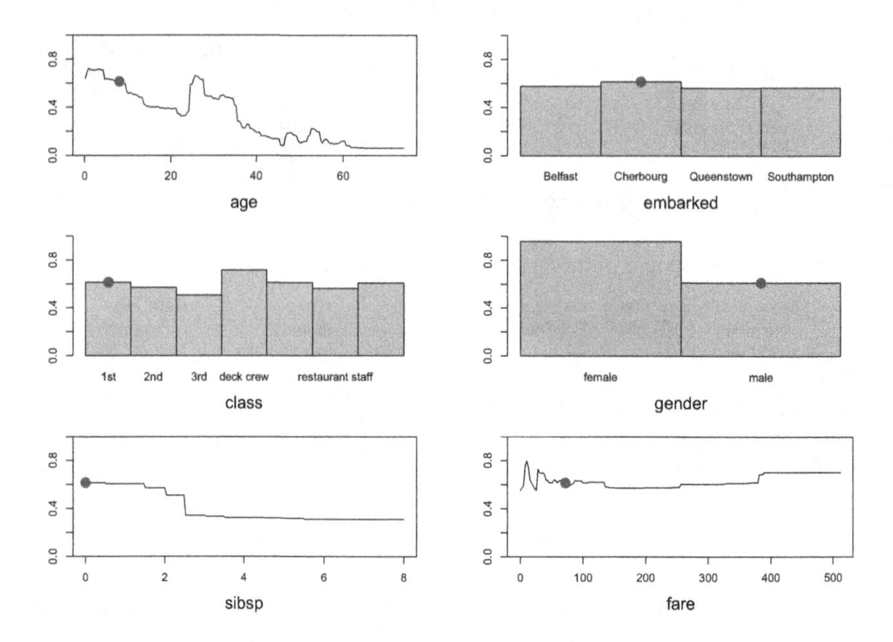

Fig. 8. Probability of survival for selected person in Titanic as a function of selected inputs.

5 Conclusion

As seen from the results in this paper, CIU provides a new alternative to LIME and Shapley values. The results confirm results of earlier research that LIME explanations tend to be less rational and provide a poor fidelity with the underlying model [2,3]. CIU and Shapley values provide quite similar results for two of the studied use cases, which can be considered to be a comforting result for both methods. However, the 'ground truth' of Shapley values and CIU differ significantly and further empirical and theoretical studies regarding these differences and their effects would be important for the XAI community as a whole. The core conclusions of the paper are the following:

1. By considering 'importance' and 'utility' as different parts of an explanation, CIU can provide more versatile explanations than LIME and Shapley values.
2. Both 'importance' and 'utility' are absolute values in the range $[0, 1]$, whereas 'influence' is a relative value that only expresses how influent different input features are compared to each other.
3. CIU is not a black box itself because CI and CU values can be 'read out' by humans from input-output graphs at least for one input feature.
4. CIU does not need access to the training data. CIU can be applied to any model f, no matter if f has been produced by machine learning or not.

CIU is intuitively a more light-weight method than Shapley values because it only modifies the values of one input feature at a time, therefore requiring a smaller number of samples. However, the number of samples remains a compromise with the estimation accuracy, which makes it difficult to properly compare calculation overhead between the methods. Furthermore, calculation speed also depends on how the method has been implemented, not only on the method itself. Therefore, such a study is left as a topic of future work.

References

1. Aas, K., Jullum, M., Løland, A.: Explaining individual predictions when features are dependent: More accurate approximations to shapley values. arXiv preprint arXiv:1903.10464 (2019)
2. Adebayo, J., Gilmer, J., Goodfellow, I., Kim, B.: Local explanation methods for deep neural networks lack sensitivity to parameter values. arXiv preprint arXiv:1810.03307 (2018)
3. Alvarez-Melis, D., Jaakkola, T.S.: On the robustness of interpretability methods. arXiv preprint arXiv:1806.08049 (2018)
4. Amiri, S.S., et al.: Data representing ground-truth explanations to evaluate XAI methods. arXiv preprint arXiv:2011.09892 (2020)
5. Barbado, A., Corcho, O.: Explanation generation for anomaly detection models applied to the fuel consumption of a vehicle fleet (2020)
6. Chen, H., Janizek, J.D., Lundberg, S., Lee, S.I.: True to the model or true to the data? arXiv preprint arXiv:2006.16234 (2020)
7. Du, M., Liu, N., Hu, X.: Techniques for interpretable machine learning. Commun. ACM **63**(1), 68–77 (2020). https://doi.org/10.1145/3359786
8. Främling, K.: Explaining results of neural networks by contextual importance and utility. In: Proceedings of the AISB 1996 Conference, Brighton, UK, 1–2 April 1996
9. Främling, K.: Modélisation et apprentissage des préférences par réseaux de neurones pour l'aide à la décision multicritère. Ph.D. thesis, INSA de Lyon, March 1996
10. Främling, K.: Decision theory meets explainable AI. In: Calvaresi, D., Najjar, A., Winikoff, M., Främling, K. (eds.) EXTRAAMAS 2020. LNCS (LNAI), vol. 12175, pp. 57–74. Springer, Cham (2020). https://doi.org/10.1007/978-3-030-51924-7_4
11. Främling, K.: Explainable AI without interpretable model. arXiv preprint arXiv:2009.13996 (2020). https://arxiv.org/abs/2009.13996
12. Främling, K., Graillot, D.: Extracting explanations from neural networks. In: ICANN 1995 Conference, Paris, France, October 1995
13. Främling, K.: Contextual importance and utility in R: the 'CIU' package. In: Proceedings of 1st Workshop on Explainable Agency in Artificial Intelligence, at 35th AAAI Conference on Artificial Intelligence, pp. 110–114 (2021)
14. Gruber, S., Kopper, P.: Introduction to local interpretable model-agnostic explanations (LIME). https://compstat-lmu.github.io/iml_methods_limitations/lime.html (2020)
15. Laugel, T., Renard, X., Lesot, M.J., Marsala, C., Detyniecki, M.: Defining locality for surrogates in post-hoc interpretablity. arXiv preprint arXiv:1806.07498 (2018)
16. Lundberg, S.M., Lee, S.I.: A unified approach to interpreting model predictions. In: Guyon, I., Luxburg, U.V., Bengio, S., Wallach, H., Fergus, R., Vishwanathan, S., Garnett, R. (eds.) Advances in Neural Information Processing Systems, vol. 30, pp. 4765–4774. Curran Associates, Inc. (2017)

17. Molnar, C.: Interpretable Machine Learning. https://christophm.github.io/interpretable-ml-book/ (2019)
18. Molnar, C., Bischl, B., Casalicchio, G.: iml: an R package for interpretable machine learning. JOSS **3**(26), 786 (2018)
19. Papenmeier, A., Englebienne, G., Seifert, C.: How model accuracy and explanation fidelity influence user trust. CoRR abs/1907.12652 (2019). http://arxiv.org/abs/1907.12652
20. Pedersen, T.L., Benesty, M.: LIME: Local Interpretable Model-Agnostic Explanations (2019). https://CRAN.R-project.org/package=lime. r package version 0.5.1
21. Ribeiro, M.T., Singh, S., Guestrin, C.: "why should i trust you?" explaining the predictions of any classifier. In: Proceedings of the 22nd ACM SIGKDD International Conference on Knowledge Discovery and Data Mining, pp. 1135–1144 (2016)
22. Shapley, L.S.: A value for n-person games. In: Contributions to the Theory of Games, vol. 2, no. 28, pp. 307–317 (1953)
23. Shortliffe, E.H., Davis, R., Axline, S.G., Buchanan, B.G., Green, C., Cohen, S.N.: Computer-based consultations in clinical therapeutics: explanation and rule acquisition capabilities of the MYCIN system. Comput. Biomed. Res. **8**(4), 303–320 (1975). https://doi.org/10.1016/0010-4809(75)90009-9
24. Swartout, W.R., Moore, J.D.: Explanation in second generation expert systems. In: David, J.M., Krivine, J.P., Simmons, R. (eds.) Second Generation Expert Systems, pp. 543–585. Springer, Heidelberg (1993). https://doi.org/10.1007/978-3-642-77927-5_24
25. Wallenius, J., Dyer, J.S., Fishburn, P.C., Steuer, R.E., Zionts, S., Deb, K.: Multiple criteria decision making, multiattribute utility theory: Recent accomplishments and what lies ahead. Manage. Sci. **54**(7), 1336–1349 (2008)
26. Yang, F., Du, M., Hu, X.: Evaluating explanation without ground truth in interpretable machine learning. arXiv preprint arXiv:1907.06831 (2019)

ciu.image: An R Package for Explaining Image Classification with Contextual Importance and Utility

Kary Främling[1,2(✉)], Samanta Knapič[2], and Avleen Malhi[2,3]

[1] Department of Computing Science, Umeå University, 901 87 Umeå, Sweden
kary.framling@umu.se

[2] Department of Computer Science, Aalto University, 02150 Espoo, Finland
{samanta.knapic,avleen.malhi}@aalto.fi

[3] Department of Computing and Informatics, Bournemouth University, Poole, UK
amalhi@bournemouth.ac.uk

Abstract. Many techniques have been proposed in recent years that attempt to explain results of image classifiers, notably for the case when the classifier is a deep neural network. This paper presents an implementation of the Contextual Importance and Utility method for explaining image classifications. It is an R package that can be used with the most usual image classification models. The paper shows results for typical benchmark images, as well as for a medical data set of gastro-enterological images. For comparison, results produced by the LIME method are included. Results show that CIU produces similar or better results than LIME with significantly shorter calculation times. However, the main purpose of this paper is to bring the existence of this package to general knowledge and use, rather than comparing with other explanation methods.

Keywords: Explainable artificial intelligence · Contextual importance and utility · Image classification · Deep neural network

1 Introduction

Contextual Importance and Utility (CIU) is a method originally developed by Kary Främling in his PhD thesis [3]. CIU was developed in a context of multiple criteria decision making (MCDM), which is a domain where different mathematical models are used as decision support systems for human decision makers. Possible mathematical models include any kind of AI system, including systems created using machine learning. CIU was designed to provide a mechanism for explaining or justifying the outcome of any such AI system in a uniform way, no matter if it is considered to be a so-called black-box model or not.

The work is partially supported by the Wallenberg AI, Autonomous Systems and Software Program (WASP) funded by the Knut and Alice Wallenberg Foundation.

D. Calvaresi et al. (Eds.): EXTRAAMAS 2021, LNAI 12688, pp. 55–62, 2021.
https://doi.org/10.1007/978-3-030-82017-6_4

As in many MCDM methods, CIU makes a difference between *feature importance* and *value utility*. The feature importance expresses to what extent an input feature (or combination of input features) can change the output value. Value utility expresses to what extent the input value contributes towards a high-utility output value. In classification tasks there is usually one output per class and the output value is a class probability value, that can be directly used as the output utility value. Both feature importance and value utility can change depending on the studied instance or context, which is why feature importance is called *Contextual Importance (CI)* and value utility is called *Contextual Utility (CU)*. CI and CU are scalars in the range [0, 1] and are absolute (non-relative) values.

The purpose of this paper is to present an R implementation of CIU for explaining image classification. This 'ciu.image' package is available at https://github.com/KaryFramling/ciu.image. It is a follow-up package to the 'ciu' R package for tabular data available at https://github.com/KaryFramling/ciu [4]. After this Introduction, Sect. 2 provides implementation details of CIU for image explanation. Section 3 gives software installation and usage instructions. Section 4 shows example results on ImageNet and medical image explanations, followed by Conclusions.

2 Contextual Importance and Utility for Images

The most recent description available about CIU is found in [4]. Only the most relevant parts for explaining image classification with CIU are included here, i.e. the basic definitions of CI and CU. CI expresses how much the output value utility can change when modifying the value(s) of one or several input features $\{i\}$ relative to the total output range:

$$CI_j(\vec{C},\{i\}) = \frac{Cmax_j(\vec{C},\{i\}) - Cmin_j(\vec{C},\{i\})}{absmax_j - absmin_j} \tag{1}$$

CU expresses to what extent the current input feature values $\vec{C}$ are favorable for a high output value utility:

$$CU_j(\vec{C},\{i\}) = \frac{y_j(\vec{C}) - Cmin_j(\vec{C},\{i\})}{Cmax_j(\vec{C},\{i\}) - Cmin_j(\vec{C},\{i\})} \tag{2}$$

Here, $y_j(\vec{C})$ is the value of output j for the current instance. $Cmin$ and $Cmax$ are the minimal and maximal output values achievable by modifying the value of the given input feature(s) with indices $\{i\}$. $absmin$ and $absmax$ give the minimal and maximal possible values for the output. In the case of explaining image classification (and for classification tasks in general) it is reasonable to use $absmin = 0$ and $absmax = 1$.

'ciu.image' segments images into so-called super-pixels using Simple Linear Iterative Clustering (SLIC) [1] in the same way as the LIME package [6]. Therefore, the input features of CIU actually consist of super-pixel values. The super-pixel values could in principle be 'anything' but what is used in practice is to have only the values present/not-present, where 'not-present' corresponds to setting the super-pixel to transparent. In practice, this leads to a rather trivial implementation of CIU for explaining image classification because calculating CI and CU values of a super-pixel requires exactly two forward-passes of the classification model, i.e. one with the original image and one with the super-pixel(s) of interest set to transparent. For an image with 50 super-pixels, for instance, only 51 forward passes are needed for calculating the CIU values of all super-pixels.

When interpreting the results, the super-pixels with the highest CI values are the ones that contribute the most to the classification result. CU can only take values zero or one in this case, where $CU = 1$ signifies that the contribution is positive and $CU = 0$ signifies that the contribution is negative. As shown in Sect. 4.1, this approach works relatively well for ImageNet classification. For the gastro-enterological images in Sect. 4.2, this approach is not sufficient because bleeding in **any** super-pixel will lead to classifying the image as 'bleeding'. However, since CIU can be calculated for any number of input features, an 'inverse' option was introduced, where all other super-pixels except the studied one are set to transparent, which efficiently identifies all the super-pixels with bleeding present.

3 Installation and Use

The package is available at https://github.com/KaryFramling/ciu.image. Installation instructions are also found there. The simplest way to install the package is to first install the 'devtools' package and then install 'ciu.image' with the command `devtools::install_github('KaryFramling/ciu.image')`. The package is loaded with the command `library(ciu.image)`. A `ciu.image` object is created by calling `ciu.image.new(model)` that uses the given predictor model. The optional parameters of `ciu.image.new` and the methods of `ciu.image` objects are explained in the package documentation and reflect the latest updates to the package. Since `ciu.image` is still a research tool, it is expected to evolve over time. Currently, the core `ciu.image` methods are the following:

1. `ciu.superpixels(imgpath, ind.outputs=1, n_superpixels=50, weight=20, n_iter=10, background = "grey", strategy = "straight")`: Return a list with fields `out.names`, `outval`, `CI`, `CU`, `cmin`, `cmax` for the requested number of outputs in `ind.outputs`, where outputs are ordered according to decreasing output value. Only `imgpath` is a compulsory parameter, for the others the default values are often appropriate.
2. `plot.image.explanation(imgpath, ind.outputs=1, threshold=0.02, show_negative=FALSE, n_superpixels=50, weight=20, n_iter=10, background="grey", strategy="straight", ciu.sp.results=NULL, title=NULL)`: Return a list of ggplot objects for the requested number of outputs in `ind.outputs`, where outputs are ordered according to decreasing output value. Most parameters are the same as for `ciu.superpixels`.

The use of 'ciu.image' typically happens as follows:

```
ciu <- ciu.image.new(model, predict_function, output.names)
plist <- ciu$plot.image.explanation(imgpath)
print(plist[[1]])
```

A complete code example is shown in Appendix 1.

4 Results

The experiments were run using Rstudio Version 1.3.1093 on a MacBook Pro, with 2,3 GHz 8-Core Intel Core i9 processor, 16 GB 2667 MHz DDR4 memory, and AMD Radeon Pro 5500M 4 GB graphics card. The LIME R package was used for producing LIME results [6]. For image classification, CIU is entirely deterministic so it always produces the same results, whereas LIME results tend to vary from one run to the other.

4.1 ImageNet Classification

Results are shown here for VGG16 and VGG19 models included in the 'keras' package, pre-trained on ImageNet images. The two images used are shown in Fig. 1. Appendix 1 shows the complete source code for producing the cat classification results shown in Fig. 2, both for CIU and LIME. For CIU, this source code is also included in the online documentation and is accessible by writing `?ciu.image.new` on the R command line. Cat calculation times are 30 s for CIU and 5 min 44 s for LIME. Dog playing guitar times are 18 s for CIU and 4 min 44 s for LIME.

(a) Cat.

(b) Dog playing guitar.

Fig. 1. Original images, with super-pixel borders shown by yellow lines. (Color figure online)

For the cat explanations shown in Fig. 2, CIU and LIME explanations are quite similar. However, CIU extracts what makes the difference between Egyptian, Tabby and Tiger cat according to the VGG16 model, whereas LIME

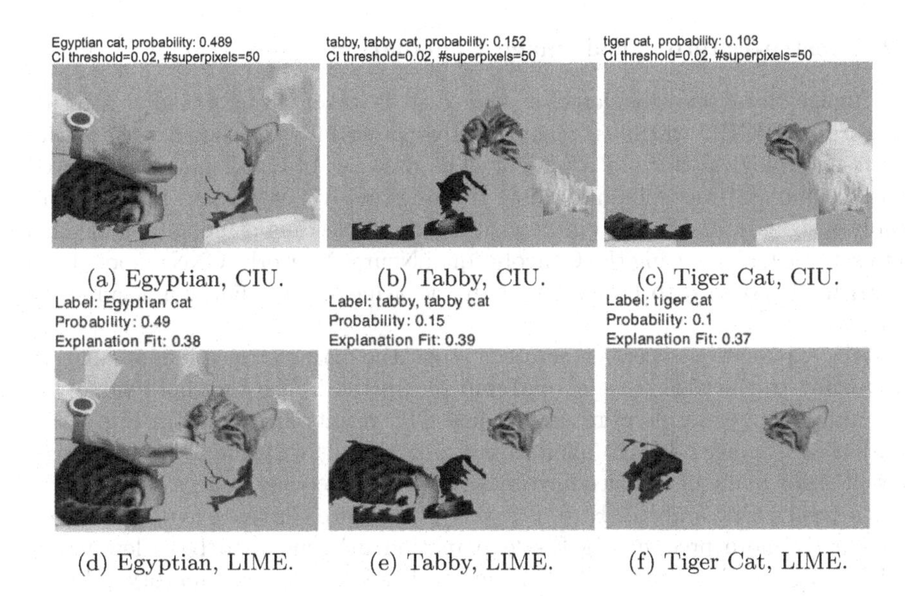

(a) Egyptian, CIU. (b) Tabby, CIU. (c) Tiger Cat, CIU.

(d) Egyptian, LIME. (e) Tabby, LIME. (f) Tiger Cat, LIME.

Fig. 2. Cat image explanation results using LIME and CIU trained with VGG16.

includes the same super-pixels for all three kinds of cat, with a smaller subset included for the two lower-probability cat types Tabby and Tiger Cat. For the guitar-playing dog image used in [7], the results are shown in Fig. 3. In this case, the interpretation of image explanations tends to be subjective and also depends on how the underlying trained model makes the classification, so it does not make much sense to declare a 'winner'. Furthermore, LIME results tend to change somewhat from one run to the other, whereas CIU results are guaranteed to be identical for every run.

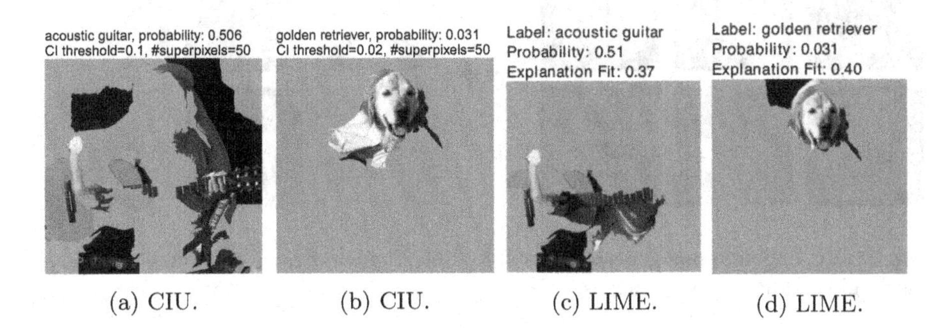

(a) CIU. (b) CIU. (c) LIME. (d) LIME.

Fig. 3. Dog playing guitar image explanation results using LIME and CIU trained with VGG19 for 'Acoustic guitar' and 'Golden retriever'.

4.2 Gastro-Enterological Image Explanation

The image data set considered in this case is taken from a Video Capsule Endoscopy (VCE), which is a non-invasive procedure to visualize the entire gastro-enterological tract of a patient. The data set of 3,295 images, retrieved from Coelho[1] [2] was split into 2,941 training and 354 validation images (randomly assigned), and it is a binary classification (bleeding or not). The medical data set was trained using the Convolutional Neural Network (CNN) model from [5], with 50 epochs with batch size of 16 and achieving a validation accuracy of 98.58%.

CIU explanations were generated using the parameter value `strategy="inverse"` to the `plot.image.explanation` method. The threshold value was 0.01 and 50 super-pixels were used. Some CIU results are shown in Fig. 4. For non-bleeding images, LIME failed to produce a result with the default settings, as well as for many of the bleeding images. For the bleeding images where LIME gave a result, CIU and LIME results were often quite similar, even though CIU was clearly more precise. LIME's 'explanation fit' tended to be below 0.001, which indicates that the fitted LIME model has low or no explanatory value.

For 'bleeding' images, the parts (super-pixels) identified by CIU were considered relevant and correct by a user panel and also corresponded to the masks of 'correct' answers available for the image set. For 'non-bleeding' images, all super-pixels that belong to the actual image should be included because they are all 'non-bleeding'. However, the black corners of the images are present in all images and therefore do not have any discriminatory effect between 'bleeding' and 'non-bleeding' images, so they have no significance for the classification. CIU indeed filters out those black areas from the explanation, as seen in Fig. 4. CIU took less than 7 s per image, whereas LIME took about 1 min 40 s per image.

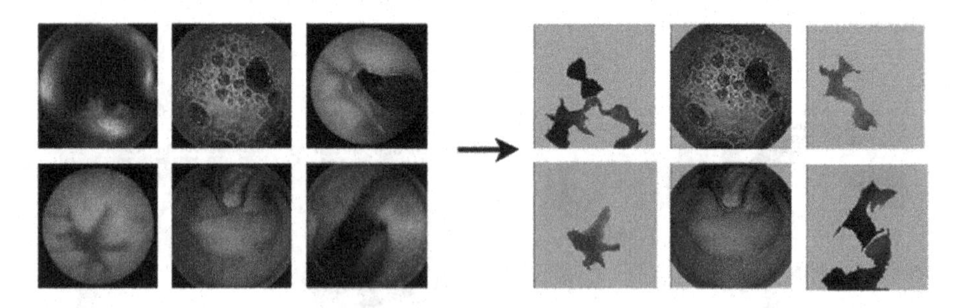

Fig. 4. CIU explanations generated for 'bleeding' (left and right columns) and 'non-bleeding' (middle column) images.

[1] https://rdm.inesctec.pt/dataset/nis-2018-003.

5 Conclusions

This CIU implementation for explaining image classification shows that CIU can produce explanations that are at least at comparable level to LIME. For explaining gastro-enterological image classification, CIU manages to produce 'makes-sense' explanations for all images, whereas LIME fails to produce explanations for several images. Moreover, CIU is orders of magnitude faster than LIME, which might in the future be used for exploiting CIU's capability to deal with super-pixel combinations in different ways, rather than only setting one super-pixel transparent, or the opposite. Therefore, CIU's performance can be expected to improve further with future research.

Appendix 1: Source Code for ImageNet Cat Results

```
library(keras)
library(lime)
library(magick)
library(ciu.image)
imgpath <- system.file('extdata', 'kitten.jpg',
  package = 'ciu.image')
# load VGG16 image classifier trained on imagenet database
model <- application_vgg16(weights = "imagenet", include_top = TRUE)
# We have to tell how images are prepared and evaluated.
vgg_predict_function <- function(model, imgpath) {
  predict(model, image_prep(imgpath))
}
# Standard preparation for imagenet, VGG16 & VGG19
image_prep <- function(x) {
  arrays <- lapply(x, function(path) {
    img <- image_load(path, target_size = c(224,224))
    x <- image_to_array(img)
    x <- array_reshape(x, c(1, dim(x)))
    x <- imagenet_preprocess_input(x)
  })
  do.call(abind::abind, c(arrays, list(along = 1)))
}
model_labels <- readRDS(system.file('extdata',
'imagenet_labels.rds', package = 'ciu.image'))
ciu <- ciu.image.new(model, vgg_predict_function,
output.names = model_labels)
# Get explanation for three topmost classes.
# Use `threshold` parameter for adjusting CI level to show.
plist <- ciu$plot.image.explanation(imgpath, c(1,2,3))
for ( i in 1:3 ) print(plist[[i]])

# These lines generate corresponding LIME explanations.
explainer <- lime(imgpath, as_classifier(model, model_labels),
image_prep)
explanation <- explain(imgpath, explainer, n_labels = 3,
n_features = 50, n_superpixels=50)
```

```
explanation <- as.data.frame(explanation)
p <- plot_image_explanation(explanation, display = 'block',
threshold = 0.01)
print(p)
```

References

1. Achanta, R., Shaji, A., Smith, K., Lucchi, A., Fua, P., Süsstrunk, S.: SLIC superpixels compared to state-of-the-art superpixel methods. IEEE Trans. Pattern Anal. Mach. Intell. **34**(11), 2274–2282 (2012). https://doi.org/10.1109/TPAMI.2012.120
2. Coelho, P., Pereira, A., Leite, A., Salgado, M., Cunha, A.: A deep learning approach for red lesions detection in video capsule endoscopies. In: Campilho, A., Karray, F., ter Haar Romeny, B. (eds.) ICIAR 2018. LNCS, vol. 10882, pp. 553–561. Springer, Cham (2018). https://doi.org/10.1007/978-3-319-93000-8_63
3. Främling, K.: Modélisation et apprentissage des préférences par réseaux de neurones pour l'aide à la décision multicritère. Ph.D. thesis, INSA de Lyon, March 1996. https://tel.archives-ouvertes.fr/tel-00825854
4. Främling, K.: Contextual importance and utility in R: the 'ciu' package. In: Proceedings of 1st Workshop on Explainable Agency in Artificial Intelligence, at 35th AAAI Conference on Artificial Intelligence, pp. 110–114 (2021)
5. Malhi, A.K., Kampik, T., Pannu, H.S., Madhikermi, M., Främling, K.: Explaining machine learning-based classifications of in-vivo gastral images. In: 2019 Digital Image Computing: Techniques and Applications, DICTA 2019, Perth, Australia, 2–4 December 2019, pp. 1–7. IEEE (2019)
6. Pedersen, T.L., Benesty, M.: LIME: Local Interpretable Model-Agnostic Explanations (2019). R package version 0.5.1. https://CRAN.R-project.org/package=lime
7. Ribeiro, M.T., Singh, S., Guestrin, C.: "Why should i trust you?": explaining the predictions of any classifier. In: Proceedings of the 22nd ACM SIGKDD International Conference on Knowledge Discovery and Data Mining, San Francisco, CA, USA, 13–17 August 2016, pp. 1135–1144 (2016)

Shallow2Deep: Restraining Neural Networks Opacity Through Neural Architecture Search

Andrea Agiollo[1,2](✉), Giovanni Ciatto[1], and Andrea Omicini[1]

[1] Alma Mater Studiorum—University of Bologna, Cesena, Italy
{andrea.agiollo,giovanni.ciatto,andrea.omicini}@unibo.it
[2] The Research Hub by Electrolux Professional S.p.A., 33170 Pordenone, PN, Italy

Abstract. Recently, the Deep Learning (DL) research community has focused on developing efficient and highly performing Neural Networks (NN). Meanwhile, the eXplainable AI (XAI) research community has focused on making Machine Learning (ML) and Deep Learning methods *interpretable* and *transparent*, seeking *explainability*. This work is a preliminary study on the applicability of Neural Architecture Search (NAS) (a sub-field of DL looking for automatic design of NN structures) in XAI. We propose *Shallow2Deep*, an evolutionary NAS algorithm that exploits local variability to restrain opacity of DL-systems through NN architectures simplification. *Shallow2Deep* effectively reduces NN complexity – therefore their opacity – while reaching state-of-the-art performances. Unlike its competitors, *Shallow2Deep* promotes variability of localised structures in NN, helping to reduce NN opacity. The proposed work analyses the role of local variability in NN architectures design, presenting experimental results that show how this feature is actually desirable.

Keywords: Neural Architecture Search · Evolutionary algorithm · Opacity · Interpretability

1 Introduction

Data-driven intelligent systems pervade modern society. The recent advancements of Machine Learning (ML) and Deep Learning (DL) are boosting the adoption of neural networks (NN) in several contexts, including, but not limited to, healthcare, finance, law, and domestic appliances. For this reason, the exploitation of DL-enabled systems in industrial applications and every-day life requires precision and efficiency. However, both features come at a price. In fact, state-of-the-art neural architectures are characterised by an ever-increasing *structural* complexity – in terms of layers, neurons, and their connections –, which is expected to make neural networks even more precise and accurate.

The structural complexity of NN, however, brings about a number of drawbacks. For instance, it makes training more eager for computational resources

D. Calvaresi et al. (Eds.): EXTRAAMAS 2021, LNAI 12688, pp. 63–82, 2021.
https://doi.org/10.1007/978-3-030-82017-6_5

and data. Furthermore, it represents a bottleneck in the engineering process of DL systems—which is commonly performed by data scientists, manually. Finally, and more importantly, it contributes to the well-known *opacity* issues making NN inner operation hard to understand for human beings—there including both expert practitioners and users. For all these reasons, research efforts devoted to the identification of *small* – i.e. structurally simpler – yet highly-performing neural architectures are gaining momentum within the DL community [17,39].

Opacity of DL systems, in particular, is a critical aspect which should be reduced and possibly avoided. As suggested by the eXplainable Artificial Intelligence (XAI) initiative [1,11], there is an urgent need for making the operation and outcomes of modern intelligent systems more human-interpretable. So far, several means have been proposed into the XAI literature to serve these purposes, following as many strategies. When it comes to DL, however, most existing techniques focus on either *(i)* easing the *inspection* of NN – via visualisation facilities –, or *(ii)* enabling their replacement with transparent models of similar performance—such as rules lists or decision trees. In other words, not enough care seems to be given to the problem of making DL models more transparent.

Arguably, a possible way to decrease NN opacity is to reduce their structural complexity, while preserving their predictive performance. Many works focusing on NN explainability can produce rules lists or decision trees equivalent models for small NN [8,40]: e.g., REFANN [36] is a rule extraction procedure tailored on neural networks having a *single* hidden layer. While such approaches can hardly be applied to complex NN, we argue that an automated procedure capable of reducing the internal structure of a NN may pave the way towards a wider adoption of algorithms that would otherwise be inapplicable.

Generally speaking, complexity reduction may bring benefits at several levels, including transparency and training time, other than the capability to extract human-readable rules or trees out of NN. However, a limiting factor along this line is that, currently, NN structures are handcrafted by human experts via a trial-and-error procedure, targeting predictive performance rather than lower structural complexity. Furthermore, experts' experience and intuition play a pivotal role in the process, making it hard to automate and reproduce. This is where Neural Architecture Search (NAS) [12,42] comes into play. The general goal of NAS is to *automate* the identification of the best NN structure for a given task. Several approaches are being explored, modelling the NAS as a search problem in the space of all possible NN architectures. To the best of our knowledge, however, no work so far focuses on controlling network structural complexity.

Accordingly, in this paper we propose *Shallow2Deep*, a novel NAS algorithm aimed at keeping NN structural complexity under control. Our algorithm allows data scientists to *automatically* and efficiently detect highly-predictive architectures for convolutional NN targeting pattern matching tasks—such as image or speech recognition. It enforces structural constraints over the searched NN architecture, limiting the NN structural complexity and therefore its opacity. In other words, *Shallow2Deep* fits NN design by providing a means to control the depth of a NN, possibly enabling, e.g., the applicability of rule-extraction algorithms in complex tasks.

Our solution differs from other NAS approaches in a number of ways. First, it promotes local *variability* in NN architectures— meaning that it supports and encourages variability in the different layers composing a NN. Then, it favours local *specialisation* of NN sub-structures—thus letting each layer of the NN specialise on different tasks, depending on their depth. Finally, it promotes progressive complexity, avoiding *overthinking*—a well-acknowledged [19] tendency of deep NN to learn too many concepts, becoming more complex than needed. We present a full operational formalisation of the *Shallow2Deep* algorithm along with a number of experiments showing its practical feasibility and versatility.

2 Background

2.1 Neural Architecture Search

Neural networks are biologically-inspired computational models, made of several elementary units (neurons) interconnected into a *directed-acyclic* graph (DAG) via *weighted* synapses. NN can be *trained* on data via backpropagation [16] and exploited into both supervised and unsupervised learning tasks such as classification, regression, and anomaly detection. The training phase makes a NN learn from data. Yet, only network synapses weights are modified in this phase, whereas its overall graph structure (topology henceforth) is not allowed to vary. It is rather assumed to be *manually* engineered by data scientists.

Convolutional Neural Networks (CNN) are particular sorts of NN whose topology consists of a cascade of *convolutional* layers. In other words, CNN can learn how to apply a number of convolution operations [30] to the data. Convolutions let the network spot relevant features into the input data, at possibly different scales. Thus, CNN are primarily used to solve complex pattern-recognition tasks, such as in computer vision or speech recognition. Yet, how many convolutions a network may learn as well as their interconnections depend on NN topology. Again, this implementation detail requires human intervention.

Traditionally, NN development workflows are deeply influenced by the choices by human experts. Network architectures represent the most relevant aspect requiring human contribution. However, as neither theorems nor methods ensure optimal results, human choices may lead to sub-optimal or inefficient solutions.

To avoid inefficiencies introduced by human errors in NN design, *neural architecture search* (NAS) has been proposed [26]. NAS automates network architecture engineering: it aims at learning a network topology that can achieve reasonably-good performances on specific tasks, by letting a search algorithm look for the best network topology among the admissible ones.

To keep the computational complexity of NAS acceptable, several approaches have been proposed in the literature. Virtually all of them try to *(i)* reduce the search space size and *(ii)* control the whole search duration by leveraging on a *greedy* or *evolutionary* search strategy.

A common means to restrict the search space is to assume the network topology to be composed by a sequence of units called *cells*. Each cell contains a number of *blocks*, connected over a DAG structure. Blocks, in turn, are groups of neurons having a predefined internal organisation—commonly corresponding to a particular mathematical operator. In CNN, for instance, blocks are commonly constrained to represent convolutional layers, each one representing different sorts of convolutional filters—e.g. 3×3, 5×5, etc. Directed connections among any two cells A and B are modelled as directed connections among the output block of A and the input block of B.

Within the scope of this paper, we denote by $\mathbb{O}$ the *operation set*, i.e. the set of all possible *sorts* of blocks for any given NAS problem. We assume $\mathbb{O}$ is always a finite-cardinality set. For instance, in the particular case of CNN, $\mathbb{O}$ may contain different sorts of convolutional layers (e.g. $\mathbb{O} = \{3 \times 3, 5 \times 5, 7 \times 7\}$).

State-of-the-art NAS approaches mostly differ in which and how many blocks and cells are exploited, how these can be connected with each others, or which (meta-)heuristic the search of the optimal topology leverages upon. For instance, a method is proposed in [22] where NN is built as a sequence of identical cells—so that only the internal structure of one cell is optimised. In [32] a *regularised* evolutionary meta-heuristic is presented introducing an age property to favour younger genotypes. The approach aims at optimising two sorts of cells – "normal" or "reduction" cells – which are concatenated in a predefined way to obtain the final NN. Similarly, a *continuous* evolutionary approach is proposed in [45] sampling the population of different generations from a super network $\mathcal{N}$ of shared parameters to search for normal and reduction cells. In [7], a probabilistic approach is applied to reduce the search memory requirements to obtain the best structures for the normal and reduction cell.

Another common approach is to fix cells operations and look for the best cell combination that can compose the NN. In [9] a multi-objective oriented algorithm is presented exploiting both evolution and reinforcement learning to search for the best composition of various predefined cells. In [38] a method is presented that explicitly incorporates model complexity into the objective function while searching for the best cell sequence. However, cells are selected from a predefined pool, and they are not optimised any further. Conversely, a differentiable NAS framework is proposed in [43], searching for the best cell placement in the NN structure. Finally, in [5] a method is presented that can search NN architectures avoiding proxies and limitations typical of other approaches—e.g., training on a smaller dataset, learning with only few blocks, training for few epochs.

Reduction in NN complexity can also be achieved via network pruning techniques [24]. However, here we focus on NAS techniques only, since network pruning is a post-hoc technique not taking into account NN structure, as it is applied *after* training—when the structure has already been fixed. For this reason, structural inefficiencies of NN can not be tackled using network pruning techniques.

To the best of our knowledge, our work represents the first attempt to produce a NAS strategy capable of identifying both a good cell structure and a global architecture, avoiding constraints on architecture design. In other words, our approach improves the NN performance over complexity ratio, allowing *Shallow2Deep* to produce smaller – yet reasonably performing – NN architectures.

2.2 eXplainable AI *vs.* Neural Networks

Neural networks and computer vision are commonly exploited behind the scenes of intelligent systems involved in several critical applications (e.g., intelligent medicine, automotive, etc.). There, NN provide intelligent systems with high-precision pattern recognition capabilities. However, as for most ML methods, NN engineering only focuses on attaining high predictive performance, whereas poor care is given when it comes to determine *why* NN provide a particular outcome. To complicate this issue, advanced applications – such as computer vision – may easily involve *deep* NN, having an intricate internal structure which makes them hard to analyse and understand—even for experts. Such intricacy is described into the literature as "opacity" [21]—a feature characterising most ML algorithms up to some extent, which are also known as "black boxes" for that very same reason [14].

Recently, the opacity issue characterising ML and, in particular, NN reached the general public attention—also because of the GDPR regulations[1]. Accordingly, safety- or privacy-critical intelligent applications leveraging on ML should be designed accounting for properties such as interpretability, reliability, and transparency [41]. In other words, human users must be able to understand the criteria behind machine-aided or -driven decisions.

The XAI community [3] is currently intercepting those needs by proposing methods aimed at tackling the opacity issues that characterise ML- and DL-powered systems. Most methods proposed so far essentially focus on *(i)* inspection or visualisation [4,48] techniques, aimed at "debugging" the inner functioning of NN; *(ii)* heatmaps [2,35], or feature relevance analyses [18,28], aimed at analysing a network behaviour w.r.t. its inputs; or *(iii)* symbolic knowledge (e.g., rules, or trees) extraction algorithms [6,49], aimed at distilling human-intelligible information out of the intricate structure of sub-symbolic predictors. In other words, following a nomenclature introduced in [10], current methods either attempt to make NN more easily interpretable or *a-posteriori* explainable.

NN opacity is deeply entangled with their structural complexity. The more NN architectures are intricate the less NN are interpretable. By keeping NN structural complexity under control, data scientists may limit their opacity. In turn, limiting NN opacity, plays a fundamental role in XAI as it may make NN both *a-priori* interpretable and *a-posteriori* explainable. Accordingly, to the best of our knowledge, our work represents the first attempt to control NN opacity – i.e. structural complexity – while automating the search for a well-performing architecture.

[1] https://eur-lex.europa.eu/eli/reg/2016/679/oj.

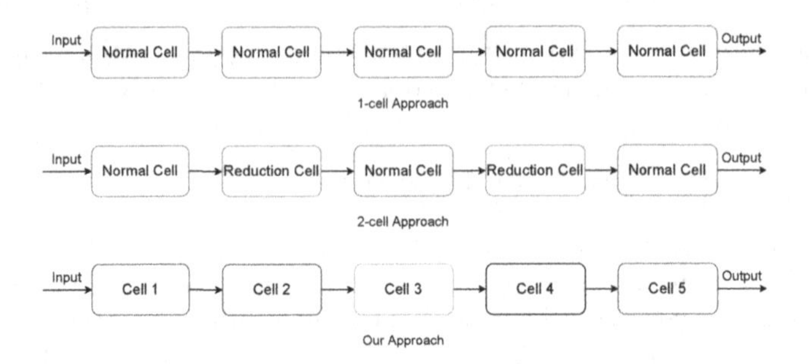

Fig. 1. *Shallow2Deep* avoids architecture design limitations, common in other NAS algorihtms. Here $C = 5$.

3 Shallow2Deep

In this section we present *Shallow2Deep*, a novel exhaustive NAS algorithm: we first present its architecture design along with the corresponding search space (Sect. 3.1), then we discuss its overall working principle, its fundamental hypotheses, and the details of its composing modules.

3.1 Architecture Design

Most popular NAS approaches blindly build a NN architecture by repeating the same *elementary* structure (cell) several times—assuming that the internal structure of a cell has been manually optimised. This sounds like a reasonable approach, considering the history of NN development and validation. Hand-crafted successful networks (e.g., VGG [37], ResNet [15], Mobilenet [17], etc.) are composed by repetitions of a certain peculiar element (e.g., convolutions, skip connections, inverted mobile bottleneck, etc.). Furthermore, repetitions lead to a reduction of the search space size, when an effective combination of the elementary cells must be automatically computed.

However, although understandable from a computational perspective, such an approach is not reasonable in terms of predictive capability. Relying on the same elementary structure at different depth levels of the network architecture may hinder the predictive performance of the resulting NN as a whole. In fact, assuming a particular elementary structure is good enough to let a network's *shallow*[2] layers perform valuable feature extraction, it is unlikely that the same structure is equally good to provide that network's *deep* layers with more sophisticated pattern matching capabilities. The different layers of a well-trained NN are expected to perform totally-different feature-mining tasks. This is why we argue that the best elementary structure for shallow and deep layers of NN are not architecturally equal. *Shallow2Deep* builds on this, providing a means to look for a good cell structure specification for both shallow and deep layers.

[2] By "shallow" (resp. "deep") layers of a NN we mean the *inner* layers close to the input (resp. output) neurons.

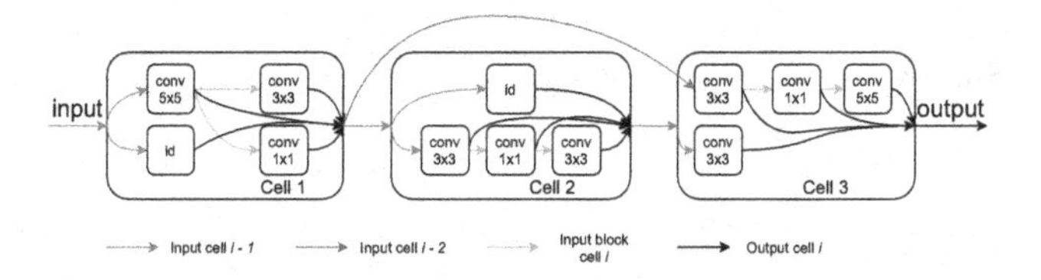

Fig. 2. Blocks in *Shallow2Deep* cells can get input from previous cell (green), the cell before that (blue) or any other block in the cell (yellow). Cell output is obtained concatenating outputs of all blocks belonging to the cell (black). Here $C = 3$ and $B = 4$.

More precisely, *Shallow2Deep* constructs NN classifiers with a fixed number C of *different* cells. Cells can differ in terms of the topology they assume and the operations they apply. We define the difference between cells w.r.t. topology and operations as *structure variability. Shallow2Deep* promotes local *structure variability* in NN architectures, avoiding design limitations.

Figure 1 highlights the main difference among state-of-the-art NAS mechanisms and our approach: *Shallow2Deep* lets each cell vary independently from each other. However, similarly to other authors, we assume cells to be ordered from shallow side to the deep side. Accordingly, the 1st cell is the closest one to the inputs, while the last one is the closest to the outputs.

In particular, *Shallow2Deep* lets each cell contain B blocks—being B a positive and finite integer. Each block represents a particular sort of NN layer. Following a convention introduced in [46], we denote by $\mathbb{O}$ the *finite* set of all possible sorts of blocks—which in turn depend on the particular task the target NN aims to solve. For instance, if the considered NN targets image recognition tasks, we let $\mathbb{O}$ contain simple convolutions, other than the identity block—e.g. some $n{\times}n$ convolutional layers (for $n = 1, 3, 5, \ldots$), plus the identity layer $f(x) = x$.

Each block of the ith cell can accept as input the output of $i - 1^{th}$ and $i - 2^{th}$ cells and of any other block in the same i^{th} cell. However, loops and cycles among blocks connections are not allowed. In other words, the blocks topology must be a DAG. This is necessary to preserve the feed-forward architecture of the NN. Furthermore, each block can provide output to any amount of other blocks.

The whole output of a cell is attained by concatenating the outputs of all blocks belonging to that cell, as in [22]. Figure 2 provides an example of an admissible topology that can be created by blocks and cells following the aforementioned rules.

3.2 Search Algorithm

Virtually all NAS algorithms proposed into the literature so far deal with reduced search spaces attained via strict architectural constraints. Conversely, our approach avoids the excessive simplification of the architectural design by allowing the internal structure of each cell to vary. A greedy search algorithm is then

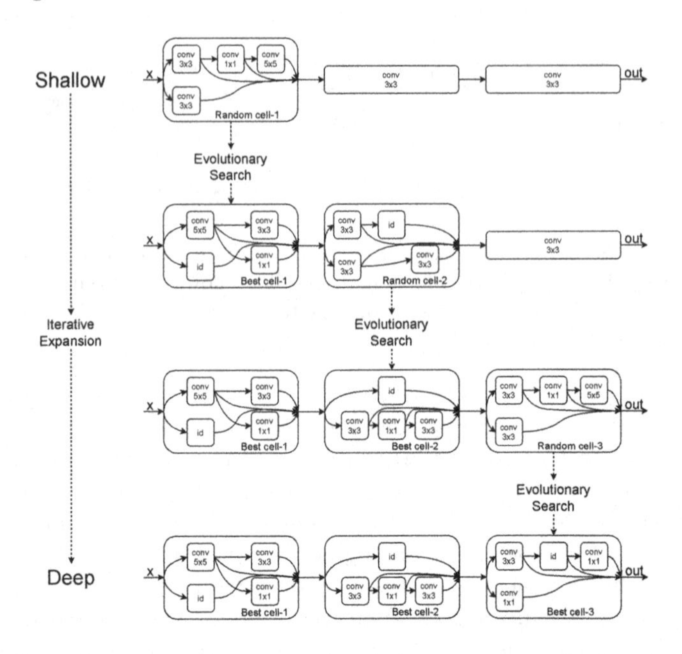

Fig. 3. *Shallow2Deep* iteratively searches for the best structure of cells going from shallow to deep ones. Shallow bests are searched using simple superstructures to increase the feature expressiveness and reduce training time. Local bests are kept fixed while deeper best are searched, reducing the complexity.

employed to automate the selection of the actual cells structures, in an iterative way. It relies on the successive search of locally-optimal cell structures proceeding from the shallower cells to the deeper ones.

As exemplified in Fig. 3, *Shallow2Deep* consists of the iterative repetition of a local search algorithm aimed at selecting the (locally) best internal structure of the ith cell. The search algorithm is repeated for all $i = 1, \ldots, C$, in such a way that the internal structure of the ith cell is only optimised *after* that $(i-1)^{th}$ one has already been optimised. In particular, here rely on an evolutionary algorithm to tackle local search. During local search, a population of NN is considered based on the structures that need to be analysed. The NN under examination are trained on a subset of the training set in order to find well behaving local structures—i.e. cell. To keep the whole process time-efficient, while optimising the ith cell, all the jth cells ($j \in \{1, \ldots, i-1\}$) are left unaffected by the training process. Moreover, to maximize the knowledge extracted at the ith cell during its discovery process, all k^{th} cells ($k \in \{i+1, \ldots, C\}$) are built as bare as possible. Following literature, we consider bare cells to be composed of a single block applying a 3×3 convolution operation [13,37]. In other words, *Shallow2Deep* greedly proceeds from the shallowest cell to the deepest one.

While further details concerning our design choices are provided in Sect. 4, some insights can be provided by the way a well-trained NN operates. The shallow layers of a NN aim to mine low level features. Complex features are extracted by deeper layers, reliably building on top of low level information. Therefore, *Shallow2Deep* searches for structures of deeper cells iteratively, building on the knowledge acquired at previous search steps.

Cell Search. The *Shallow2Deep* algorithm relies on a local search of the best performing structure for each cell of the NN. The task can be accomplished through a variety of different search algorithms, from reinforcement learning to evolutionary algorithms [45,50]. In *Shallow2Deep* we exploit *evolutionary* (a.k.a. genetic) heuristic algorithms.

Evolutionary algorithms are a family of population-based metaheuristic optimization algorithms inspired by biological evolution. They commonly rely on a set of predefined stochastic mechanisms – namely, generation, mutation, selection, mating, fitness, etc. – which let the algorithm randomly explore a vast search space in a smart way. Technically, these algorithms attempt to solve an optimisation problem by generating population of random solutions for the problem at hand, and by simulating evolution for a predefined amount of iterations—a.k.a. *generations*. Solutions are more or less likely, to survive among generations depending on their *fitness*—i.e. a measure of the quality of a particular solution w.r.t. the problem at hand. To prevent the search to step into local optima, evolutionary algorithms may exploit a number of strategies to introduce more randomness in the precess, such as mutations—meaning that solutions may randomly mutate while stepping through generations.

We choose to rely on evolutionary algorithms because of their *(i)* flexibility, *(ii)* support to *space pruning* [25] – a feature that we plan to support in the future –, other than *(iii)* the many successful works on NAS leveraging on evolutionary approaches as well (cf. [22,23,46]). In particular, our evolutionary algorithm is inspired to regularised evolution proposed in [32]. However, we avoid regularisation through aging and introduce a randomised approach to explore untouched areas of the search space.

As any other evolutionary approach, our algorithm mimics biological evolution by letting a *population* of N randomly-generated NN step through a number ν of generations. More in details, the number of generations (i.e., ν) represents the maximum amount of iterations that the evolutionary algorithm should perform before returning the final solution. While transitioning between generations, NN may probabilistically *mutate*, other than being allowed to survive depending on their *fitness*. Accordingly, while the *mutation* mechanism lets the algorithm *randomly* explore different internal structures for the ith cell, the *fitness measure* lets the algorithm assess how good a particular internal structure of the ith cell actually is. The *Shallow2Deep* algorithm can then go on with its iteration and focus on the $(i+1)^{th}$ cell. Once reached the νth generation, the best fitting NN is used to determine the final interal structure to be chosen for the ith cell.

Accordingly, in the remainder of this section, we delve into the details of how mutation and fitness actually work in the particular case of *Shallow2Deep*.

Algorithm Stub. We denote by P_n the nth generation of the population. Similarly, we denote by P_0 the initial population, which is randomly generated.

The population size is kept fixed to throughout the local search procedure, as it is commonly done for evolutionary algorithms. In other words, for all $i \in \{1, \dots, \nu\}$, the population P_n is such that $|P_n| = N$ and all the architectures of all networks in P_n conform to the constraints described in Sect. 3.1.

Then, our evolutionary algorithm refines the population through 3 steps which are repeated at every generation. These steps are:

train—where all the NN in P_n are trained on (a subset of) the data set;
selection—where the NN which are not among the top-m fittest ones are removed from P_n;
incubation—where P_n is enriched with new NN – attained via mutation – aimed at replacing the ones cutted off by the selection step.

Shallow2Deep assumes the available data to be partitioned into 3 parts, namely the *training*, *validation*, and *test* sets. While the train step only leverages on the training set, the selection step evaluates the fitness measure of each network against the validation set. The test can then be used to assess the performance of the final network architecture output by *Shallow2Deep*.

Concerning the incubation step, it is aimed at helping *Shallow2Deep* both from a performance-maximisation and search-space-exploration-speed perspective. More precisely, it aims at generating new NN following two criteria:

- c networks are attained by mutating as many individuals in P_n through the application of *mutation* transformation;
- $r = N - m - c$ networks are randomly generated from the search space.

Best behaving structures mutation helps performance maximisation, enhancing the focus on those evolutionary paths that have proven to be strong in recent history of the population. Partially randomising incubation helps search space exploration as it allows the evolution to look for points in the space farther apart from previously beaten evolutionary paths.

Once all the three steps have been completed for generation n, and a new population has been attained, P_{n+1} and the evolutionary search can proceed with generation $n + 1$. The process is repeated ν times, after which the best performing local structure is considered as found.

Fitness Measure. Fitness is measured on the *validation* set using the most adequate performance measure for the task at hand. Accordingly, in case the to-be-defined network targets classification tasks, accuracy or F1-score measures may be used. Conversely, in the case of regression tasks, MSE, MAE, or R^2 measures may be exploited instead.

In the particular case of image recognition tasks, classification accuracy is an adequate choice. More complex performance metrics may consider also FLOPS [39] and latency [38]. However, these are left for future works.

Mutation. The *mutation* transformation is applied to some NN – referred as the *parent* – in order to attain new architectures—called *children*. It only focuses

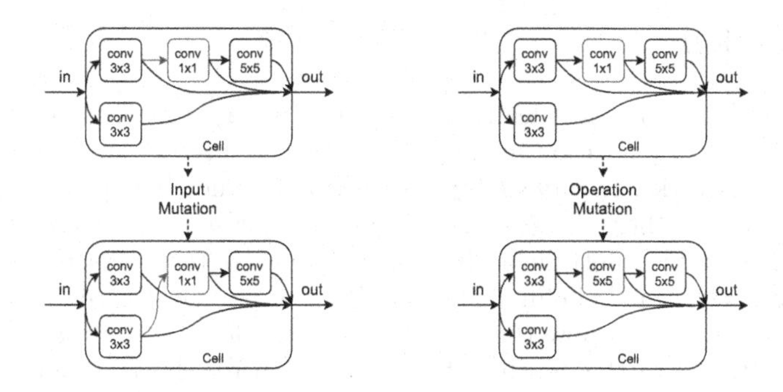

Fig. 4. Mutation operations available in randomized evolution. When input operation is applied, previous input block is linked with cell output if it has remained pendent, avoiding block removal.

on the internal structure of the ith cell of the parent network, possibly affecting some of its blocks. In particular, we rely on two possible mutations that can be applied to the blocks of a cell (graphically depicted in Fig. 4):

input mutation—a block B of the ith cell is selected at random, it is detached from its previous input, and the output of either another block B' in the same cell or of the jth cell as whole, with $j \in \{i-1, i-2\}$, is used as the new input of B—provided that the new connection does not introduce a loop or a cycle;

operation mutation—a block B of type $o \in \mathbb{O}$ is randomly selected from the ith, and its type is changed to some other $o' \in \mathbb{O}$ such that $o \neq o'$.

Greedy Assemble. *Shallow2Deep* requires several NN to be actually trained behind the scenes of its operation. This is true, in particular, for the evolutionary algorithm described above. In fact, while mostly focusing on one cell at a time, the algorithm must still train at least $N \cdot \nu$ networks – only differing for the content of the ith cell –, C times.

To keep the computational effort feasible, a number of strategies are in place. For istance, while performing the ith evolutionary search, *Shallow2Deep* leaves all cells of index j s.t. $1 \leq j < i$ unaffected, and does not re-train them anymore, as they have already been explored and trained in previous iterations. Dually, the algorithm always assumes all cells of index j s.t. $i < j \leq C$ to only contain a single block. In this way, the whole NN shallowness is preserved. In the particular case of image recognition tasks, that block may for instance consist of a 3×3 convolutional layer. Accordingly, during the ith evolutionary search, only cells whose index is at least equal to i are actually trained over data, and all cells whose index is greather than i have a very minimal structure.

In other words, once the ith local search is completed, the m best performing structures for the ith cell are fixed, and never retrained anymore. As part of the subsequent iterations of *Shallow2Deep*, the network architecture is deepened to produce deeper and more complex NN.

4 Discussion

Global NN architectures are ideally composed by different local structures whose role depends on their position in the NN. Following this idea, unlike most common NAS frameworks, *Shallow2Deep* framework does not rely on the replication of the same cell. Rather, *Shallow2Deep* exploits a *progressive* search of the best cells at each possible depth level, from the shallowest to the deepest ones. We here discuss the rationale behind *Shallow2Deep* progressive search.

It is well understood how the complexity of the features extracted by some NN is proportional to the depth of the layer which recognises them [29,47]. In fact, while layers that are closer to the input are appointed to extract basic features – such as edges, corners, borders, etc., in image-recognition tasks –, deeper layers aim at recognising more complex features—such as combination of shapes, combination of textures, etc. Accordingly, shallow networks are better suited to tackle simple tasks [13] where only simple features are involved. Conversely, the more complex a to-be-recognised feature is, the deeper a layer capable to recognise it must be. This happens because the recognition of a complex feature in a NN relies on the composition of more basic features extracted by shallow layers. Consequently, the lower is a feature complexity, the shallower can be the NN able to learn it. We call this phenomenon *depth-complexity proportionality* assumption.

There exists a tight link between features complexity and network depth that allows us to propose reasonable shortcuts for exhaustive architecture search methods. *Shallow2Deep* is designed on the assumption that simple features learnt by shallow networks perform reliably for deeper networks as well. Indeed, deeper NN achieve more flexible recognition capabilities than shallower ones [27,31]. Moreover, deeper NN may attain higher generalisation capabilities [34], being capable of adapting to the features that shallow NN have learnt to recognise.

Accordingly, we argue that NN built from the sequential repetition of the same local structure cannot achieve the astounding results that characterise state-of-the-art NN. Conversely, we believe it is possible to search for reliable shallow architectures and expand them in successive iterations, as done by *Shallow2Deep*. The more simple concepts are reliably learnt by shallow networks, the easier it will be to learn complex notions from their combinations. We call this phenomenon *knowledge greediness* assumption.

The progressive global assemble of *Shallow2Deep* exploits both knowledge greediness and depth-complexity proportionality assumptions to boost the overall time complexity and performance.

In particular, depth-complexity proportionality justifies the deepening of the NN architecture in successive iterations, which in turns supports the trick exploited by *Shallow2Deep* to speed up the local search phase. Indeed, population training in the evolutionary local search is the most expensive and time consuming process. Training shallower networks requires less time to complete, as the parameters to optimize are much less.

Conversely, knowledge greediness justifies *Shallow2Deep*'s strategy of iteratively expanding the depth of the NN architecture under consideration. While this certainly raises NN training time, it also lets deeper architectures rely on previously trained cells. In particular, to boost the overall search, *Shallow2Deep* fixes the parameters of shallower cells, avoiding their re-training. This idea traces back to the well-established idea of re-using pre-trained feature extractors in object detection mechanisms [20,33]. Indeed, once a network is deepened and its deeper cell structure is determined, the predictive performance of the overall network does not degrade, even if shallow layers are kept fixed.

5 Experiments

In this section we first present *Shallow2Deep* implementation and the best NN architecture obtained with it (see Sect. 5.1). In Sect. 5.2 we then compare obtained architecture with state-of-the-art models that leverage on the same operation set $\mathbb{O}$. We also analyse if *Shallow2Deep* local structures could be reused through repetition in a NN model to obtain better performance/complexity ratio. We make publicly available our implementation of *Shallow2Deep*.[3]

5.1 *Shallow2Deep* Architecture

In order to demonstrate the validity of our approach we run *Shallow2Deep* on MNIST fashion [44]. We define $\mathbb{O}$ to be the set of available operations that can be selected for each block of a cell. Similar to [46], in *Shallow2Deep* $\mathbb{O}$ contains simple convolutions and identity (*1× 1 conv*, *3× 3 conv*, *5× 5 conv*, *identity*). Consider now *Shallow2Deep* search space $\mathbb{S}$—i.e. the space that contains obtainable cells through local search. The search space cardinality – i.e. the number of obtainable cells – is $|\mathbb{S}| = (B+1)! \cdot |\mathbb{O}|^B$. Let now $\mathbb{N}$ be the search space for the overall NN—i.e. the set of obtainable NN architectures. Remembering that *Shallow2Deep* does not rely on cell repetition, the amount of NN architectures available during the overall architecture search is $|\mathbb{N}| = |\mathbb{S}|^C = \left((B+1)! \cdot |\mathbb{O}|^B\right)^C$. For our experiments we set $B = 3$ and $C = 4$, obtaining $|\mathbb{S}| = 1.54 \cdot 10^3$ and $|\mathbb{N}| = 5.57 \cdot 10^{12}$. The amount of possible NN architectures is huge, but it does not reflect the computational complexity. Indeed, thanks to its increasing depth approach, *Shallow2Deep* is capable of searching a space of size $|\mathbb{S}|^C$, while having complexity that is only proportional to $C \cdot |\mathbb{S}|$

For each cell we search for the best structure using the randomised evolution algorithm proposed in Sect. 3.2. We fix the number of generations of the evolutionary algorithm to be $\nu = 5$ for each cell and the population size to be $|P| = 50$. During *incubation* we fixed the number of surviving best models to be $m = 10$, the number of models obtained through mutations to be $c = 20$ and the number of random models added to each generation to be $r = 20$. Each model is trained for 10 epochs using learning rate $learning\_rate = 0.01$.

To show the effectiveness of *Shallow2Deep* search, we study the behaviour of the NN population against the number of generations of the overall algorithm.

[3] https://github.com/AndAgio/Shallow2Deep.

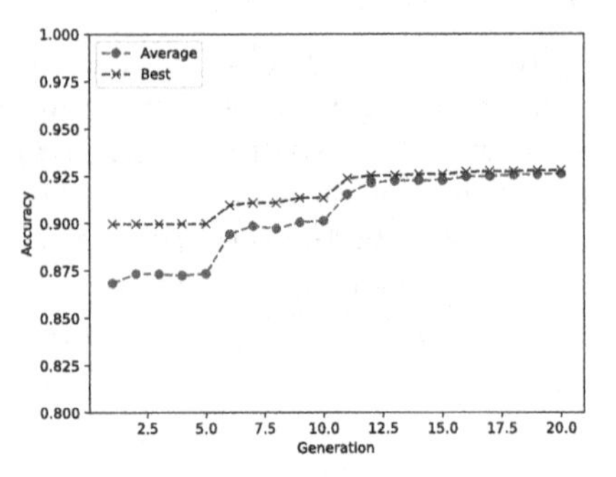

Fig. 5. Performance – i.e. classification accuracy – of NN architectures considered by *Shallow2Deep* for each generation. We consider both the average performance and the accuracy of the best model in each generation.

In *Shallow2Deep*, the user can select the number of cells C that compose the NN and ν, the number of generations that the local search takes. *Shallow2Deep* iteratively searches each of the C cells for ν generations. Therefore, the overall search of the NN architecture takes $C \cdot \nu$ generations to complete. We study the average performance – i.e. classification accuracy – of the population of NN for each of the $C \cdot \nu$ generations. We also study the accuracy of the best NN in the population for each of *Shallow2Deep* $C \cdot \nu$ generations.

Figure 5 shows the behaviour of average and best NN performance against *Shallow2Deep* generations. The classification accuracy increases with the number of generations considered, showing the success of *Shallow2Deep* search. Accuracy increments are limited since even 1st generation NN reach reasonable performances. This is due to the mild complexity of the classification task over the MNIST fashion dataset. Biggest increments in the NN accuracy are found in generations where the cell index i is increased—i.e. local search shifts to the next cell. This behaviour is expected as the increasing complexity – i.e. depth – of the NN extends its reasoning capabilities. It is also interesting to notice that this behaviour is more evident for smaller cell index i, while it becomes more attenuated for values of i close to C. In our experiments, performance reaches stability for $i = C$—i.e. there exists a negligible difference between accuracy of NN with $i = C - 1$ and $i = C$. Stabilisation of accuracy can be considered a signal that the NN is reaching a complexity limit. Surpassing this limit would increase concepts complexity while not bringing any gain in performance, introducing possible overthinking issues [19]. Therefore, *Shallow2Deep* represents a tool to automatically identify the NN complexity sweetspot over a certain task.

Figure 6 shows the architecture of the NN obtained running *Shallow2Deep* on the MNIST fashion dataset.

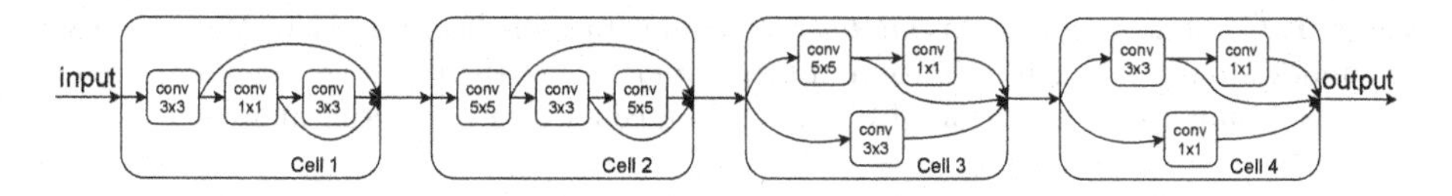

Fig. 6. NN architecture discovered by *Shallow2Deep* algorithm.

From its architecture, we point out that *Shallow2Deep* NN identifies sequential operations – i.e. blocks connected in a sequential manner inside a cell – at shallower stages of the NN—i.e. cells 1 and 2. Going deeper in the NN architecture – i.e. cells 3 and 4 –, *Shallow2Deep* building procedure identifies cells composed of parallel branches of convolutional operations. If confirmed in future investigations, this concept might give some insights on the learning process of NN. It is possible that sequential operations at shallow sections of NN help the model to learn simple concepts at the basis of their reasoning—i.e. edges, corners, simple shapes, etc. Parallel operations approach may, instead, be useful for the NN learning process when complex concepts need to be extracted—i.e. combination of shapes, combination of textures, etc. Therefore, deeper investigation of this result may be interesting.

5.2 *Shallow2Deep vs.* State-of-the-Art

We now compare the performances obtained by *Shallow2Deep* NN against state-of-the-art models that apply the same basic operations – i.e. convolutions and identity – like VGG [37] and ResNet [15]. In order to make the comparison fair, we retrain the *Shallow2Deep* NN, VGG, and ResNet on the MNIST fashion dataset from scratch. Training parameters are the same for every model considered—i.e. 60 epochs and $learning\_rate = 0.01$. Moreover, to study the effects of cell structure variability in NN architectures, we consider NN models built from the repetition of single cells found by *Shallow2Deep* local search. In other words, we select *Shallow2Deep* cell i – i.e. the cell discovered during i^{th} local search step – and we build the NN model composed of 4 cells having the same structure of cell i. We name these NN architectures $Shallow2Deep_i$.

Table 1 shows the performance over the test set T– i.e., the average accuracy and its standard deviation over 20 training runs –, the footprint – i.e. number of weights of the NN (expressed in millions, denoted by M) – of *Shallow2Deep* and state-of-the-art NN. *Shallow2Deep* NN with its variants reach state-of-the-art performances over the MNIST fashion classification dataset. NN obtained using *Shallow2Deep* are the most efficient if we consider the accuracy/footprint trade-off—i.e. division between reached accuracy and number of parameters. More in details, *Shallow2Deep* NN reaches accuracy comparable with VGG (only 0.4% less), while requiring a third of the parameters. Performances obtained by the ResNet NN over the dataset under examination are possibly due to overthinking issues. ResNet model complexity – i.e. model footprint – is higher than necessary for the selected task, which brings it to learn too many or too complex concepts, decreasing overall performances.

Table 1. Comparison between *Shallow2Deep* and state-of-the-art models. We consider also models built through repetition of single *Shallow2Deep* cells—e.g. $Shallow2Deep_1$ is the NN built from repetition of *Shallow2Deep* cell 1 in a sequential manner.

Model name	Accuracy ± std (%)	Parameters (M)
Shallow2Deep	93.26 ± 0.18	0.251
$Shallow2Deep_1$	92.87 ± 0.17	0.165
$Shallow2Deep_2$	93.31 ± 0.14	0.491
$Shallow2Deep_3$	92.73 ± 0.14	0.377
$Shallow2Deep_4$	92.28 ± 0.10	0.118
VGG	93.66 ± 0.18	0.746
ResNet	92.77 ± 0.09	1.626

We also analyse the effects of cell structure variability in NN architectures. Base *Shallow2Deep* NN version intrinsically express high level of cell structure variability, while its variants – e.g. $Shallow2Deep_i$ – do not. It is possible to notice that *Shallow2Deep* NN outperforms 3 of its $Shallow2Deep_i$ variants out of 4 in terms of absolute performances. Moreover, *Shallow2Deep* NN outperforms all of its $Shallow2Deep_i$ variants when the accuracy/footprint trade-off is considered. Therefore, we can safely state that cell structure variability allows NN models to reach higher performances while being complexity-constrained.

6 Conclusion

In this work we propose *Shallow2Deep*, a novel NAS approach that limits NN complexity and promotes local variability in their architectures. *Shallow2Deep* relies on successive searches of local optima and NN expansions – i.e. depth increment – to produce well performing NN models.

We show that *Shallow2Deep* can effectively achieve NN complexity reduction, while reaching performances comparable to the state-of-the-art. Complexity reduction is tightly linked with NN opacity. Along this line, we also discuss why *Shallow2Deep* enables the application of explainability techniques to unprecedented scenarios, by providing a means to control NN structural design.

To the best of our knowledge, the proposed work represents the first approach to design an automatic tool to produce efficient NN architectures, while identifying complexity limits of NN models and helping designers to avoid overthinking issues or unnecessary opacity increments. In particular, this work represents a first approach to the analysis of structure variability influence on NN model performances, as *Shallow2Deep* promotes local variability. Along this path, our experimental analysis demonstrates how variability over local structures that compose NN is a desirable feature to obtain small and well performing models. This idea is in contrast with previously-proposed NN design approaches that neglect local structure variability, opening new possibilities for future research.

Acknowledgements. This paper has been partially supported by *(i)* the H2020 project "StairwAI" (G.A. 101017142), and *(ii)* the CHIST-ERA IV project "EXPECTATION" (G.A. CHIST-ERA-19-XAI-005).

References

1. Adadi, A., Berrada, M.: Peeking inside the black-box: a survey on explainable artificial intelligence (XAI). IEEE Access **6**, 52138–52160 (2018). https://doi.org/10.1109/ACCESS.2018.2870052
2. Bach, S., Binder, A., Montavon, G., Klauschen, F., Müller, K.R., Samek, W.: On pixel-wise explanations for non-linear classifier decisions by layer-wise relevance propagation. PLoS ONE **10**(7), 1–46 (2015). https://doi.org/10.1371/journal.pone.0130140
3. Arrieta, A.B., et al.: Explainable explainable artificial intelligence (XAI): concepts, taxonomies, opportunities and challenges toward responsible AI. Inf. Fusion **58**(December 2019), 82–115 (2020). https://doi.org/10.1016/j.inffus.2019.12.012
4. Bau, D., Zhou, B., Khosla, A., Oliva, A., Torralba, A.: Network dissection: quantifying interpretability of deep visual representations. In: 2017 IEEE Conference on Computer Vision and Pattern Recognition, CVPR 2017, Honolulu, HI, USA, 21–26 July 2017, pp. 3319–3327. IEEE Computer Society (2017). https://doi.org/10.1109/CVPR.2017.354
5. Cai, H., Zhu, L., Han, S.: ProxylessNAS: direct neural architecture search on target task and hardware. In: 7th International Conference on Learning Representations, ICLR 2019, New Orleans, LA, USA, 6–9 May 2019. OpenReview.net (2019). https://openreview.net/forum?id=HylVB3AqYm
6. Calegari, R., Ciatto, G., Omicini, A.: On the integration of symbolic and subsymbolic techniques for XAI: a survey. Intelligenza Artificiale **14**(1), 7–32 (2020). https://doi.org/10.3233/IA-190036
7. Casale, F.P., Gordon, J., Fusi, N.: Probabilistic neural architecture search. CoRR abs/1902.05116 (2019). http://arxiv.org/abs/1902.05116
8. Chen, S., Bateni, S., Grandhi, S., Li, X., Liu, C., Yang, W.: DENAS: automated rule generation by knowledge extraction from neural networks. In: Devanbu, P., Cohen, M.B., Zimmermann, T. (eds.) ESEC/FSE 2020: 28th ACM Joint European Software Engineering Conference and Symposium on the Foundations of Software Engineering, Virtual Event, USA, 8–13 November 2020, pp. 813–825. ACM (2020). https://doi.org/10.1145/3368089.3409733
9. Chu, X., Zhang, B., Xu, R.: Multi-objective reinforced evolution in mobile neural architecture search. In: Bartoli, A., Fusiello, A. (eds.) ECCV 2020. LNCS, vol. 12538, pp. 99–113. Springer, Cham (2020). https://doi.org/10.1007/978-3-030-66823-5_6
10. Ciatto, G., Schumacher, M.I., Omicini, A., Calvaresi, D.: Agent-based explanations in AI: towards an abstract framework. In: Calvaresi, D., Najjar, A., Winikoff, M., Främling, K. (eds.) EXTRAAMAS 2020. LNCS (LNAI), vol. 12175, pp. 3–20. Springer, Cham (2020). https://doi.org/10.1007/978-3-030-51924-7_1
11. Dosilovic, F.K., Brcic, M., Hlupic, N.: Explainable artificial intelligence: a survey. In: Skala, K., et al. (eds.) 41st International Convention on Information and Communication Technology, Electronics and Microelectronics, MIPRO 2018, Opatija, Croatia, 21–25 May 2018, pp. 210–215. IEEE (2018). https://doi.org/10.23919/MIPRO.2018.8400040

12. Elsken, T., Metzen, J.H., Hutter, F.: Neural architecture search: a survey. J. Mach. Learn. Res. **20**, 55:1–55:21 (2019). http://jmlr.org/papers/v20/18-598.html
13. Golovko, V., Egor, M., Brich, A., Sachenko, A.: A shallow convolutional neural network for accurate handwritten digits classification. In: Krasnoproshin, V.V., Ablameyko, S.V. (eds.) PRIP 2016. CCIS, vol. 673, pp. 77–85. Springer, Cham (2017). https://doi.org/10.1007/978-3-319-54220-1_8
14. Guidotti, R., Monreale, A., Ruggieri, S., Turini, F., Giannotti, F., Pedreschi, D.: A survey of methods for explaining black box models. ACM Computi. Surv. **51**(5) (2018). https://doi.org/10.1145/3236009
15. He, K., Zhang, X., Ren, S., Sun, J.: Deep residual learning for image recognition. In: 2016 IEEE Conference on Computer Vision and Pattern Recognition, CVPR 2016, Las Vegas, NV, USA, 27–30 June 2016, pp. 770–778. IEEE Computer Society (2016). https://doi.org/10.1109/CVPR.2016.90
16. Hecht-Nielsen, R.: Theory of the backpropagation neural network. Neural Netw. **1**(Supplement-1), 445–448 (1988). https://doi.org/10.1016/0893-6080(88)90469-8
17. Howard, A.G., et al.: MobileNets: efficient convolutional neural networks for mobile vision applications. CoRR abs/1704.04861 (2017). http://arxiv.org/abs/1704.04861
18. Janzing, D., Minorics, L., Blöbaum, P.: Feature relevance quantification in explainable AI: a causal problem. In: Chiappa, S., Calandra, R. (eds.) The 23rd International Conference on Artificial Intelligence and Statistics (AISTATS 2020). Proceedings of Machine Learning Research, vol. 108, pp. 2907–2916 (2020). http://proceedings.mlr.press/v108/janzing20a.html
19. Kaya, Y., Hong, S., Dumitras, T.: Shallow-deep networks: understanding and mitigating network overthinking. In: Chaudhuri, K., Salakhutdinov, R. (eds.) 36th International Conference on Machine Learning, ICML 2019, 9–15 June 2019, Long Beach, CA, USA. Proceedings of Machine Learning Research, vol. 97, pp. 3301–3310 (2019). http://proceedings.mlr.press/v97/kaya19a.html
20. Li, J., Liang, X., Shen, S., Xu, T., Feng, J., Yan, S.: Scale-aware fast R-CNN for pedestrian detection. IEEE Trans. Multimedia **20**(4), 985–996 (2018). https://doi.org/10.1109/TMM.2017.2759508
21. Lipton, Z.C.: The mythos of model interpretability. Queue **16**(3), 31–57 (2018). https://doi.org/10.1145/3236386.3241340
22. Liu, C., et al.: Progressive neural architecture search. In: Ferrari, V., Hebert, M., Sminchisescu, C., Weiss, Y. (eds.) ECCV 2018. LNCS, vol. 11205, pp. 19–35. Springer, Cham (2018). https://doi.org/10.1007/978-3-030-01246-5_2
23. Liu, H., Simonyan, K., Yang, Y.: DARTS: differentiable architecture search. In: 7th International Conference on Learning Representations, ICLR 2019, New Orleans, LA, USA, 6–9 May 2019. OpenReview.net (2019). https://openreview.net/forum?id=S1eYHoC5FX
24. Liu, J., Tripathi, S., Kurup, U., Shah, M.: Pruning algorithms to accelerate convolutional neural networks for edge applications: a survey. CoRR abs/2005.04275 (2020). https://arxiv.org/abs/2005.04275
25. Luo, R., Tan, X., Wang, R., Qin, T., Chen, E., Liu, T.: Neural architecture search with GBDT. CoRR abs/2007.04785 (2020). https://arxiv.org/abs/2007.04785
26. Miller, G.F., Todd, P.M., Hegde, S.U.: Designing neural networks using genetic algorithms. In: Schaffer, J.D. (ed.) 3rd International Conference on Genetic Algorithms, Fairfax, VA, USA, pp. 379–384. Morgan Kaufmann, June 1989

27. Nam, H., Han, B.: Learning multi-domain convolutional neural networks for visual tracking. In: 2016 IEEE Conference on Computer Vision and Pattern Recognition, CVPR 2016, Las Vegas, NV, USA, 27–30 June 2016, pp. 4293–4302. IEEE Computer Society (2016). https://doi.org/10.1109/CVPR.2016.465
28. Nguyen, A.M., Dosovitskiy, A., Yosinski, J., Brox, T., Clune, J.: Synthesizing the preferred inputs for neurons in neural networks via deep generator networks. In: Lee, D.D., Sugiyama, M., von Luxburg, U., Guyon, I., Garnett, R. (eds.) Advances in Neural Information Processing Systems 29: Annual Conference on Neural Information Processing Systems 2016, Barcelona, Spain, 5–10 December 2016, pp. 3387–3395 (2016). https://proceedings.neurips.cc/paper/2016/hash/5d79099fcdf499f12b79770834c0164a-Abstract.html
29. Nguyen, A.M., Yosinski, J., Clune, J.: Multifaceted feature visualization: uncovering the different types of features learned by each neuron in deep neural networks. CoRR abs/1602.03616 (2016). http://arxiv.org/abs/1602.03616
30. O'Shea, K., Nash, R.: An introduction to convolutional neural networks. CoRR abs/1511.08458 (2015). http://arxiv.org/abs/1511.08458
31. Peng, S., Ji, F., Lin, Z., Cui, S., Chen, H., Zhang, Y.: MTSS: learn from multiple domain teachers and become a multi-domain dialogue expert. In: AAAI Conference on Artificial Intelligence (AAAI-20 Technical Tracks 5), vol. 34, pp. 8608–8615. AAAI Press (2020). https://doi.org/10.1609/aaai.v34i05.6384
32. Real, E., Aggarwal, A., Huang, Y., Le, Q.V.: Regularized evolution for image classifier architecture search. In: AAAI Conference on Artificial Intelligence (AAAI-19, IAAI-19, EAAI-20), vol. 33, pp. 4780–4789. AAAI Press (2019). https://doi.org/10.1609/aaai.v33i01.33014780
33. Ren, S., He, K., Girshick, R.B., Zhang, X., Sun, J.: Object detection networks on convolutional feature maps. IEEE Trans. Pattern Anal. Mach. Intell. **39**(7), 1476–1481 (2017). https://doi.org/10.1109/TPAMI.2016.2601099
34. Rolnick, D., Tegmark, M.: The power of deeper networks for expressing natural functions. In: 6th International Conference on Learning Representations, ICLR 2018, Vancouver, BC, Canada, 30 April– 3 May 2018, Conference Track Proceedings. OpenReview.net (2018). https://openreview.net/forum?id=SyProzZAW
35. Samek, W., Binder, A., Montavon, G., Lapuschkin, S., Müller, K.: Evaluating the visualization of what a deep neural network has learned. IEEE Trans. Neural Networks Learn. Syst. **28**(11), 2660–2673 (2017). https://doi.org/10.1109/TNNLS.2016.2599820
36. Setiono, R., Leow, W.K., Zurada, J.M.: Extraction of rules from artificial neural networks for nonlinear regression. IEEE Trans. Neural Netw. **13**(3), 564–577 (2002). https://doi.org/10.1109/TNN.2002.1000125
37. Simonyan, K., Zisserman, A.: Very deep convolutional networks for large-scale image recognition. In: Bengio, Y., LeCun, Y. (eds.) 3rd International Conference on Learning Representations, ICLR 2015, San Diego, CA, USA, 7–9 May 2015, Conference Track Proceedings (2015). http://arxiv.org/abs/1409.1556
38. Tan, M., et al.: MnasNet: platform-aware neural architecture search for mobile. In: IEEE Conference on Computer Vision and Pattern Recognition, CVPR 2019, Long Beach, CA, USA, 16–20 June 2019, pp. 2820–2828. Computer Vision Foundation/IEEE (2019). https://doi.org/10.1109/CVPR.2019.00293
39. Tan, M., Le, Q.V.: EfficientNet: rethinking model scaling for convolutional neural networks. CoRR abs/1905.11946 (2019). http://arxiv.org/abs/1905.11946
40. Thrun, S.: Extracting rules from artificial neural networks with distributed representations. In: 7th International Conference on Neural Information Processing Systems (NIPS 1994), pp. 505–512. MIT Press (1994)

41. Tjoa, E., Guan, C.: A survey on explainable artificial intelligence (XAI): towards medical XAI. CoRR abs/1907.07374 (2019). http://arxiv.org/abs/1907.07374
42. Wistuba, M., Rawat, A., Pedapati, T.: A survey on neural architecture search. CoRR abs/1905.01392 (2019). http://arxiv.org/abs/1905.01392
43. Wu, B., et al.: FBNet: hardware-aware efficient convnet design via differentiable neural architecture search. In: IEEE Conference on Computer Vision and Pattern Recognition, CVPR 2019, Long Beach, CA, USA, 16–20 June 2019, pp. 10734–10742. Computer Vision Foundation/IEEE (2019). https://doi.org/10.1109/CVPR.2019.01099
44. Xiao, H., Rasul, K., Vollgraf, R.: Fashion-MNIST: a novel image dataset for benchmarking machine learning algorithms. CoRR abs/1708.07747 (2017). http://arxiv.org/abs/1708.07747
45. Yang, Z., et al.: CARS: continuous evolution for efficient neural architecture search. In: 2020 IEEE/CVF Conference on Computer Vision and Pattern Recognition, CVPR 2020, Seattle, WA, USA, 13–19 June 2020, pp. 1826–1835. IEEE (2020). https://doi.org/10.1109/CVPR42600.2020.00190
46. Ying, C., Klein, A., Christiansen, E., Real, E., Murphy, K., Hutter, F.: NAS-Bench-101: towards reproducible neural architecture search. In: Chaudhuri, K., Salakhutdinov, R. (eds.) Proceedings of the 36th International Conference on Machine Learning, ICML 2019, Long Beach, California, USA, 9–15 June 2019. Proceedings of Machine Learning Research, vol. 97, pp. 7105–7114. PMLR (2019). http://proceedings.mlr.press/v97/ying19a.html
47. Yosinski, J., Clune, J., Nguyen, A.M., Fuchs, T.J., Lipson, H.: Understanding neural networks through deep visualization. CoRR abs/1506.06579 (2015). http://arxiv.org/abs/1506.06579
48. Zhang, Q., Wu, Y.N., Zhu, S.: Interpretable convolutional neural networks. In: 2018 IEEE Conference on Computer Vision and Pattern Recognition, CVPR 2018, Salt Lake City, UT, USA, 18–22 June 2018. pp. 8827–8836. IEEE Computer Society (2018). https://doi.org/10.1109/CVPR.2018.00920
49. Zhang, Q., Yang, Y., Ma, H., Wu, Y.N.: Interpreting CNNs via decision trees. In: IEEE Conference on Computer Vision and Pattern Recognition, CVPR 2019, Long Beach, CA, USA, 16–20 June 2019, pp. 6261–6270. Computer Vision Foundation/IEEE (2019). https://doi.org/10.1109/CVPR.2019.00642
50. Zoph, B., Le, Q.V.: Neural architecture search with reinforcement learning. In: 5th International Conference on Learning Representations (ICLR 2017). Toulon, France, 24–26 April 2017. https://openreview.net/forum?id=r1Ue8Hcxg

Visual Explanations for DNNs with Contextual Importance

Sule Anjomshoae(✉), Lili Jiang, and Kary Främling

Department of Computing Science, Umeå University, Umeå, Sweden
{sulea,lili.jiang,kary.framling}@cs.umu.se

Abstract. Autonomous agents and robots with vision capabilities powered by machine learning algorithms such as Deep Neural Networks (DNNs) are taking place in many industrial environments. While DNNs have improved the accuracy in many prediction tasks, it is shown that even modest disturbances in their input produce erroneous results. Such errors have to be detected and dealt with for making the deployment of DNNs secure in real-world applications. Several explanation methods have been proposed to understand the inner workings of these models. In this paper, we present how Contextual Importance (CI) can make DNN results more explainable in an image classification task without peeking inside the network. We produce explanations for individual classifications by perturbing an input image through over-segmentation and evaluating the effect on a prediction score. Then the output highlights the most contributing segments for a prediction. Results are compared with two explanation methods, namely mask perturbation and LIME. The results for the MNIST hand-written digit dataset produced by the three methods show that CI provides better visual explainability.

Keywords: Deep learning · Explainable artificial intelligence · Image classification · Contextual importance

1 Introduction

Deep Neural Networks (DNNs) have improved the accuracy of prediction tasks in many applications including object recognition and natural language processing. However, DNNs inability to show their reasoning is hindering the use of these models in safety-critical systems such as in autonomous driving and medical domains. Moreover, their susceptibility to adversarial inputs (i.e., image and audio data is modified in a subtle way that is undetectable to humans) easily leads to incorrect predictions. The existing work on adversarial attacks on DNNs shows graffiti and art stickers cause to misclassify a turn right sign as a stop sign, and a stop sign as a 45-speed limit sign [10]. Explanations for such cases can help to evaluate the model and identify the patterns which the model has learned during training. This is a reason for the recent increase in research about explainable black-box algorithms as one means of working toward more robust and interpretable DNNs [2,6,24].

D. Calvaresi et al. (Eds.): EXTRAAMAS 2021, LNAI 12688, pp. 83–96, 2021.
https://doi.org/10.1007/978-3-030-82017-6_6

In this work, we present the concept of contextual importance to provide visual justifications for image classification on a standard DNN. The DNN's convolutional layer applies sliding filters to capture shift-invariant patterns and learn robust features to make predictions. Given the predicted class index and the probability, we investigate the most important features contributing to the prediction using a perturbation method. In this way, we present the visual representation of contextual importance value to justify the image classification results. Our main contributions are summarized as follows:

- We introduce contextual importance explanations for image classification tasks that can be applied to any CNN-based network without requiring alteration to the model.
- We present contrastive explanations that highlight class-discriminating features for multiple class predictions.
- We show examples of visual explanations on visibly distorted and noisy images.
- We present explanations in high confidence cases for incorrect predictions to help diagnose features contributing to misclassication.

2 Related Work

Researchers have been focusing on integrating explanation facilities into computer vision algorithms [20,21,23]. Generally, these explanation methods can be categorized as interpretable models and post-hoc explanations. Interpretable models focus on the internal functioning of the models; they analyze the interaction between neurons and what each neuron has learned. Post-hoc methods explain instance-specific predictions on the basis of how each feature influences the final outcome. In general, both approaches can have some limitations and strengths. The results of interpretable models are directly explainable without requiring another model to generate explanations. However, they restraint the model to increase comprehensibility, as a result, these may oversimplify the problem at hand. In contrast, the post-hoc explanations are not restricting the model but they may be limited in their approximate nature [8].

Several interpretable models have been proposed to make the complex blackbox models more understandable. Some of the techniques for understanding and diagnosing DNNs include gradients, visual analytics, and decomposition methods. Gradient methods highlight the unit changes and emphasize the important features or regions in an image. In this way, it is possible to find the prototypes that have the highest probability to be predicted as a certain class of a trained DNN. Considering this, Li *et al.*, proposed an interpretable neural network architecture whose predictions are based on the similarity of an input to a small set of prototypes learned during training [14]. This approach is able to produce artificial images that maximize a neuron activation or class confidence value. Nguyen *et al.*, presented the activation maximization method which synthesizes an image based on high activation on a neuron, then reveals the features learned by each neuron in an interpretable way [17]. Some works proposed visual analysis by clustering important neurons based on the features and the interactions between them [12,15]. Furthermore, decomposition methods such as layer-wise relevance methods are presented to analyze which pixels are contributing to what extent in a classification result [5]. Yoo *et al.*, proposed a regularization technique using forward and backward interacted-activation by defining a sum of layer-wise differences between neuron activations. This computation between forward and backward directions provide some kind of interpretability for DNNs [22].

On the other hand, post-hoc explanations are mostly proposed based on saliency maps and gradient visualizations methods in computer vision tasks. Saliency maps are being created in various ways. One way is to visualize by going backward through the inverted network from an output of interest. It highlights the discriminative features of the image with respect to the given class [21]. Another method uses class activation mapping with the gradients of a target input in the final convolutional layer to produce a rough localization map highlighting the important regions [19]. This method is further developed for explaining occurrences of multiple object instances in a single image [7]. In another work, class activation mapping is combined with the global average pooling layer to visualize class-specific image regions for revealing the attention map of DNNs on an image [25].

Moreover, some methods suggested generating explanations by approximating a black-box model by a simple model locally with the perturbations of the original instance. Then, they present the super-pixels with the highest positive weights as an explanation [18]. In our approach, contextual importance measures the influence of an individual feature on a prediction result by perturbing inputs without transforming the model into an interpretable one. The effects are then visualized to explain the outcome based on the main contributing group of pixels (i.e., interpretable regions). The visualization shows the relative importance of each region considering the whole image for the current prediction. Contextual importance explains the prediction results by analysing the effect on the output of the model, therefore remains faithful to the original model.

3 Method

The DNN (Deep Neural Network) architecture studied here is composed of three convolutional layers each followed by batch normalization, ReLu, max pooling, and finally fully connected layer as seen in Fig. 1. A convolution layer applies a sliding convolutional filter to an input image. The weighted sum of the input pixels within the window produces an output pixel at each point allowing the convolutional layer to learn visual patterns and features. Batch normalization reduces the sensitivity to network initialization and speeds up the training of the convolutional layer. ReLU activation function performs a threshold operation for each element by setting negative values to zero. Following this, the max-pooling layer performs down-sampling by dividing the input into sets of non-overlapping rectangles, and outputs the maximum for each region. Finally, a high-level reasoning is made through the fully connected layer in the network. This hierarchical architecture allows DNNs to extract increasingly abstract features from the first layer to the final layer.

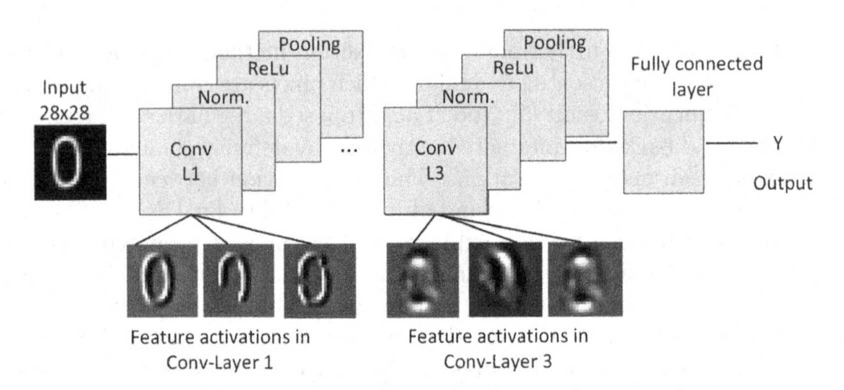

Fig. 1. DNN architecture for the recognition of the hand-written digit dataset

3.1 Contextual Importance

Contextual importance (*CI*) method is one of the earlier studies to address the post-hoc explainability for predictions made by black-box algorithms, specifically for a tabular data type [11]. However, the *CI*'s potential for image explanations has yet to be investigated. In this study, we propose generating visual explanations for DNNs through the contextual importance. This concept originates from the idea that the set of input features forms the context, and given this, the importance of feature is dependent on other feature values. Therefore, contextual importance indicates the degree of significance of a feature value (or set of feature values) when changes are made to that particular value(s) while the rest of the inputs remain constant. This is a model-agnostic approach which makes it possible to explain the outcomes of both linear and non-linear learning models as presented in [3,4].

Feature importance usually signifies a measure for how much one feature affects the outcome when taking into account the whole dataset. Building an explanation method is simple in the case of a purely linear model, where every feature's importance is constant and irrelevant to the other feature values. For linear models such as the weighted sum, the weight directly expresses the importance of each feature and the combined importance of several features corresponds to the sum of the weights. Neural networks are mainly useful for tasks where linear models are not sufficiently expressive. When speaking about post-hoc explanations of non-linear methods, the feature importance might be specific for the current set of input values.

When dealing with non-linear models, it becomes non-trivial how to define feature importance. It seems like most current model-agnostic methods use the local gradient as an indicator of the contextual importance [9]. The definition of contextual importance is based on a different principle. Rather than observing how much an output value changes with fixed amount of small perturbations to the current input value of the studied feature(s), we study how much perturbations over the whole range of possible feature values affects the output range. *CI* is then the ratio of the observed output range and the greatest possible output range. If the observed output range is greater for one feature than another, then the former is more important. Contextual importance is formally defined as:

$$CI_j(\vec{C},\{i\}) = \frac{Cmax_j(\vec{C},\{i\}) - Cmin_j(\vec{C},\{i\})}{absmax_j - absmin_j} \tag{1}$$

where $\vec{C}$ is the vector of input values that defines the context, $Cmax_j(\vec{C},\{i\})$ is the maximal value of output j observed when modifying the values of inputs $\{i\}$ and keeping the values of the other inputs at those specified by $\vec{C}$. Correspondingly, $Cmin_j(\vec{C},\{i\})$ is the minimal value of output j observed. $absmin_j$ and $absmax_j$ are typically the minimal and maximal values of output j in the training set, which signifies 0 and 1 for classification tasks. The estimation of $Cmax_j(\vec{C},\{i\})$ and $Cmin_j(\vec{C},\{i\})$ can be performed with Monte-Carlo simulation or more efficient sampling methods regardless of the black-box model or the learned function. For image case, we create samples through perturbing the image. This results in getting only one probability value for each perturbed sample, then the range for minimal and maximal values corresponds to probability value of when a region is on the scene and when it is absent. Therefore *CI* is defined as;

$$CI_j(C,\{i\}) = \frac{out_j(C) - out_j(C,\{i\})}{absmax_j - absmin_j} \tag{2}$$

where $out_j(C)$ is the value of the output j for the context C. $out_j(C,\{i\})$ is the prediction score for the perturbed sample. In this way, $CI_j(C,\{i\})$ expresses where the value of $out_j(C) - out_j(C,\{i\})$ is located in the $[absmax_j, absmin_j]$ range.

3.2 Contextual Importance Explanations for DNNs

DNNs learn to detect the most defining features such as color and edges in their first convolutional layer since this layer receives weighted connections from the input layer. The activation of each node is a weighted sum of pixel intensity values that are passed over to an activation function. So, the set of incoming weights to a node is measuring what that node is about. This can be easily observed in the first layer (see Fig. 1). Then the prediction is made following the last convolutional feature-map where the parts with the highest gradient are the most important for the prediction. Generally, the saliency maps created based on the final layer or another intermediate layer feature-map loses details significantly. We propose identifying interpretable regions through an over-segmentation method which is the process of segmenting the object(s) from the background and fracturing an image into subcomponents [1]. As a result, this increases the chances that boundaries of importance are extracted.

Then, we measure the contextual importance by masking each subcomponent at a time and present the results as visual evidence for a prediction. We note that "interpretable regions" are referred to as "features" throughout the paper.

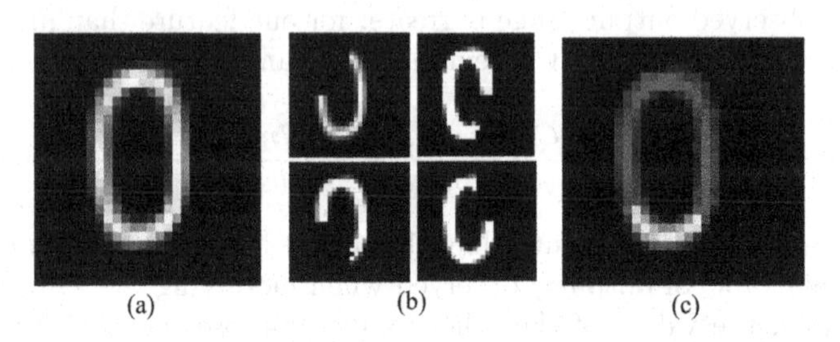

Fig. 2. Contextual importance explanations a)Input image b)Perturbed samples obtained through over-segmentation c)Visualization of the contextual importance

This process is consisting of four steps as outlined in Algorithm 1. Given an input C, we get the prediction class $out_j(C)$ (line 1). We then find samples $(C, \{i\}) \in C$ by filtering each region one after another from C where $(C, \{i\})$ is representative of the input, which is shown in Fig. 2(b). Model runs for each perturbed sample for prediction on output index j (line 2). We find the most important features by simply observing how the prediction score drops for the each interpretable samples $(C, \{i\})$ when they are not on the scene. Then, we concatenate prediction values from the interpretable samples for $out_j(C) - out_j(C, \{i\})$. We compute $CI_j(C, \{i\})$ and visualize the results (line 3 and line 4) as seen in Fig. 2(c). The CI values higher than the threshold (0.01) are represented in color. In this way, we identify which features were activated the most from the perturbed inputs. Our `MATLAB` implementation of this algorithm is available in the `GitHub` repository.[1]

Algorithm 1. Explanations for DNN with CI

Given: Context C that specifies input image, output index j, $absmin_j$, $absmax_j$, model f.
Require: Sample set $(C, \{i\})$ contains perturbed samples.

1: **Run** f with C to get $out_j(C)$
2: **Run** f on set $(C, \{i\})$ and get $out_j(C, \{i\})$
3: **Calculate** $CI_j(C, \{i\})$ using (2)
4: **Return** $CI_j(C, \{i\})$

[1] https://github.com/shulemsi/MNIST_CNN_CI_Explanations.

4 Experimental Results

We apply the proposed method to the MNIST hand-written digit dataset, which has 60,000 training and 10,000 test samples [13]. We provide region-wise explanations derived from the image context as supporting evidence for the class predictions.

The results of contextual importance are illustrated in Fig. 3. The proposed explanation method shows the contributing features to the predicted class based on the degree of contextual importance (i.e., those regions with higher CI value than the threshold). The second row in Fig. 3 shows the over-segmentation clusters, which are used as interpretable regions to perturb the image. The pixel location of high-significance features is rendered in color in the last row. We also report the prediction scores for the input image and the highlighted region, which shows the difference when the low-importance regions are left out. We observe that omitting the features with low importance slightly increased the confidence for the predicted class, which also justifies why the model predicts a certain class.

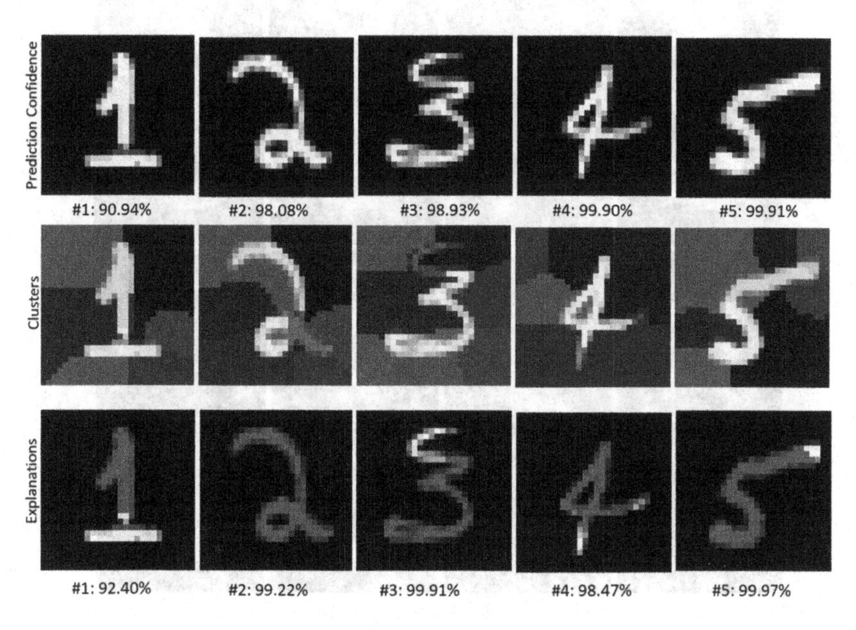

Fig. 3. Contextual importance explanations. The first line is the input images with the probability score for the predicted class. The second line is the over-segmentation clusters for generating perturbed samples. The last line shows features with the highest contextual importance and the prediction score for the explanations.

4.1 Visual Comparisons

To evaluate the proposed explanation method, we provide a visual comparison with two methods namely mask perturbation [23] and Local Interpretable Model-agnostic Explanations (LIME) [18]. Both methods produce explanations based on how the prediction score varies when the features are altered. With this exercise, we look for the immediate intelligibility and explicitness in visual explanations. Figure 4 shows the comparison results. The masking method analyses how sensitive the model's prediction on a class by occluding different parts of an image with a gray square, then provides a pixel-level explanation where the results are presented as heatmaps. Parts with high impacts on the output are highlighted with bright colors; conversely, low impacts are shown in dark colors.

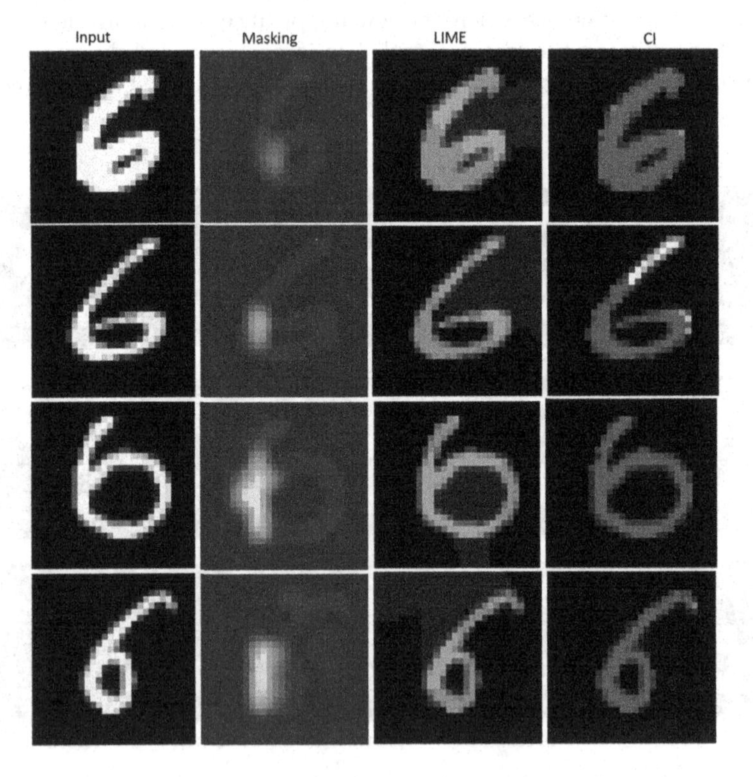

Fig. 4. Visual comparison of explanation methods. The masking method provides a saliency map based on the prediction scores. LIME highlights the relevant super-pixels for the prediction. Contextual importance provides a visual representation of the importance value as the most contributing features to the outcome.

LIME provides a region-based explanation by discovering the relevant segments. Segments with the highest positive weights are presented as an explanation (in red). Yet, the parts of the background are shown with positive relevance along with non-zero values. The masking method is implemented on CNN results and LIME uses the random forest algorithm for the hand-written digit classification task. While both methods give important intuitions, neither is precise in demonstrating the features leading to a prediction. Masking method gives significant insight into identifying the minimal region which has the most impact on the output, however their results are not immediately intelligible. Morever, the masking method could be computationally expensive when handling an instance with high dimensions since they need to sequentially perturb the image. LIME results in losing significant parts of the object. Given that humans intuitively pinpoint the most typical features of an class on an image, the contextual importance could be considered better for humans to understand and evaluate the predictions.

4.2 Explanations on Contrastive Cases

Humans generally explain the cause of an event relative to some other event that did not occur. Thus explanations are usually in the form of "Why this?- and not that", with a contrasting case that did not occur, even it is implicit in the question [16]. It is common for humans to state and request contrastive facts to distinguish between similar examples instead of giving complete explanations (i.e., listing all the causes of an event). Contrastive explanations could be more intuitive and valuable particularly for multivariate datasets since the cognitive load of complete explanations could be high in those cases.

As the contextual importance can be computed for every function that outputs a prediction value for all classes, this makes it possible to explain why a certain class $out_j(C)$ is more likely than another (i.e., here $out_c(C)$ indicates the contrastive case). For this, feature importance over the perturbation variable $(C, \{i\})$ is computed to explain the model's prediction results for the contrasting case $out_c(C)$ and compare it with the initial case. Thus, we are not only interested in visualizing the present features but also looking for missing features. Given an input C with prediction class $out_c(C)$, we find predictions for interpretable perturbations for the index of the contrastive case. We find the importance of the features that are contributing to prediction class $out_c(C)$ using the same equation (Eq. 2). The visualization demonstrates the most critical features contributing to different classes (see Fig. 5).

Thereby contrastive explanations are generated in the form of either;

- The C is predicted as class $out_j(C)$, because features $(C, \{i\})$ have high importance, which are not typical of class $out_c(C)$.
- The C is not predicted as class $out_j(C)$, because features $(C, \{i\})$ has no importance, which are typical of class $out_c(C)$.

Here, an example is presented for number 7 and a set of features that distinguish it from number 2, which has the second high probability. Figure 5 illustrates the visualizations of contextual importance values as well as the absent features in number 2, which distinguishes it from 7. Our comparisons with LIME show

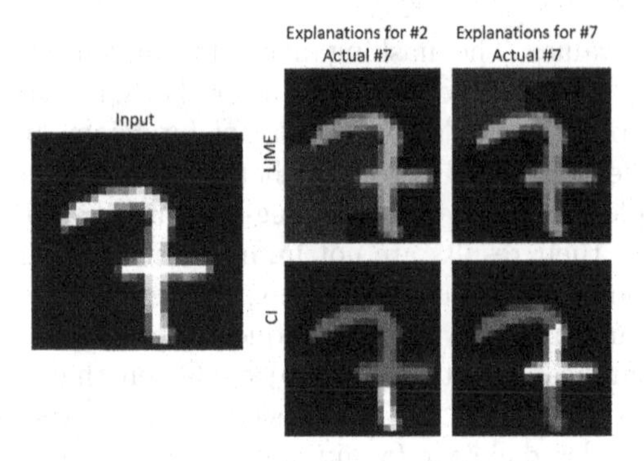

Fig. 5. CI and LIME comparisons for contrastive cases.

varying explanation results for both the actual class and the contrastive case. The important features are highlighted in order to identify and distinguish the two classes. LIME shows background as the positive relevance besides the non-zero pixels. The reason for this could be the result of the clustering algorithm. Moreover, it is very often that LIME produces the same explanations for different class predictions. Figure 6 demonstrates this for two different forms of the same digit. LIME resulted in identical explanations for number 5 and actual class number 6, which is not very explanatory to know why this image is labeled as either of the two classes.

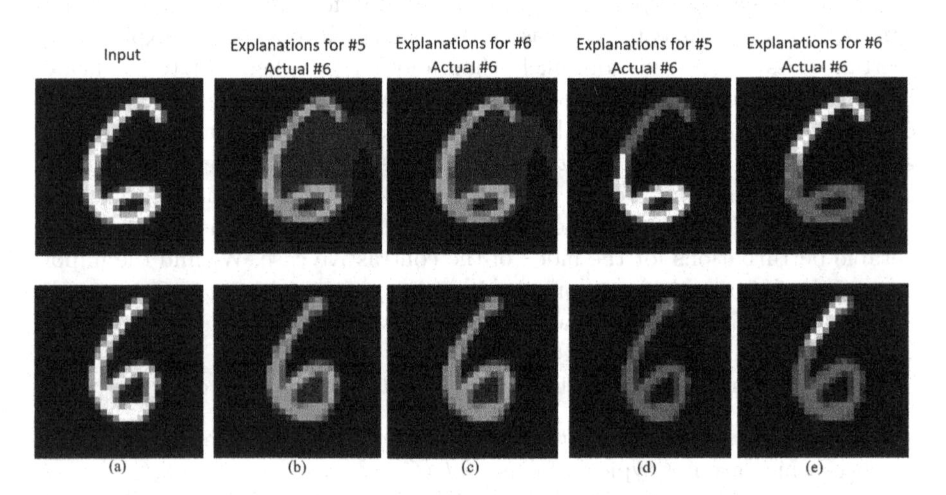

Fig. 6. Contrastive explanations (a) Input image number 6 (b) LIME explanations for 5 (c) LIME explanations for 6 (d) Contextual importance explanations for 5 (e) Contextual importance explanations for 6.

4.3 Explanations on Distorted Images

Even minor alterations in an image can produce completely different classification results. Handwritten text is likely to have varying noises such as underlines, neighboring characters, or stray marks. Here, we tested the model with different images of the numbers that were not in the training set to observe whether the model can extract enough common features from distorted pictures to identify a number. Finally, the explanations are generated to check whether they are having invariance to these visual differences.

We also tested the contextual importance for the consistency of the explanations by adding noise to the input image and evaluating how it affects the explanations. The degree of noise resistance and distortion invariance is experimented through contextual importance. The distorted images are tested to verify if the model predicts the correct label. The importance over the perturbation variable $(C, \{i\})$ is computed for the output index j, then the contextual importance values are visualized for each class. We are interested in visualizing whether such noises would be shown as relevant features and if the explanations are robust to such variations.

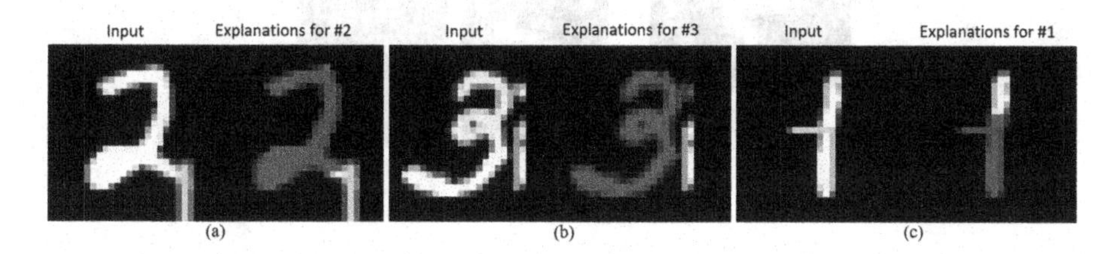

Fig. 7. Contextual importance explanations for partially distorted and noisy images (a) Explanations for 2 (98.80%) (b) Explanations for 3 (99.95%) (c) Explanations for 1 (98.21%).

The explanations for the distorted numbers are shown in Fig. 7. Despite the noise and distortion, these samples were correctly recognized by the model with high confidence. For the given examples, DNN is able to extract the salient features from the cluttered image and contextual importance provided robust explanations under the noisy conditions. Although the completely invariant classification of complex shapes in the physical world is still a challenging task, contextual importance could provide a solution to the problem of robustness concerning partial distortions. We also note that such explanations depend on the outcome of over-segmentation clusters.

4.4 Explanations on Misclassification

In this section, we show explanations for a misclassification case to identify the features causing a wrong prediction. The contextual importance for the perturbation variable $(C, \{i\})$ is computed for the predicted class and for the true class.

Then, both classes are visualized to demonstrate the relevant features for the incorrect class $out_j(C)$ and analyse why the model predicts as $out_j(C)$ instead $out_c(C)$. We present LIME and contextual importance explanations for example of true class number 9 and predicted class number 4 in Fig. 8. This input example classified as number 4 by 51% and the actual class is 9 by 19% probability. LIME gives explanations for class number 4 (12%) while it failed to find any relevant features for the actual class (6%). The contextual importance features represented for number 4 (by 99%) are resembles to explanations produced by LIME as shown in Fig. 8. Contextual importance for ground truth (9) (by 26%) highlights all the pixels as important. Although the prediction score is low for the ground truth, it still provides discriminative features for each class.

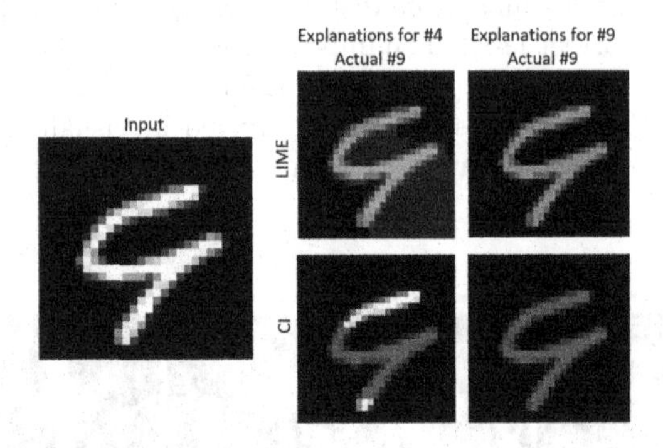

Fig. 8. CI and LIME comparisons on misclassification

This kind of result gives no direct explanation about why a model makes a wrong prediction, however, they could potentially enhance trust as it helps to identify the contributing features and evaluate whether the model is performing in an arbitrary way. Hence, understanding the features learned by a model will give an opportunity to improve the dataset and correct the model.

5 Conclusion

In this paper, we proposed contextual importance to provide visual explanations for DNN predictions. The method presents region-wise explanations for image classification results by visualizing the contribution of each region based on the degree of importance. The visual comparison results show that the contextual importance provides explicit visual justification for the DNN predictions. The results also demonstrate that the region with the high importance gives a class score close to the initial prediction score. This suggests that the proposed method is able to extract the most relevant features for the prediction to justify an outcome. The idea is further supported by providing explanations for wrong predictions to investigate the regions contributing to misclassification. The results indicate that contrastive explanations offer a way to analyze the misclassified examples and identify class discriminative features. Being limited to only a hand-written digit dataset, this study lacks the implementation of contextual importance for the multi-object classification explanations, which is an important consideration for our future work. Another interesting research direction could be exploring contextual importance for text-based explanations in a multi-object classification task.

Acknowledgements. This work was supported by the Wallenberg AI, Autonomous Systems and Software Program (WASP) funded by the Knut and Alice Wallenberg Foundation.

References

1. Achanta, R., Shaji, A., Smith, K., Lucchi, A., Fua, P., Süsstrunk, S.: SLIC superpixels compared to state-of-the-art superpixel methods. IEEE Trans. Pattern Anal. Mach. Intell. **34**(11), 2274–2282 (2012)
2. Alvarez-Melis, D., Jaakkola, T.S.: Towards robust interpretability with self-explaining neural networks. In: Proceedings of the 32nd International Conference on Neural Information Processing Systems, pp. 7786–7795. Curran Associates Inc. (2018)
3. Anjomshoae, S., Främling, K., Najjar, A.: Explanations of black-box model predictions by contextual importance and utility. In: Calvaresi, D., Najjar, A., Schumacher, M., Främling, K. (eds.) EXTRAAMAS 2019. LNCS (LNAI), vol. 11763, pp. 95–109. Springer, Cham (2019). https://doi.org/10.1007/978-3-030-30391-4_6
4. Anjomshoae, S., Kampik, T., Främling, K.: Py-CIU: a python library for explaining machine learning predictions using contextual importance and utility. In: IJCAI-PRICAI 2020 Workshop on Explainable Artificial Intelligence (XAI) (2020)
5. Bach, S., Binder, A., Montavon, G., Klauschen, F., Müller, K.R., Samek, W.: On pixel-wise explanations for non-linear classifier decisions by layer-wise relevance propagation. PLoS one **10**(7), e0130140 (2015)
6. Bodria, F., Giannotti, F., Guidotti, R., Naretto, F., Pedreschi, D., Rinzivillo, S.: Benchmarking and survey of explanation methods for black box models. arXiv preprint arXiv:2102.13076 (2021)
7. Chattopadhay, A., Sarkar, A., Howlader, P., Balasubramanian, V.N.: Grad-CAM++: generalized gradient-based visual explanations for deep convolutional networks. In: 2018 IEEE Winter Conference on Applications of Computer Vision (WACV), pp. 839–847. IEEE (2018)
8. Du, M., Liu, N., Hu, X.: Techniques for interpretable machine learning. Commun. ACM **63**(1), 68–77 (2019)
9. Du, M., Liu, N., Hu, X.: Techniques for interpretable machine learning. Commun. ACM **63**(1), 68–77 (2019). https://doi.org/10.1145/3359786
10. Eykholt, K., et al.: Robust physical-world attacks on deep learning visual classification. In: Proceedings of the IEEE Conference on Computer Vision and Pattern Recognition, pp. 1625–1634 (2018)
11. Främling, K.: Modélisation et apprentissage des préférences par réseaux de neurones pour l'aide à la décision multicritère. Ph.D. thesis, Institut National de Sciences Appliquées de Lyon, Ecole Nationale Supérieure des Mines de Saint-Etienne, France (1996)

12. Hohman, F., Park, H., Robinson, C., Chau, D.H.: Summit: scaling deep learning interpretability by visualizing activation and attribution summarizations. arXiv preprint arXiv:1904.02323 (2019)
13. LeCun, Y., Bottou, L., Bengio, Y., Haffner, P.: Gradient-based learning applied to document recognition. Proc. IEEE **86**(11), 2278–2324 (1998)
14. Li, O., Liu, H., Chen, C., Rudin, C.: Deep learning for case-based reasoning through prototypes: a neural network that explains its predictions. In: Thirty-Second AAAI Conference on Artificial Intelligence (2018)
15. Liu, M., Shi, J., Li, Z., Li, C., Zhu, J., Liu, S.: Towards better analysis of deep convolutional neural networks. IEEE Trans. Visual Comput. Graphics **23**(1), 91–100 (2016)
16. Miller, T.: Explanation in artificial intelligence: insights from the social sciences. Artif. Intell. **267**, 1–38 (2019)
17. Nguyen, A., Dosovitskiy, A., Yosinski, J., Brox, T., Clune, J.: Synthesizing the preferred inputs for neurons in neural networks via deep generator networks. In: Advances in Neural Information Processing Systems, pp. 3387–3395 (2016)
18. Ribeiro, M.T., Singh, S., Guestrin, C.: Why should i trust you?: Explaining the predictions of any classifier. In: Proceedings of the 22nd ACM SIGKDD International Conference on Knowledge Discovery And Data Mining, pp. 1135–1144. ACM (2016)
19. Selvaraju, R.R., Cogswell, M., Das, A., Vedantam, R., Parikh, D., Batra, D.: Grad-CAM: visual explanations from deep networks via gradient-based localization. In: Proceedings of the IEEE International Conference on Computer Vision, pp. 618–626 (2017)
20. Shrikumar, A., Greenside, P., Kundaje, A.: Learning important features through propagating activation differences. In: Proceedings of the 34th International Conference on Machine Learning, vol. 70, pp. 3145–3153 (2017). JMLR.org
21. Simonyan, K., Vedaldi, A., Zisserman, A.: Deep inside convolutional networks: visualising image classification models and saliency maps. arXiv preprint arXiv:1312.6034 (2013)
22. Yoo, C.H., Kim, N., Kang, J.W.: Relevance regularization of convolutional neural network for interpretable classification. Network **50**, 5 (2019)
23. Zeiler, M.D., Fergus, R.: Visualizing and understanding convolutional networks. In: Fleet, D., Pajdla, T., Schiele, B., Tuytelaars, T. (eds.) ECCV 2014. LNCS, vol. 8689, pp. 818–833. Springer, Cham (2014). https://doi.org/10.1007/978-3-319-10590-1_53
24. Zhang, Q., Nian Wu, Y., Zhu, S.C.: Interpretable convolutional neural networks. In: Proceedings of the IEEE Conference on Computer Vision and Pattern Recognition, pp. 8827–8836 (2018)
25. Zhou, B., Khosla, A., Lapedriza, A., Oliva, A., Torralba, A.: Learning deep features for discriminative localization. In: Proceedings of the IEEE Conference on Computer Vision and Pattern Recognition, pp. 2921–2929 (2016)

Towards Explainable Recommendations of Resource Allocation Mechanisms in On-Demand Transport Fleets

Alaa Daoud[1(✉)], Hiba Alqasir[2], Yazan Mualla[3], Amro Najjar[4], Gauthier Picard[5], and Flavien Balbo[1]

[1] Mines Saint-Étienne, CNRS, UMR 6158, LIMOS - Institut Henri Fayol, Saint-Étienne, France
{alaa.daoud,flavien.balbo}@emse.fr

[2] Université Lyon, UJM-Saint-Etienne, CNRS, Institut Optique Graduate School, Laboratoire Hubert Curien UMR 5516, Saint-Étienne, France
h.alqasir@univ-st-etienne.fr

[3] CIAD, Univ. Bourgogne Franche-Comté, UTBM, 90010 Belfort, France
yazan.mualla@utbm.fr

[4] University of Luxembourg, Esch-sur-Alzette, Luxembourg
amro.najjar@uni.lu

[5] ONERA/DTIS, Université de Toulouse, Toulouse, France
gauthier.picard@onera.fr

Abstract. Multi-agent systems can be considered a natural paradigm when modeling various transportation systems, whose management involves solving hard, dynamic, and distributed allocation problems. Such problems have been studied for decades, and various solutions have been proposed. However, even the most straightforward resource allocation mechanisms lead to debates on efficiency vs. fairness, business quality vs. passenger's user experience, or performance vs. robustness. We aim to design an analytical tool that functions as a recommendation system for on-demand transport (ODT) authorities. This tool recommends specific allocation mechanisms that match the authority's objectives and preferences to solve allocation problems for particular contextual scenarios. The paper emphasizes the need for transparency and explainability of resource allocation decisions in ODT systems to be understandable by humans and move toward a more controllable resource allocation. We propose in this preliminary work a multi-agent architecture and general implementation guidelines towards meeting these requirements.

Keywords: Multi-agent systems · Explainable Artificial Intelligence · Intelligent transport systems · Resource allocation

1 Introduction

Today's transport systems are constructed of complex, large-scale interactions in a dynamic environment. In on-demand transport (ODT) systems, a fleet of vehicles is distributed in an urban area to meet potential requests to transfer people

D. Calvaresi et al. (Eds.): EXTRAAMAS 2021, LNAI 12688, pp. 97–115, 2021.
https://doi.org/10.1007/978-3-030-82017-6_7

or goods between origin and destination locations. Agent-based and multi-agent systems provide a suitable scheme to model such complexity. In multi-agent models of ODT systems, vehicles are represented by agents that are mobile in their spatial environment and may have communication abilities. The system's spatial environment consists of a network of roads, facilities, and urban infrastructure artifacts. The agents may have the possibility to communicate with each other and with other system entities to share information and coordinate their actions [34].

The allocation problems are major issues in the management of ODT systems. They have been studied for decades, and various solutions have been proposed. However, even the most straightforward cases of resource allocation lead to debates on efficiency versus fairness [25], business quality versus passengers' experience [36], and performance versus robustness [22].

We are interested in building an analytical tool that functions as a recommendation system for resource allocation methods for ODT scenarios. There are several stakeholders involved in ODT systems including passengers, drivers, service providers, etc.). What we mean by the term *User* in this document is a human user representing the transport authority and looking for the best solution method to solve the problem regarding the authority's preferences and the actual context parameters.

This potential tool takes as input the set of parameters for the scenario (vehicle fleet properties and request distribution model), user's objective function, and preferences, in addition to the environment model (road network and traffic model).

This system simulates the problem scenario and its solutions with different classes of AI methods, then produces to the user the recommended solution model (the solution method and its tuned parameters) that produce results matching the user objective and preferences for the input scenario.

In future Artificial Intelligence (AI) systems, it is vital to guarantee a smooth human-agent interaction, as it is not straightforward for humans to understand the agent's state of mind, and explainability is an indispensable ingredient for such interaction [28]. Recent works in the literature highlighted explainability as one of the cornerstones for building trustworthily responsible and acceptable AI systems [24,32,35]. Consequently, the emerging research field of eXplainable Artificial Intelligence (XAI) gained momentum both in academia and industry [3,6,19]. XAI is allowing, through explanations, users to understand, trust, and effectively manage the next generation of AI solutions [20].

Providing users with some form of control over the recommendation process can be realized by allowing them to tell the system what they like or by engaging them in adjusting the recommendation profile to synthesize recommendations from different sources [40]. High-quality explanations allow a better understanding of the results and help the user to make the right decisions. Reliable answers increase confidence in the system, while explanations that reflect system inaccuracies allow the user to modify the system's reasoning or control the weighting parameter that reorganizes or regenerates recommendations.

2 About the Need for Explainability

The human perspective is what differentiates ODT from most routing and transport problems; in addition to the technical factors, the quality of the service is influenced by human satisfaction factors, including the stability of service quality, service availability, wait-time, information privacy, passengers' special constraints, and preferences [9].

The following examples show that global system decisions may not fit all stakeholders' preferences: a decision may make some people dissatisfied.

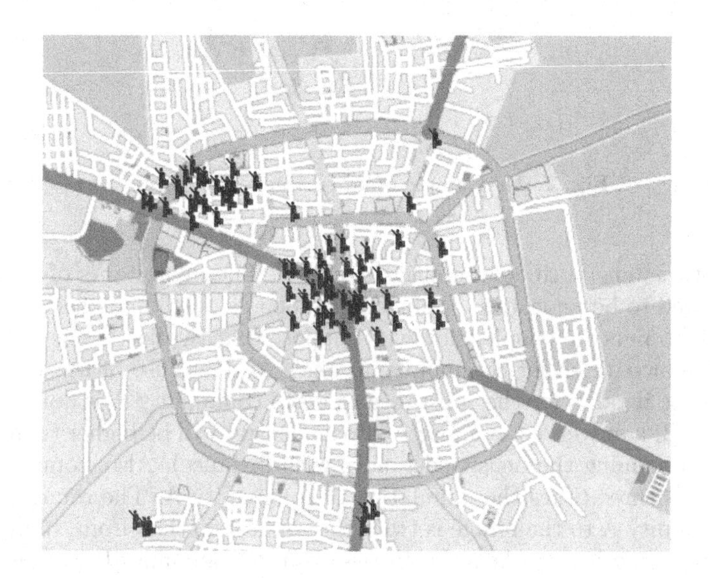

Fig. 1. Passenger request distribution at rush hours.

Scenario 1: Dial-a-ride in Rush Hours. At rush hours, taxi-ride demand is usually concentrated at specific parts of the city, e.g., city center and train stations, as seen in Fig. 1. The objective of the transport authority is to maximize the number of satisfied requests while reducing operational costs. An efficient allocation mechanism will dispatch as many vehicles as possible to the crowded areas to serve passengers, prioritizing the requests whose destinations are near other crowded areas. As a consequence, in this example, most of the vehicles move back and forth between the two areas, which reduces the chance of far passengers and makes them wait for a long time for being served, regardless of the urgency level of their requests that may be higher than those who do their ordinary work-home trips from the city center.

Scenario 2: Emergency Management ODT. The example of Fig. 2, introduced by [1] represents a disaster management situation. However, this kind of emergency transport can be modeled as an ODT system [4,37,41]. In this example, a failure in facility X leads to a leak of toxic substances. The leakage grows over

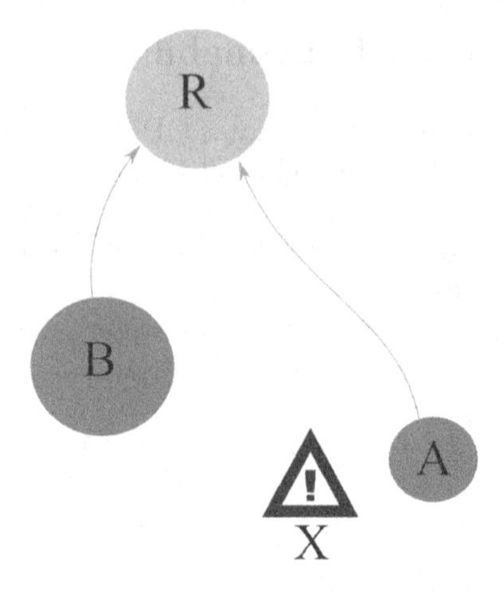

Fig. 2. An example of an emergency scenario.

time and threatens both communities A and B. The inhabitants of these communities need to be relocated to refuge R as soon as possible. A fleet of shuttles is available to relocate people. However, suppose that the fleet's size is not large enough to evacuate either community in one hour. Because of the wind direction, the time it takes for the substance to reach community A is double that of community B; however, community A's toxic density will be higher than in community B (assuming the density degrades with distance). Also, community B's population is three times the population of community A. The round-trip time from community A to the refuge is twice the round-trip time from community B to the refuge. In other words, a shuttle assigned to community B can carry twice the number of evacuees compared to the same shuttle assigned to community A. If the goal is to maximize the number of evacuees moved to the refuge within one hour, the answer would be to assign the entire fleet to community B since the round trip time is shorter for this community. However, if the goal is to evacuate high-risk individuals as quickly as possible, the answer would be to assign the entire fleet to community A. While both of these responses seem correct for the corresponding objective, neither seems fair.

Providing explanations for the system decision may increase people's satisfaction [5], and maintain the AI system's acceptability. When a recommendation mechanism is too complicated for lay users, the system may need to justify why the recommendation has been made [12,39]. The EU General Data Protection Regulation introduces a right of explanation for citizens to obtain "meaningful information about the logic involved" for automated decisions [17]. Generating explanations of autonomous decisions in multi-agent environments is even more difficult than providing explanations in other contexts [23]. In addition to identifying the technical reasons that led to the decision, it is necessary to convey the agents' preferences. It is necessary to decide what to reveal about other agents' preferences to increase user satisfaction while considering other agents' privacy, and how those features led to the final decision.

To provide useful explanations, it is necessary to identify the features of the context and decisions relevant to a specific user. Given these features, other relevant agents' preferences should be identified, and any relevant statements that touch on important concepts such as fairness should be generated. Using these features, preferences, and concepts, various explanations could be generated using subsets of them. The selected subset should be transferred in a certain communication form. The personalization of explanations could also be used at this stage since explanations are subjective and depend on multiple factors [38]. As to personalize explanations, there is a need to build a user, or mental model [21] that influences the generation of explanations.

In our resource allocation scenario in vehicle fleets, the allocation process can provide a set of constraints that lead to the proposed allocation. It will be necessary to identify the relevant constraints and generalize statements related to other agents' preferences and general system constraints related to fairness [26]. Then, we can use user satisfaction models to choose the best constraints, and generalized statements to present.

3 AV-OLRA Metamodel

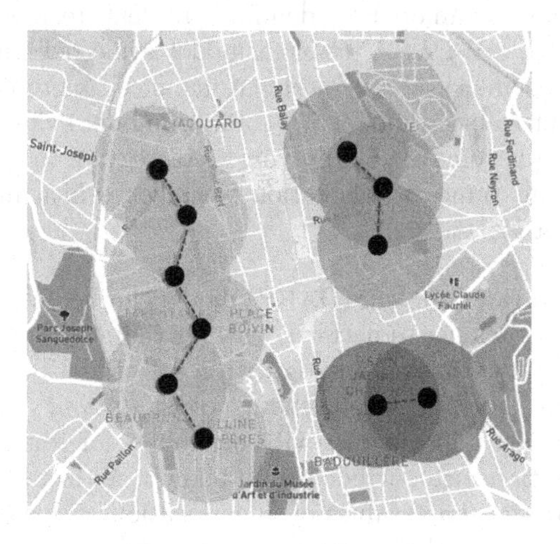

Fig. 3. AV-OLRA's dynamic composition of connected sets.

In previous work, we presented *AV-OLRA* a metamodel for resource allocation in autonomous vehicle fleets [11]. In this model, an autonomous vehicle is any vehicle that can make autonomous decisions, and interact with other entities in the surrounding environments, besides its self-driving capabilities. We consider vehicles communicate locally within limited ranges, and can pass transitive messages.

Connectivity between two vehicles is achieved if the distance between them is less than or equals their communication range. However, as the vehicles' communication range is limited, and to maximize their connectivity, two vehicles can be connected by transitivity if both are connected to another vehicle. We define the concept of *connected set* (CS) as a dynamic set of entities that can communicate with each other either directly or by transitive message passing. CSs are composed, split, and merged at run-time based on vehicles' movement as shown in Fig. 3. When the communication range is long enough, all vehicles in some urban area can communicate globally i.e. all the vehicles belong to one CS.

Considering the transport requests as dynamic resources that can be consumed by or allocated to vehicles, the *AV-OLRA* metamodel is formulated as:

$$AV\text{-}OLRA := (\mathcal{R}, \mathcal{V}, \mathcal{G}, \mathcal{T}) \tag{1}$$

where $\mathcal{R}$ defines a dynamic set of resources that occur to be available for a specific time window at the time of execution, representing passengers' requests; the set of consumers $\mathcal{V}$ represent a fleet of m autonomous vehicles that are mobile and can only communicate within a limited range; $\mathcal{G}$ is a directed graph representing the urban infrastructure network that defines the problem spatial environment, with $\mathcal{N}$ the set of nodes, and $\mathcal{E}$ the set of edges, $e_{ij} \in \mathcal{E}$ is the edge between the nodes i and j, ω is a valuation function that associates each edge $e \in \mathcal{E}$ with the value ω_e based on a temporal distance measure (e.g., average driving time in minutes), which will be used to calculate the operational costs of vehicle trips; $\mathcal{T}$ defines the temporal dimension of the problem as a discrete-time horizon.

Instantiating this metamodel by defining the feature model of these components results in an *AV-OLRA* problem model while defining these features' exact values leads to an *AV-OLRA* problem instance. A problem model can be solved with different solution models. A solution model defines the strategy by which the allocation is computed. Applying a strategy X to a problem instance I results in assigning values to allocation variables, which means achieving a feasible solution if it exists.

Example. The dial-a-ride problem (DARP) model can be defined in some urban area u by defining the u's urban network features (number of nodes, edges, facilities, etc.). The fleet's vehicles are taxis with a set of attributes for capacity as the number of seats, average speed, energy consumption, and communication range. The requests are trip requests with attributes for some passengers, pick-up and drop-off locations, time-window, and budget. An instance of this problem model is defined by the exact values of node locations, edge distances, number of vehicles in the fleet with their capacities speed, range, and initial locations, in addition to the set of passenger requests and the time slot in which the scenario takes place (the time horizon of the problem instance). Here the allocation variables are the vehicles' schedules; for each vehicle, we have a schedule as a list of couples $(location, time)$ defining the locations that the vehicle needs to visit (for pick-up or delivery) and their potential visit time.

4 Explainable MAS for AV-OLRA Recommendation

In this section, we introduce *EX-AV-OLRA*, an extension to *AV-OLRA* metamodel with explainability-related components. We present a multi-agent model for an explainable recommendation system that realizes the *EX-AV-OLRA* model. *EX-AV-OLRA* model is formulated as:

$$EX\text{-}AV\text{-}OLRA := (\mathcal{R}, \mathcal{V}, \mathcal{G}, \mathcal{T}, \mathcal{X}) \tag{2}$$

where $(\mathcal{R}, \mathcal{V}, \mathcal{G}, \mathcal{T})$ define an *AV-OLRA* and $\mathcal{X}$ defines the explaining mechanism.

We aim to design a recommender system in which a human user sets the scenario parameters to create an AV-OLRA instance, setting objective and utility preferences. The system's output is a recommendation to use the solution method that is the best match to user preferences, supported with multi-level explanations of why particular methods are recommended and why others are discouraged.

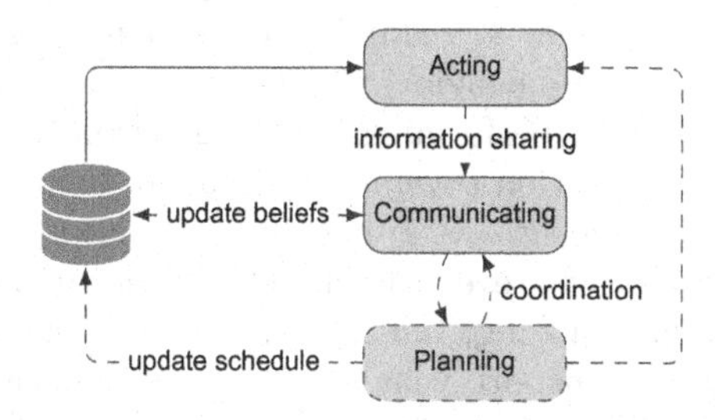

Fig. 4. Generic AV agent behavior in *AV-OLRA* (dashed components are generic, to be implemented for any specific strategy).

The generic multi-agent model of *AV-OLRA* consists basically of *Autonomous Vehicle (AV)* agents who are mobile in their spatial environment to serve trip requests and may communicate within a limited range with other agents and surrounding artifacts. We can distinguish three different *sub-behaviors* (*acting*, *communicating*, and *planning*) shown in Fig. 4.

The multi-agent model for the explainable recommender system extends the previous model. An additional agent type *Monitor Agent (MA)* plays the role of proxy for *AV*s to produce human-readable personalized explanations for the recommended methods. Unlike the inter-*AV*s' limited-range communication model, the *MA* can interact with *AV*s globally (see Fig. 5).

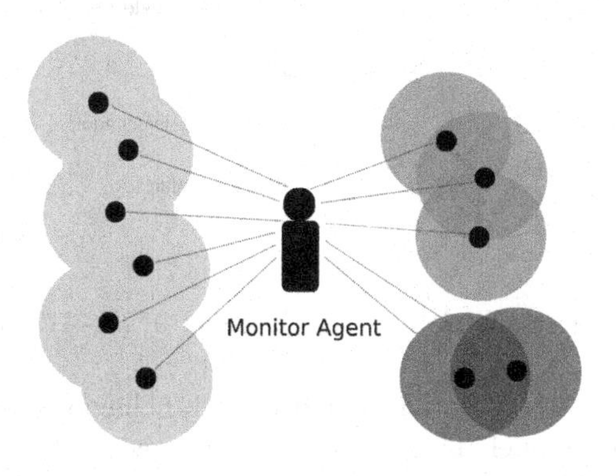

Fig. 5. *MA* and *AV* agents interaction.

This interaction is only to monitor the performance of AV agents and logging the explanations of their actions during the simulation. The MA's role is to aggregate the explanations of AVs' actions during the simulation to enable building explainable recommendations. In practice, only the AVs are deployed. Their communication constraints in real world scenarios should be taken into account during the simulation. To do so, this global interaction means should never be used for communication between AVs.

4.1 AV Agents' Behavior

An AV agent repeatedly performs the following actions representing the behavior of a vehicle in the system: 1. read the received messages and update the context (*communicating* sub-behavior), 2. choose the locations to visit (*planning* sub-behavior), 3. act by performing a driving action (*acting* sub-behavior), 4. broadcast context information (*communicating* sub-behavior).

The *acting* and *communicating* sub-behaviors are always the same whatever the problem instance and whatever the chosen solution model. The AV agent can perform four actions (moving, waiting-for/marauding requests, picking-up, and dropping off) as a transport vehicle. As a communicating agent, an AV can join/leave a connected set and send, receive, or broadcast messages. The communication behavior depends mainly on the value of the communication range, which is an attribute of the scenario. The *planning* sub-behavior represents how an AV obtains its dynamic schedule. This behavior depends on the allocation mechanism, which is specific to each coordination mechanism that defines the solution model. A *coordination mechanism* is defined by three components $\langle DA, AC, AM \rangle$, where DA denotes the level of decision autonomy which is either centralized (C) or decentralized (D); AC denotes the agents' cooperativeness level with (S) or without (N) sharing of schedule information, and AM is the chosen allocation mechanism (e.g. "Auctions", "Greedy", "DCOP", etc.).

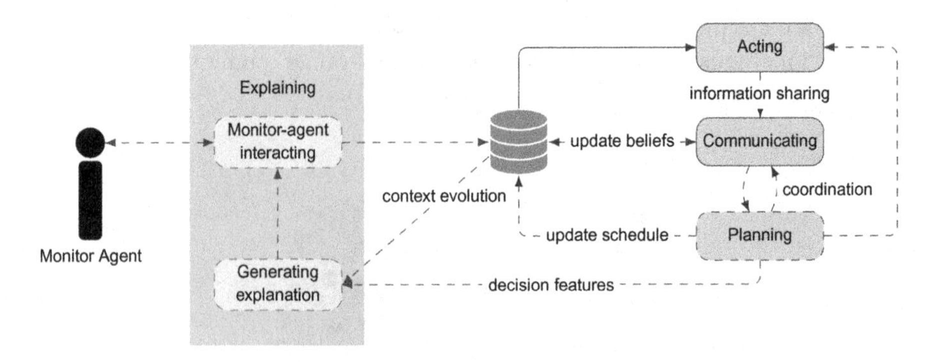

Fig. 6. Explainable AV agent behavior in *EX-AV-OLRA*

In this work, we propose to add another sub-behavior (the *explaining* sub-behavior) to the AV model. This sub-behavior consists of two phases *generating explanation* and *monitor-agent interacting* as shown in Fig. 6.

Table 1. Examples of solution models and what should be explained.

Solution	*DA*	*AC*	*AM*	Explanation examples
Selfish	*D*	*N*	Greedy	Why prioritizing a specific request?
Dispatching	*C*	*S*	MILP	Which constraints are violated?
Market	*D*	*S*	Auctions	How winner determination computed? Why accepting some trade options?
Cooperative	*D*	*S*	DCOP	What are individual costs and utilities?

Generating explanation phase is triggered whenever a decision is taken (in *planning* sub-behavior). The *AV* gathers all information related to the taken decision (the leading constraints, context information, potential improvement in the solution quality, etc.), in addition to the changed decision variables and their values. This information, together with the contextual data gathered in the previous steps from the agent belief base, are used to generate an understandable explanation for the taken decision. When the explanation is generated, the agent moves to the *monitor-agent interacting* phase. In the *monitor-agent interacting* phase, the generated explanations are sent to the monitor agent and stored in the *AV* belief base. To reflect the behavior of *AV*s in real world scenarios, the *MA* should never play the role of communication mediator between *AV*s.

The set of possibly explainable actions and decisions depend basically on the chosen solution model. Table 1. Lists some examples of solution models in line with their possibly explainable decisions.

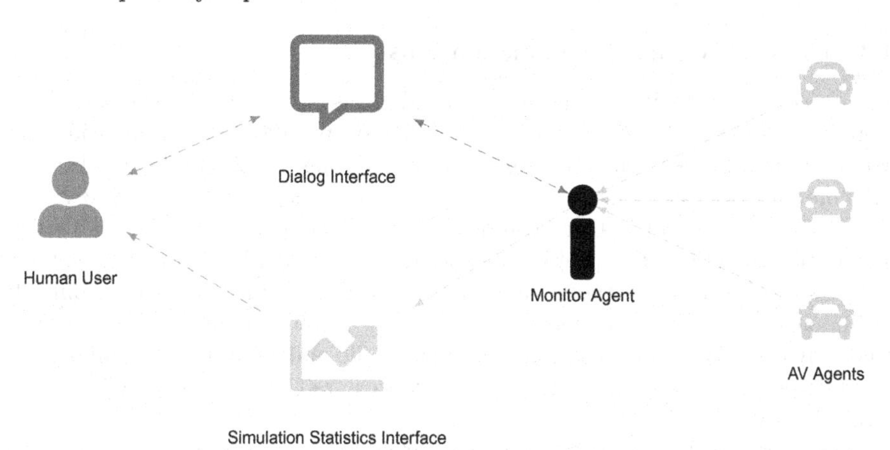

Fig. 7. The interacting behavior of the *Monitor Agent.*

4.2 *Monitor Agent*'s Behavior

The role of the *MA* is to be a proxy between *AV*s and the user. It interacts with the user via dialogues to build a user profile that simulates the user preferences and objectives. *MA* could be formed in a group of agents for fault tolerance and backup reasons and to avoid having a bottleneck in the model. Additionally, members inside this group may execute different explanations behaviors and interact/cooperate to provide the explanation to the human. The most important point is to have one interface with the human user to avoid overwhelming her/him with many interfaces. We can look at the monitor agent as the personal assistant of the human that could be embedded in his/her smartphone for example. Therefore, and even with a group of*MA*s, the interface with the human is preferably unified through one agent as a representative of the group.

MA gathers the statistics of decisions and their explanations from *AV*s. It aggregates these explanations in several abstraction levels. Following a similar approach of [33], the *MA* builds a multilevel explanation tree. The leaves of this tree correspond to particular agent's actions explanations. The root corresponds to the global abstract explanation for the final recommendation, and intermediate levels correspond to explanations for the evolution of evaluation metrics. At the end of the simulation scenario, it ranks the different solution methods based on their matching to the user profile providing a summary explanation for the ranking decision. The user could ask for a detailed explanation –to handle this, the *MA* defines a new granularity for selecting the right level of explanation that is communicated to the user. While the user is asking for more details, *MA* proceeds from the root to leaves gradually, providing at every step the corresponding level of explanations. It stops when the user stops asking or reaching the leaves representing the atomic details that can not be expanded. The next section discusses how an *MA* computes its recommendation.

4.3 Computing the Recommendations

The objective of *MA* is to assign values to its decision variable by the end of the scenario execution. *MA* has three sets of variables: profiles, recommendation and explanation variables. The recommendation variables are the ranking values for the different candidate methods. The explanation variables aggregate individual *AV*s' explanations and *MA*'s reasoning on the evolution of the evaluation metrics during the execution. The profile variables define a model based on the available features of allocation methods that match the user-defined features profile. If we manage to get such a model, then making recommendations for a user is relatively easy. We need to look at the user profile and compute its similarity to the different candidate methods. The candidates are then ranked based on their similarity value.

The user profile u in the set of user profiles U is represented by a vector of n features $u = [u_1, ..., u_n]$ defines the user's preferred values for the different evaluation metrics. Given that the system implements k solution methods that are potential candidates, this set of candidate methods is represented by $M = \{m_1, ..., m_k\}$; $m \in M$ is the feature vector of the candidate $m = [f_1^m, ..., f_n^m]$ we define a distance $dist$ function to calculate the n-dimensional euclidean distance between feature vectors:

$$dist : U \times M \rightarrow \mathbb{R}^+$$

$$dist(x, y) = \sqrt{\sum_{i=1}^{n}(x_i - y_i)^2}$$

A perfect-match method m' to user profile u if exists, will have $dist(u, m') = 0$ otherwise the following similarity function will be used to rank the candidate methods:

$$sim : U \times M \rightarrow [0, 1]$$

In its simplest form, sim function is the inverse of $dist$.

$$sim(x, y) = \frac{1}{dist(x, y)}$$

so that the highest recommended method m' to user u is the one with higher value of $sim(u, m')$.

4.4 Creating and Communicating the Explanations

As seen before, we need explanations that are scalable for multiple levels, we can distinguish two types of actions to be explained: the AV's individual decisions and the aggregated decision by MA.

The classical approach in XAI is the straightforward design of interpretable models on the original data to reveal the logic behind actions proposed by the system. State-of-the-art interpretable models, including decision trees, rules, and linear models; are considered to be understandable and readable by humans [27]. This applies to the individual decisions of AVs in our model, every AV is an autonomous agent having predefined interpretable behavior, and can justify his decisions with their technical and social reasons (based on its believes of itself and the context).

Another XAI approach is the post hoc interpretability, given the decisions made by the system, the problem consists of reconstructing an explanation to make the system intelligible without exposing or modifying the underlying model internally. The generation of explanations is an epistemic and computational action, carried out on-demand according to the current state of a model, and meta-knowledge on the functionalities of the system. It is intended to produce a trustworthy model based on features or exemplars. This applies to the aggregation of decisions made by MA, to the statistics-based matching, and to the recommendations.

An explanation can indeed be in any type of interaction. The advantage of human-like interaction is that it provides to the user higher levels of satisfaction, trust, confidence, and willingness to use autonomous systems. For this reason, many techniques have been developed to generate natural language (NL) descriptions of agent behavior and detected outliers or anomalies. This entails answering questions such as, why did an agent choose a particular action? Or what training data were most responsible for that choice? The internal state and action representations of the system are translated into NL by techniques such as recurrent neural networks [15], rationale generation [16], adding explanations to a supervised training set such that a model learns to output a prediction as well as an explanation [8].

5 ExDARP: A Use Case of Explainable Decisions in Decentralized DARP

In DARP scenarios, a fleet of vehicles is distributed through an urban network to satisfy passenger dynamic trip requests. A human *User* sets the scenario parameters, including fleet size, vehicle characteristics, request distribution model in addition to the user-defined objective, and utility preferences.

The *User* of the system is a representative of the transport authority; he wants to know the best solution method for solving the problem regarding the authority's preferences and the actual context parameters. The *Monitor Agent* uses this information to build the user profile and the scenario profile, which is then passed to the simulator. The system runs simulations with several solution methods; at each simulation tick, a snapshot of the problem context is solved with the different solution methods, explanations for local decisions are logged, and statistics for evaluation metrics are computed. At the end of the scenario execution, the results from different solution methods are compared, assessed in line with the statistics and user preferences.

Solution methods with the highest match to user preferences are recommended to the *User* (abstractly, macro-level) explaining the features of these methods – e.g., greedy method favors closer requests with short distances, which means lower operational cost. The *User* can also ask the system to monitor why some solution method is not suitable for his scenario – e.g., centralized dispatching requires continuous communication between vehicles and the dispatching portal, this consumes bandwidth in dynamic settings, making it unsuitable for scenarios with limited communication. – or, centralized dispatching requires exponential execution time to reach an optimal solution, it is unsuitable for emergency response scenarios. If the *User* requires fine-grained details, he can trace the evolution of evaluation metrics and ask for explanations for remarkable spots – e.g., a valley in the QoS chart followed by a peak can be explained as follows: at that time slot, 70% of vehicles were carrying passengers on the route to their far destinations, so for a while, only a low number of requests is satisfied, which means low values of QoS for a while; when these long trips ended one by one, the number of satisfied requests increases rapidly causing the peak in QoS. The *User* can continue asking for finer-grained details until reaching an explanation for individual vehicle actions.

Figure 8 illustrates a simple instance of DARP in a part of an urban network where the two vehicles V_1 and V_2 located in A and B are available and aware of a passenger request d_1 to travel from C to H which is announced at t_1. Considering the symmetric weights on edges represents the edge crossing operational cost in terms of the average time that a vehicle needs to move between its two ends. In the absence of central authority, vehicle agents should act autonomously and make decentralized decisions to find a solution in which the request is allocated to one of them. One of the common decentralized solution methods to such allocation problems is the market-based allocation [2,10,13] in which coordinating vehicle trips is handled by market mechanisms such as auctions and trade offers. Both V_1 and V_2 can serve d_1 without violating its constraints. To decide who

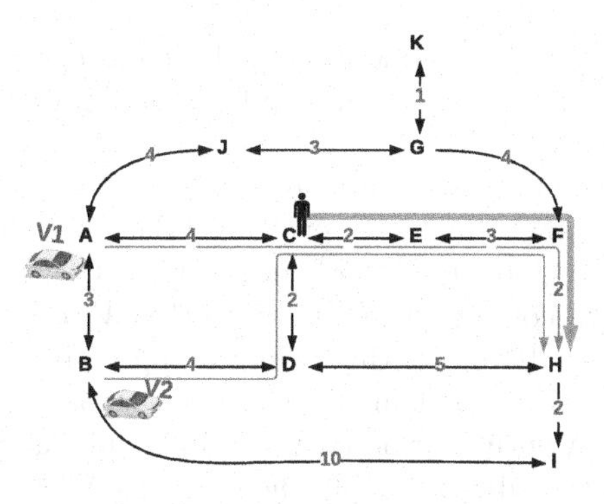

Fig. 8. A simple instance of DARP

serve it, vehicles instantiate an auction on d_1 where V_1's offer Bid_{V_1} is to serve d_1 with total operational cost of 11, while V_2 offers to serve it with 13. V_1 can explain its offer as "Serving d_1 costs me 11 time units because: Following the shortest paths, reaching C (the pick-up location of d_1) from my current location A requires 4 time units, and reaching H (the drop-off location of d_1) from C requires at least 7 time units". In the same way V_2 can explain: "Serving d_1 costs me 13 time units because: Following the shortest paths, reaching C (the pick-up location of d_1) from my current location B requires 6 time units, and reaching H (the drop-off location of d_1) from C requires at least 7 time units".

V_1 wins the auction on d_1, Explaining "how winner determination computed?" (see line 3 in Table 1.) can be: "V_1's offer has a lower cost than V_2's. The lower the cost is, the better the QoB achieved." Or, "V_1 reaches pick-up location 2 time units earlier than V_2 which means lower waiting time so better QoS".

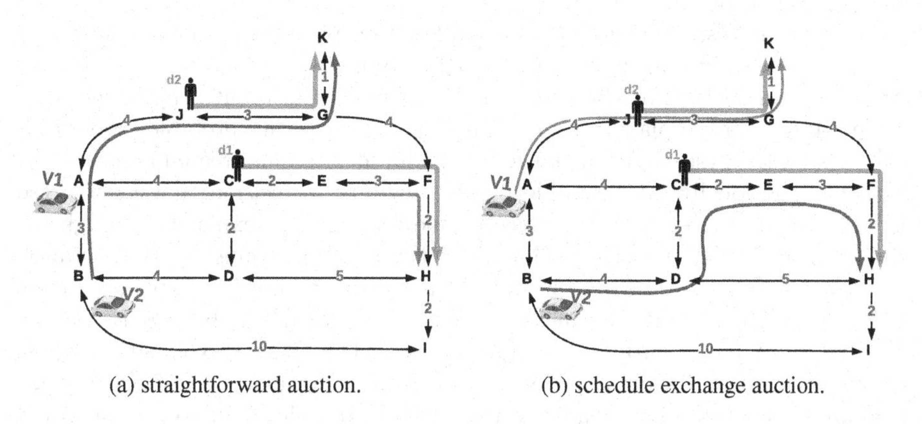

(a) straightforward auction.

(b) schedule exchange auction.

Fig. 9. Combinatorial auctions.

After d_1 is added to V_1's schedule, a new request d_2 for a passenger trip from J to K specifying the pick-up time-window at J as $tw_{d_2} = [5, 15]$ is announced. V_2 can offer to serve it with 11 time unit (pick-up in 7 time units and delivery after 4), as shown in Fig. 9(a). While V_1 cannot offer a reasonable bid, it explains: "I am committed to d_1, the earliest pick-up I can offer for d_2 is t_{27} which violates d_2's time-window constraint".

Some run-time optimization protocols such as ORNInA [10] and the sliding-horizon method [2] allow vehicles to exchange their scheduled requests if this exchange improves the global quality of the solution. V_2 can offer to pull d_1 from V_1's schedule for extra cost of 2 time units, so that V_1 can bid for d_2 with only 8 time units without violating its time-window constraint Fig. 9(b).

This exchange should be explained because it increases d_1's waiting time by 2 time units. In ORNInA optimization protocol, V_1 should accept V_2's pull bid, and then an explanation to "why accepting the exchange?" would be: "Abandoning d_1 in favor of V_2 decreases the global operational cost value by 1. It also decreases the accumulated waiting time by 1"

6 Related Work

Basically, one can distinguish between post hoc interpretability and the design of the explanation. In the former, the task is to create an explanation of the decision made by the system. While in the latter, the task is to design an interpretable model with its explanations.

Post-hoc interpretability approaches are divided into three categories depending on the motivation for having an explanation: model explanation, outcome explanation, and model inspection. The model explanation problem is to understand the reasoning of the system as a whole, while, the outcome explanation problem is to provide an explanation of the output of an intelligent system on a given input instance. Finally, the model inspection problem lies between the two previous problems [19].

We are interested in the outcome explanation problem because in this type it is not necessary to explain all the logic behind the system, only explaining why a certain decision has been returned for a particular input. Approaches that solve the outcome explanation problem yield a locally interpretable model that can explain the system output for a specific input in human-readable terms. For example, the locally interpretable model might be a decision tree constructed from a neighborhood of the instance in question, and an explanation might be the path of the decision tree followed by the attribute values of the instance. The proposed agnostic solutions to the problem of outcome explanation are generalizable by definition. Thus, in some cases, they can also be used for diverse data types. A recent proposal is LORE (LOcal Rule-based Explanations) [18], which overcomes previous solutions in terms of performance and clarity of the explanations. LORE first relies on a local interpretable predictor learned on a synthetic neighborhood generated by a genetic algorithm. Then, it generates an explanation which comprises (i) a decision rule, which explains the reasons for the decision; (ii) a set of counterfactual rules, suggesting changes in the functionality of the instance that lead to a different outcome. Another agnostic explanation method is MAME (Model Agnostic Multilevel Explanations) [33], which takes as input a post-hoc local explainability technique and an unlabeled dataset. Then generates multiple explanations for each of the examples corresponding to different levels of cohesion between explanations of the examples. As mentioned earlier, this type of methods could be applied to explain the aggregation of decisions made by *MA*.

The work of [7] argues that explainable planning can be designed as a wrapper around an existing planning system that uses the existing planner to answer challenging questions. To do so, they presented a prototype framework for explainable planning as a service.

In the task allocation domain, [42] introduces *AlgoCrowd* that offers efficient and explainable AI task allocation optimization designed to emphasize on fair treatment of workers, whiling reducing managers' workload to find suitable workers for tasks. they aimed at guaranteeing that workers of similar capability and productivity receive equitable incomes in the long run. This objective is translated into minimizing the workers' regret when their incomes are compared to similar peers.

In the domain of passenger transport, Ehmke and Horstmannshoff [14] propose the idea of personalized creation of multi-modal travel itineraries. They demonstrate the potential and limitations of "black box" mathematical optimization and discuss how to include more complex passenger preferences. Using solution sampling, they present a simple idea to identify the solution space's characteristics and allow travelers to restrict or improve their preferences interactively. These works consider central "black-box" AI tools and provide explanations produced by another external or wrapper AI tool. We consider in this paper a decentralized "transparent-box" multi-agent model.

In the "black-box" models, the user is unaware about the technical and logical reasons of the decisons so that another explaining system aims at reasoning on these decisions and provide explanations for them. On the other hand, in the "transparent-box" models, the user is aware about the system's logic and agent behaviors, so we can say the system is technically interpretable for every individual decision but the aggregation of these decisions to achieve the global objectives still need to be explained. Therefor, In our decentralized transparent-box multi-agent model, the required explanation should not only focus on technical reasons but also it is necessary to convey the user's preferences. It is necessary to decide what to reveal about agents' preferences to increase user satisfaction, trust, and control while considering how those features led to the final decision.

Agent-based approaches have been employed in the literature to provide explanations to humans in intelligent transport systems. For example, Mualla et al. proposed a Human-Agent Explainability Architecture (HAExA) to formulate context-aware explanations for remote robots represented as agents [28,31]. Considering that the human understandability of AI is subjective, they conducted empirical human-computer interaction studies employing Agent-based Simulation (ABS) [29]. The experiment scenario was about investigating the role of XAI in the communication between Unmanned Aerial Vehicles (UAVs) and humans in the context of package delivery in a smart city [30]. Their results showed that a balance between the simplicity of explanations and adequacy of the information contained in the explanations is needed. One interesting research direction, which we considered in this paper, is the importance of user-aware explanations [38]. Additionally, The results showed that ABS offers a test-bed environment to conduct human studies that facilitate the explanations reception by the human and visualize the behavior of the remote robots, represented as agents.

7 Conclusion

In this work, we explore the direction of explaining planning decisions in multi-agent resource allocation for ODT scenarios. Because there exists a huge variety of methods for resource allocation, the choice between these methods cannot be considered a straightforward decision. Moreover, these cannot be only seen as technical issues. The need for matching human satisfaction and controllable decisions requires these decisions to be transparent and self-explainable.

We aimed at conceptually designing a multi-agent model for an explainable recommendation of resource allocation mechanisms that match the user preferences and objectives in solving ODT scenarios, in particular contexts. We defined the system's main components and how the explanations are generated and aggregated in multiple levels of abstraction, to scale for the user-defined level. We defined some general guidelines and assumptions for generating the explanations and communicating them to human users. We illustrated this proposal through some case study examples.

However, to bring this model into reality many open questions should be addressed, including: How generic is this model to explain the variety of allocation mechanisms? How to define the right levels of detail that match the user needs, cognitive capacities, and competencies? How can we assess the quality of explanation and could that be automatic? We propose to deploy a proof-of-concept implementation of this model to address these challenges.

References

1. Aalami, S., Kattan, L.: Fair dynamic resource allocation in transit-based evacuation planning. Transp. Res. Part C Emerg. Technol. **94**, 307–322 (2018)
2. Agatz, N.A., Erera, A.L., Savelsbergh, M.W., Wang, X.: Dynamic ride-sharing: a simulation study in metro Atlanta. Transp. Res. Part B **9**(45), 1450–1464 (2011)
3. Anjomshoae, S., Najjar, A., Calvaresi, D., Främling, K.: Explainable agents and robots: results from a systematic literature review. In: Proceedings of 18th International Conference on Autonomous Agents and MultiAgent Systems, pp. 1078–1088. International Foundation for Autonomous Agents and Multiagent Systems (2019)
4. Borowski, E., Stathopoulos, A.: On-demand ridesourcing for urban emergency evacuation events: an exploration of message content, emotionality, and intersectionality. Int. J. Disaster Risk Reduction **44**, 101406. ISSN 2212-4209 (2020). https://doi.org/10.1016/j.ijdrr.2019.101406, https://www.sciencedirect.com/science/article/pii/S221242091930799X

5. Bradley, G.L., Sparks, B.A.: Dealing with service failures: the use of explanations. J. Travel Tourism Mark. **26**(2), 129–143 (2009)
6. Calvaresi, D., Mualla, Y., Najjar, A., Galland, S., Schumacher, M.: Explainable multi-agent systems through blockchain technology. In: Calvaresi, D., Najjar, A., Schumacher, M., Främling, K. (eds.) EXTRAAMAS 2019. LNCS (LNAI), vol. 11763, pp. 41–58. Springer, Cham (2019). https://doi.org/10.1007/978-3-030-30391-4_3
7. Cashmore, M., Collins, A., Krarup, B., Krivic, S., Magazzeni, D., Smith, D.: Towards Explainable AI Planning as a Service. arXiv:1908.05059 (2019)
8. Codella, N.C., et al.: Teaching meaningful explanations. arXiv preprint arXiv:1805.11648 (2018)
9. Cordeau, J.-F., Laporte, G.: The dial-a-ride problem: models and algorithms. Ann. Oper. Res. **153**(1), 29–46 (2007)
10. Daoud, A., Balbo, F., Gianessi, P., Picard, G.: Ornina: a decentralized, auction-based multi-agent coordination in ODT systems. In: AI Communications, pp. 1–17 (2020)
11. Daoud, A., Balbo, F., Gianessi, P., Picard, G.: A generic multi-agent model for resource allocation strategies in online on-demand transport with autonomous vehicles. In: Proceedings of the 20th International Conference on Autonomous Agents and MultiAgent Systems, AAMAS 2021, pp. 1489–1491. International Foundation for Autonomous Agents and Multiagent Systems. ISBN 978-1-45-038307-3 (2021)
12. di Sciascio, C., Brusilovsky, P., Veas, E.: A study on user-controllable social exploratory search. In: 23rd International Conference on Intelligent User Interfaces, pp. 353–364 (2018)
13. Egan, M., Jakob, M.: Market mechanism design for profitable on-demand transport services. Transp. Res. Part B Methodol. **89**, 178–195 (2016)
14. Ehmke, J., Horstmannshoff, T.: Position Paper: Explainable Search of Multimodal Itineraries. In: Modellierung (2020)
15. Ehsan, U., Harrison, B., Chan, L., Riedl, M.O.: Rationalization: a neural machine translation approach to generating natural language explanations. In: Proceedings of the 2018 AAAI/ACM Conference on AI, Ethics, and Society, pp. 81–87 (2018)
16. Ehsan, U., Tambwekar, P., Chan, L., Harrison, B., Riedl, M.O.: Automated rationale generation: a technique for explainable AI and its effects on human perceptions. In: Proceedings of the 24th International Conference on Intelligent User Interfaces, pp. 263–274 (2019)
17. Goodman, B., Flaxman, S.: European union regulations on algorithmic decision-making and a "right to explanation". AI Mag. **38**(3), 50–57 (2017)
18. Guidotti, R., Monreale, A., Ruggieri, S., Pedreschi, D., Turini, F., Giannotti, F.: Local rule-based explanations of black box decision systems. arXiv preprint arXiv:1805.10820 (2018)
19. Guidotti, R., Monreale, A., Ruggieri, S., Turini, F., Giannotti, F., Pedreschi, D.: A survey of methods for explaining black box models. ACM Comput. Surv. (CSUR) **51**(5), 93 (2018)
20. Gunning, D.: Explainable artificial intelligence (XAI). Defense Adv. Res. Projects Agency (DARPA) Web **2**(2), 47 (2017)
21. Hoffman, R.R., Mueller, S.T., Klein, G., Litman, J.: Metrics for explainable AI: Challenges and prospects. arXiv preprint arXiv:1812.04608 (2018)
22. Jin, Y., Sendhoff, B.: Trade-off between performance and robustness: an evolutionary multiobjective approach. In: Fonseca, C.M., Fleming, P.J., Zitzler, E., Thiele, L., Deb, K. (eds.) EMO 2003. LNCS, vol. 2632, pp. 237–251. Springer, Heidelberg (2003). https://doi.org/10.1007/3-540-36970-8_17

23. Kraus, S., et al.: AI for explaining decisions in multi-agent environments. In: Proceedings of the AAAI Conference on Artificial Intelligence, vol. 34, no. 09, pp. 13534–13538 (2020)
24. Lipton, Z.C.: The mythos of model interpretability: in machine learning, the concept of interpretability is both important and slippery. Queue **16**(3), 31–57 (2018)
25. Liu, Y., Li, Z., Liu, J., Patel, H.: A double standard model for allocating limited emergency medical service vehicle resources ensuring service reliability. Transp. Res. Part C Emerg. Technol. **69**, 120–133 (2016)
26. Ludwig, J., Kalton, A., Stottler, R.: Explaining complex scheduling decisions. In: IUI Workshops (2018)
27. Molnar, C., Casalicchio, G., Bischl, B.: Interpretable machine learning-a brief history, state-of-the-art and challenges. arXiv preprint arXiv:2010.09337 (2020)
28. Mualla, Y.: Explaining the Behavior of Remote Robots to Humans: An Agent-based Approach. Ph.D. thesis, University of Burgundy - Franche-Comté, Belfort, France (2020). http://www.theses.fr/2020UBFCA023
29. Mualla, Y., Bai, W., Galland, S., Nicolle, C.: Comparison of agent-based simulation frameworks for unmanned aerial transportation applications. Procedia Comput. Sci. **130**, 791–796 (2018)
30. Mualla, Y., Najjar, A., Kampik, T., Tchappi, I., Galland, S., Nicolle, C.: Towards explainability for a civilian UAV fleet management using an agent-based approach. In: 1st Workshop on Explainable AI in Automated Driving: A User-Centered Interaction Approach, Utrecht, Netherland. arXiv preprint arXiv:1909.10090 (2019)
31. Mualla., Y., Tchappi, I.H., Najjar, A., Kampik, T., Galland, S., Nicolle, C.: Human-agent explainability: An experimental case study on the filtering of explanations. In: Proceedings of the 12th International Conference on Agents and Artificial Intelligence - Volume 1: HAMT. INSTICC, pp. 378–385. SciTePress. ISBN 978-9-89-758395-7 (2020). https://doi.org/10.5220/0009382903780385
32. Preece, A.: Asking 'Why' in AI: explainability of intelligent systems-perspectives and challenges. Intell. Syst. Acc. Finan. Manage. **25**(2), 63–72 (2018)
33. Ramamurthy, K.N., Vinzamuri, B., Zhang, Y., Dhurandhar, A.: Model agnostic multilevel explanations. arXiv preprint arXiv:2003.06005 (2020)
34. Ronald, N., Thompson, R., Winter, S.: Simulating demand-responsive transportation: a review of agent-based approaches. Transp. Rev. **35**(4), 404–421 (2015)
35. Rosenfeld, A., Richardson, A.: Explainability in human-agent systems. In: Autonomous Agents and Multi-Agent Systems, pp. 1–33 (2019)
36. Schilde, M., Doerner, K.F., Hartl, R.F.: Metaheuristics for the dynamic stochastic dial-a-ride problem with expected return transports. Comput. Oper. Res. **38**(12), 1719–1730 (2011)
37. Schofer, J.L.: T. C. R. Program, and N. R. C. U. S. T. R. Board. Resource Requirements for Demand-responsive Transportation Services. Transportation Research Board. ISBN 978-0-30-908778-0 (2003). Google-Books-ID: RG9wnNBKCy4C
38. Singh, R., et al.: Directive explanations for actionable explainability in machine learning applications. arXiv preprint arXiv:2102.02671 (2021)
39. Tintarev, N., Masthoff, J.: Designing and evaluating explanations for recommender systems. In: Ricci, F., Rokach, L., Shapira, B., Kantor, P.B. (eds.) Recommender Systems Handbook, pp. 479–510. Springer, Boston (2011). https://doi.org/10.1007/978-0-387-85820-3_15
40. Tsai, C.-H.: Controllability and explainability in a hybrid social recommender system. Ph.D. thesis, University of Pittsburgh (2020)

41. Yankov, D.: Discrete Event System Modeling of Demand Responsive Transportation Systems Operating in Real Time. Graduate Theses and Dissertations (2008). https://scholarcommons.usf.edu/etd/575
42. Yu, H., et al.: Fair and explainable dynamic engagement of crowd workers. In: Proceedings of the Twenty-Eighth International Joint Conference on Artificial Intelligence. IJCAI-19, pp. 6575–6577. International Joint Conferences on Artificial Intelligence Organization (2019). https://doi.org/10.24963/ijcai.2019/961

XAI Vision, Understanding, Deployment and Evaluation

A Two-Dimensional Explanation Framework to Classify AI as Incomprehensible, Interpretable, or Understandable

Ruben S. Verhagen[1(✉)], Mark A. Neerincx[1,2], and Myrthe L. Tielman[1]

[1] Delft University of Technology, Van Mourik Broekmanweg 6, 2628 XE Delft, The Netherlands
{R.S.Verhagen,M.A.Neerincx,M.L.Tielman}@tudelft.nl

[2] TNO, Kampweg 55, 3769 DE Soesterberg, The Netherlands

Abstract. Because of recent and rapid developments in Artificial Intelligence (AI), humans and AI-systems increasingly work together in human-agent teams. However, in order to effectively leverage the capabilities of both, AI-systems need to be understandable to their human teammates. The branch of eXplainable AI (XAI) aspires to make AI-systems more understandable to humans, potentially improving human-agent teamwork. Unfortunately, XAI literature suffers from a lack of agreement regarding the definitions of and relations between the four key XAI-concepts: transparency, interpretability, explainability, and understandability. Inspired by both XAI and social sciences literature, we present a two-dimensional framework that defines and relates these concepts in a concise and coherent way, yielding a classification of three types of AI-systems: incomprehensible, interpretable, and understandable. We also discuss how the established relationships can be used to guide future research into XAI, and how the framework could be used during the development of AI-systems as part of human-AI teams.

Keywords: Explainable AI · Human-agent teaming · Transparency · Interpretability · Understandability · Explainability

1 Introduction

Rapid developments in the field of Artificial Intelligence (AI) have resulted in the design and adoption of intelligent systems/agents (A/IS) working together with humans. For such human-AI teams to work effectively and efficiently, it is crucial that AI-systems are understandable and predictable to their human teammates [22–24]. The eXplainable Artificial Intelligence (XAI) community aims to make AI more understandable, however, there is a lack of clear definitions and relationships between key concepts in XAI. The objective of this paper is to identify similarities, differences and inconsistencies in the description and usage

D. Calvaresi et al. (Eds.): EXTRAAMAS 2021, LNAI 12688, pp. 119–138, 2021.
https://doi.org/10.1007/978-3-030-82017-6_8

of these concepts, and to establish a framework in which the concepts can be unambiguously defined and related to each other.

Autonomous and intelligent systems/agents (A/IS) are characterized by their abilities to *sense* their environment, *reason* about their observations and goals, and consequently make *decisions* and *act* within their environment in a goal-driven manner [9]. Thanks to these capabilities, A/IS often outperform humans with respect to handling complex problems and rapid and rational decision-making. Consequently, the adoption domains of A/IS range from applications in healthcare to military defense. On the other hand, humans still surpass A/IS regarding the handling of uncertainty and unexpected situations. In an attempt to assemble their diversity in skills and leverage the unique abilities of both, A/IS and humans are increasingly paired to create human-agent teams (HATs).

Several factors are crucial for and determine the success of human-agent teams. Some of the most cited involve mutual trust and understanding; shared mental models and common ground; observability, predictability and directability; transparency and explainability; and teaming intelligence [22–24,33]. Unfortunately, many of these factors are lacking in contemporary human-agent teams. For example, most A/IS demonstrate extremely limited directability and often possess only rudimentary teaming intelligence (i.e., the knowledge, skills, and strategies necessary to effectively team) [23]. Furthermore, A/IS often demonstrate poor transparency and explainability, making it hard for human teammates to properly understand their inner workings, behavior, and decision-making [3,26,30]. This, in turn, negatively affects factors like mutual trust and understanding, eventually resulting in decreased global team performance [22,23].

To understand the behavior of A/IS, humans attribute A/IS behavior by assigning particular mental states (i.e., Theory of Mind) that explain the behavior [3,14,28–30]. Such mental states involve beliefs, desires/goals, emotions, and intentions. For example, humans trying to understand a robot entering a burning house can do so by attributing it to the goal to save a victim. A/IS capable of self-explaining their behavior and actions based on the reasons for the underlying intentions (e.g., beliefs, goals, emotions) help human teammates to build this ToM of the A/IS. This, in turn, will result in better understanding of the capabilities and limits of the A/IS and eventually better human-agent collaboration [3].

Explainable AI (XAI) methods, techniques, and research emerged as a means of making AI-systems more *understandable* to humans [16]. This relatively new community is characterized by the distinction between data-driven - and goal-driven XAI [3] (or perceptual vs. cognitive XAI [31]). Data-driven XAI is about *explaining* and *understanding* the decisions and inner workings of "black-box" machine learning algorithms given certain input data [3,15]. In contrast, goal-driven XAI/explainable agency refers to building goal-driven A/IS (e.g., robots) *explaining* their actions and reasons leading to their decisions to lay users [3,25].

Although fundamentally different branches, both data- and goal-driven XAI are characterized by the same fundamental issue: a lack of consensus with regards to the definition of and relations between key XAI concepts. Furthermore, provided definitions often suffer from a high level of ambiguity because they frequently refer to related notions. For example, the concepts of *transparency*, *interpretability*, *explainability*, and *understandability* are all frequently used in XAI literature, but often interchangeably, differently, with recourse to each other, or without even being defined. Without establishing clear distinctions and relations between these notions, the resulting ambiguity significantly hampers the comprehensibility of research centered around these concepts. We argue that prior to implementing, manipulating, or investigating these key concepts it is fundamental to first define and relate them. Only in this way, we can truly know what exactly we are trying to develop and evaluate.

To address the lack of agreement concerning the definition of and relations between key XAI notions, we propose a two-dimensional explanation framework that establishes clear concept definitions and relationships between them. This framework is based on both XAI and social sciences literature, and focuses primarily on A/IS disclosing and clarifying causes underlying their behavior and reasoning to human teammates (i.e., goal-driven XAI). Our framework explicitly addresses the lack of consensus and ambiguity problem by establishing clear distinctions and relations between system *transparency*, *interpretability*, *explainability*, and *understandability*. More specifically, the framework discriminates between system *interpretability* and *understandability* as passive and subjective characteristics concerning user knowledge of the system, versus system *transparency* and *explainability* as active and objective characteristics involved with disclosing and clarifying relevant information. Ultimately, these definitions result in the classification of three types of AI-systems: *incomprehensible*, *interpretable*, and *understandable* systems. We argue *transparency* can make *incomprehensible* systems *interpretable*, and *explainability* can make *interpretable* systems *understandable*. Adopting our distinctive concept definitions and mutual relationships can benefit XAI community by clarifying what kind of systems can be developed, and how we can evaluate them.

The remainder of the paper is structured as follows. In Sect. 2 we demonstrate the terminology problem by providing an overview of literature defining the key concepts. Next, we present our two-dimensional framework in Sect. 3. In Sect. 4 we discuss how the framework can be used to guide future XAI research, be applied in practice, and other relevant future directions. Finally, we conclude our paper in Sect. 5.

2 Background

Several works introduced or defined key XAI concepts such as *interpretability*, *explainability*, *transparency*, and *understandability*. However, the lack of consensus on the exact meanings and relations between these notions remains a prevalent issue. This section aims to highlight the problem and discuss relevant and significant prior contributions, before proposing our framework attempting to establish clear distinctions and relations between the concepts. First, we

Table 1. Several definitions for key XAI concepts, illustrating their ambiguity and relatedness.

Concept	Definition
Explainability	How well humans can understand AI-system decisions [30,37]
Interpretability	To explain or present in understandable terms to humans [4,11] How well humans can understand AI-system decisions [30,37]
Transparency	Representing system states in a way that is open to scrutiny, analysis, interpretation, and understanding by humans [1] Characteristic of model to be understandable for humans [4] Capacity of method to explain how a system works, even when behaving unexpectedly [37]
Understandability	To make a human understand how a model works, without any need for explaining its internal structure [4] Measuring how well humans understand model decisions [4] Capacity of a method of explainability to make a model understandable by end users [37]

demonstrate the lack of consensus problem and ambiguity of several proposed definitions. Next, we discuss some definitions, distinctions, and classifications that influenced our work. Finally, we discuss a framework that might help to unambiguously define and relate XAI concepts.

2.1 Problem

Unambiguously defining and relating XAI concepts is challenging. A small survey of available definitions in the literature demonstrates it is particularly hard to do so without recourse to related concepts (Table 1). Table 1 clearly demonstrates the ambiguity and relatedness of the defined concepts, and fails to provide any clear distinctions between them. For example, all of these concepts are defined at least once as *how understandable the AI-system is to humans.*

2.2 Transparency

Turilli and Floridi [36] introduce a clear definition for *transparency* which influenced our work. They suggest *transparency* refers to forms of *information visibility* and the possibility of *accessing* information, intentions, or behaviors that have been intentionally revealed through a process of *disclosure*. This disclosed information (i.e., made explicit and openly available) can then be exploited by potential users to support their own decision-making process.

Despite considering *transparency* and *explainability* as synonyms, Walmsley's [38] discussion of *transparency* influenced our work. Walmsley [38] divides the notion of *transparency* into two major categories: outward - vs. functional *transparency*. Outward *transparency* concerns the relationship between the AI-system and externals, such as developers and users. This includes *transparency* about development reasons, design choices, values driving the system developers, and capabilities and limitations of the system. In contrast, functional *transparency* concerns the inner workings of the system. This includes *transparency* about how and why the system behaves in general (type functional *transparency*[1]), or came up with certain decisions or actions (token functional *transparency*[2]).

2.3 Related Work

Ciatto et al. [8] propose an abstract and formal framework for XAI that, in contrast to most work, introduces a clear distinction between *interpretation* and *explanation*. The framework stresses the objective nature of *explanation*, in contrast with the subjective nature of *interpretation*. The act of *interpreting* some object X is defined as the activity performed by an agent A assigning a subjective meaning to X. Furthermore, Ciatto et al. [8] argue an object X is interpretable for an agent A if it is easy for A to assign a subjective meaning to X (i.e., A requires low computational or cognitive effort to *understand* X). The authors stress the subjective nature of *interpretations*, as agents assign them to objects according to their State of Mind and background knowledge.

In contrast, *explaining* is defined as the epistemic and computational activity of producing a more *interpretable* object X' out of a less interpretable one X, performed by agent A. They argue this activity can be considered objective because it does not depend on the agent's perceptions and State of Mind. Consider, for example, decision tree extraction (the *explaining* activity) from a neural network (object X) to produce a decision tree (the *explanation*/object X'). In the end, the effectiveness of the explanations always remains a subjective aspect.

This framework differs from ours in a few ways. In particular, Ciatto et al. [8] provide a formal framework focused on data-driven XAI, whereas we provide more general definitions in a goal-driven XAI context. In contrast, the intentions of the paper and provided definitions are similar to our work. We also define *interpretability* as a subjective system characteristic reflecting user knowledge about a system, and *explainability* as an epistemic and computational activity aimed at increasing user knowledge about the system.

Barredo Arrieta et al. [4] provide a brief clarification of the distinctions and similarities between *transparency*, *interpretability*, *explainability*, and *understandability*. So this part of their work is very similar in its intents to our work, despite focusing on data-driven XAI instead of goal-driven XAI. However, we argue that their attempt at clarifying the distinctions and similarities between the concepts fails to resolve any ambiguity. For example, the authors first argue *interpretability* is a passive model characteristic referring to the level at which a given model makes sense for a human, but later as the ability to explain or provide the meaning in *understandable* terms to a human.

[1] Also referred to as global explanations in XAI literature.

[2] Also referred to as local explanations in XAI literature.

In summary, Barredo Arrieta et al. [4] define *interpretability* (i.e., their first definition), *understandability*, and *transparency* as passive model characteristics reflecting human knowledge and understanding of a model. In contrast, they define *explainability* as an active model characteristic, denoting any action taken by a model with the intent of clarifying or detailing its internal functions. Unlike Barredo Arrieta et al. [4], we consider *transparency* as an active system characteristic concerned with disclosing information to generate knowledge about system elements. Similar to them, we also define *interpretability* and *understandability* as passive characteristics reflecting system knowledge and understanding, and *explainability* as actively clarifying or detailing system elements.

Rosenfeld and Richardson [32] formally define *explainability* and its relationship to *interpretability* and *transparency*, in the case of a ML-based classification algorithm. The authors define *explainability* as the ability for the human user to *understand* the algorithm's logic. This ability to *understand* is achieved from the *explanation*, which they define as the human-centric objective for the user to *understand* the algorithm, using an *interpretation*. *Interpretation/interpretability* is defined as a function mapping data, data schemes, outputs, and algorithms to some representation of the algorithm's internal logic. Furthermore, the authors argue an *interpretation* is *transparent* when the connection between the *interpretation* and algorithm is *understandable* to the human, and when the logic within the *interpretation* is similar to that of the algorithm.

All in all, the work of Rosenfeld and Richardson [32] differs from our work in several ways. First of all, they focus on data-driven XAI and provide formal definitions, whereas our work focuses on goal-driven XAI and provides more general definitions. More importantly, the provided definitions differ from our view. Rosenfeld and Richardson [32] consider *explainability* as passive and subjective, defining it as the ability to *understand*. In contrast, we consider *explainability* as an active system characteristic, and argue their definition of *explainability* reflects *understandability* instead. In addition, the authors consider *interpretability* as active and objective, defining it as providing representations of an algorithm's internal logic. However, we consider *interpretability* as passive and subjective, reflecting user knowledge and understanding of a system/algorithm, and argue their definition of *interpretability* reflects *explainability* instead.

Sanneman and Shah [34] propose an interesting situation awareness-based levels of XAI framework. This framework argues AI-systems part of human-AI teams should explain what the system did or decided (XAI for Perception), why the system did this (XAI for Comprehension), and what the system might do next (XAI for Projection). The authors argue XAI for Comprehension should provide information about causality in the system, aimed at supporting user comprehension of the system's behavior. Examples include explanations linking behavior to the system's goals, constraints, or rules.

This framework broadly aligns with ours, but includes a few differences as well. First of all, we agree with their distinction between providing information for perception and comprehension. However, whereas Sanneman and Shah [34] define both of them as explanations, we refer to XAI for Perception as *transparency*/disclosing information, and XAI for Comprehension as *explainability*/clarifying disclosed information. We argue XAI for Projection can be defined as both *transparency* and *explainability*, depending on whether the system discloses next actions (i.e., *transparency*) or also clarifies them (i.e., *explainability*). Furthermore, the framework only focuses on *explaining* AI-system behavior like actions or decisions. However, we argue it is also possible and sometimes even necessary to explain system elements like goals, knowledge, development reasons, or design choices. By doing so, human users can build more complete mental models of the AI-system. Therefore, our framework also incorporates disclosing and clarifying other relevant system elements like goals or knowledge.

Doran et al. [10] introduce an interesting distinction between *opaque*, *interpretable*, and *comprehensible* AI-systems that influenced our work. They define *opaque* AI-systems as systems where the mechanisms mapping inputs to outputs are invisible to users. Consequently, the reasoning of the system is not observable or understandable for users. In contrast, *interpretable* AI-systems are characterized as systems where users cannot only *see*, but also *study* and *understand* how inputs are mapped to outputs. The authors argue that *interpretable* systems imply *transparency* about the underlying system mechanisms. Finally, they define *comprehensible* AI-systems as systems emitting symbols (e.g., words or visualizations) along with their output to allow users to *relate* properties of the input to their corresponding output. According to this classification, *interpretable* systems can be inspected to be understood (i.e., letting users draw *explanations* by themselves), while *comprehensible* systems explicitly provide a symbolic *explanation* of their functioning [8].

This classification of AI-systems is quite similar to the one provided in our work. However, whereas Doran et al. [10] focus on data-driven XAI and argue the notions of *interpretation* and *comprehension* are separate, we focus on goal-driven XAI and argue *understanding/comprehension* implies *interpretation.* More specifically, we claim *transparency* can make *incomprehensible* systems *interpretable*, and *explainability* can make these *interpretable* systems *understandable.* We will explain our definitions, relationships, and classification in detail in the next section.

3 A Two-Dimensional Framework to Classify AI

In this section we present and discuss our two-dimensional explanation framework providing clear distinctions and relations between key XAI concepts (Fig. 1). In short, our framework makes a distinction between *incomprehensible*, *interpretable*, and *understandable* AI-systems, and argues system *transparency* can make *incomprehensible* systems *interpretable*, whereas *explainability* can make *interpretable* systems *understandable.* In the following sections, we will explain and illustrate our framework by introducing our definitions of the concepts *transparency* and *explainability* (Sect. 3.1), and *interpretability* and *understandability* (Sect. 3.2). After that, we illustrate and discuss our framework based on the example of a search and rescue human-agent teaming scenario where a

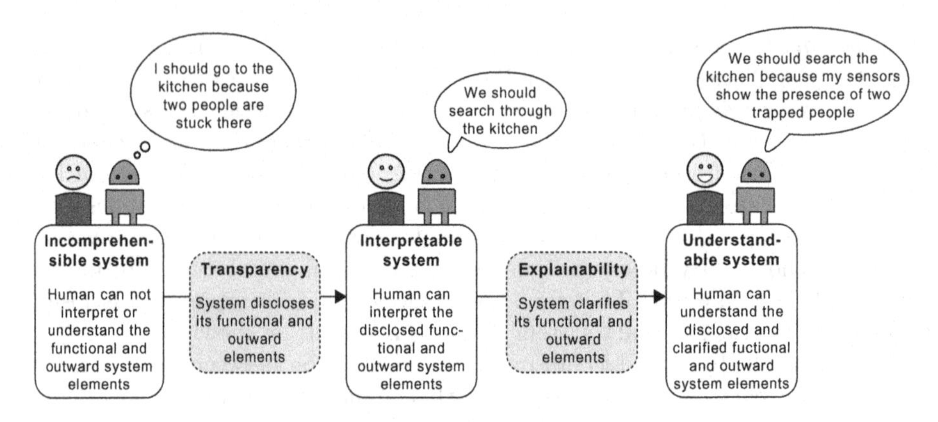

Fig. 1. Two-dimensional explanation framework providing distinctive definitions and relationships between key XAI concepts.

human collaborates with a goal-driven A/IS (Sect. 3.3). Finally, we extend our framework to include some other relevant factors enabled by system *transparency* and *explainability* in Sect. 3.4.

3.1 Transparency vs. Explainability

Whereas most prior work strongly ties or even equates system *explainability* to *interpretability* (e.g., [30, 37]), we consider them fundamentally different. Instead, we strongly tie system *transparency* to *explainability*. However, we also argue for a major distinction between these two notions. Inspired by [1] and [36], we define system *transparency* as "*disclosing* the relevant outward and functional system elements to users, enabling them to access, analyze, and exploit this disclosed information". Here, functional system elements concern elements like goals, knowledge, beliefs, decisions, and actions. In contrast, outward elements concern aspects like development reasons, intended users, and design choices.

System *transparency* can answer "*what*"-questions [30] requiring descriptive answers concerning the system elements. Consider, for example, a goal-driven autonomous and intelligent agent collaborating with a human teammate to save victims after an earthquake. According to our definition, system *transparency* is both an *active* [4] and *objective* [8] system characteristic achieved by, for example, disclosing the goal to save all injured children first by collaborating with trained firefighters. By doing so, the human teammate can gain knowledge about these system elements (here a goal and intended users respectively), without necessarily always knowing the relations between them.

The disclosure of relevant elements can be considered *active* in the sense that it is an epistemic and computational *activity* aimed at increasing user knowledge, and *objective* because this activity itself does *not depend on the human's perceptions or State of Mind.* Put differently, the computational implementation of *transparency* is independent of the human user's perceptions and State of Mind, and thus reproducible in principle [8]. However, the exact effectiveness and content of the disclosed information is a subjective aspect, reflected by measures of *interpretability* and *understandability*.

Inspired by [4,8], and [34] we define system *explainability* as "*clarifying* disclosed system elements by providing information about causality and establishing relations with other system elements, making it easier for users to *understand*, analyze, and exploit this information". *Explainability* can answer "*how*"- and "*why*"-questions [30] requiring clarifying answers concerning the system elements and how they relate and depend on each other. For example, system *explainability* can involve clarifying the disclosed goal to save all children first by linking it to the norm that children are most vulnerable, or that it will not give safety instructions because it assumes the user is a firefighter and familiar with these. Just as *transparency*, we characterize system *explainability* as an *active* [4] and *objective* [8] system characteristic aimed at increasing user knowledge and where the epistemic and computational activity itself does not depend on the human's perceptions or State of Mind.

In summary, the main difference between system *transparency* and *explainability* boils down to *disclosing* vs. *clarifying*. *Transparency* aims to provide descriptive answers providing knowledge about system elements. In contrast, *explainability* aims to ease understanding by clarifying the relations between system elements. Both are considered *active* and *objective* system characteristics, since they are epistemic and computational activities aimed at increasing user knowledge without depending on user's perceptions or State of Mind. We define *transparency* and *explainability* from a system-centric point of view as methods for sharing information, hence the categorization as active and objective/independent from the user. However, we argue that the subjective aspect concerning the effectiveness and content of the shared information also plays a crucial role, as reflected by measures of *interpretability* and *understandability*.

3.2 Interpretability vs. Understandability

In contrast to *transparency* and *explainability*, we define system *interpretability* and *understandability* as *passive* and *subjective* characteristics reflecting user knowledge of the system and depending on the user's State of Mind and background knowledge. In addition, we argue *transparency* makes system *interpretable*, whereas *explainability* makes *interpretable* systems *understandable*. Although we strongly tie *interpretability* to *understandability*, we argue for a major distinction between these two notions as well.

Inspired by [4,8,10] and [36], we define system *interpretability* as "the level at which the system's users can assign subjective meanings, draw explanations, and gain knowledge by accessing, analyzing, and exploiting disclosed outward and functional system elements". Our definition implies *interpretability* is both a *passive* [4] and *subjective* [8] system characteristic. *Passive* in the sense that *interpretability* reflects a degree of user knowledge about system elements, opposite to actively sharing information to generate knowledge (i.e., *transparency*).

Furthermore, *interpretability* can be considered *subjective* in the sense that it is highly dependent on the user's State of Mind and background knowledge [8].

Consider, again, the example of the goal-driven A/IS collaborating with a human to save victims after an earthquake. Disclosing its goal to save all children first enables human users to gain knowledge and assign subjective meanings or draw explanations by themselves (i.e., *interpret*). However, without clarifying the disclosed goal and relating it to other system elements (i.e., *explainability*), these interpretations can vary considerably. For example, the human could draw the conclusion that the system knows/beliefs the area contains a lot of children but only few elderly or adults.

On the other hand, we define system *understandability* as "the level at which the system's users have knowledge of disclosed and clarified outward and functional system elements, and the relationships and dependencies between them". *Understandability* involves knowing *how* and *why* the system reasons and functions, based on *explanations* clarifying and relating disclosed system elements. For example, clarifying the goal to save all children first because they are most vulnerable provides the user with knowledge about the relationship between the goal and a specific norm.

In summary, the main difference between system *interpretability* and *understandability* boils down to a difference in cognitive effort required to have knowledge of the system elements [8]. More specifically, we argue *interpretability* requires more cognitive effort because it implies inferring the meaning of and relations between disclosed information without explicit knowledge of this meaning and relations themselves. In contrast, *understandability* requires less cognitive effort because it implies knowing the meaning of and relations between disclosed and clarified information (facilitated by *explanations*). Both are considered *passive* [4] and *subjective* [8] system characteristics, since they reflect a degree of *user knowledge* about the system *depending on the user's State of Mind and background knowledge.* So we define *interpretability* and *understandability* from a user-centric point of view reflecting the subjective effectiveness of the *transparency* and *explainability* content. Here, *transparency* and *explainability* will be most effective when their content is tailored to the user's State of Mind and background knowledge.

3.3 Two-Dimensional Framework to Classify AI

Our framework (Fig. 1) distinguishes between three types of AI-systems (*incomprehensible*, *interpretable* and *understandable*) and establishes relations between them by integrating the defined concepts of Sect. 3.1 and Sect. 3.2. We will illustrate our framework in the context of a search and rescue human-agent teaming scenario, where a human collaborates with a goal-driven A/IS.

When collaborating with *incomprehensible* systems, humans can not *interpret* or *understand* the system elements because they are not disclosed and clarified. For example, without disclosing and clarifying its decision to search through the kitchen because it perceived stuck people, a human will not be able to interpret or understand the system's behavior. Our framework argues *transparency* can turn *incomprehensible* systems into *interpretable* ones. By disclosing its relevant functional and outward system elements (i.e., *transparency*), the human can access and exploit this information to assign subjective meanings and gain knowledge (i.e., *interpret*). Consider, for example, an A/IS disclosing the decision to search through the kitchen of a collapsed house to its human teammate. By doing so, the human can utilize this information to interpret that the A/IS perceived something urgent in the kitchen. Furthermore, we argue *explainability* can turn *interpretable* systems into *understandable* systems. By clarifying the disclosed system elements and relations between them (i.e., *explainability*), the human can more easily exploit this information to gain knowledge and build a mental model of the system (i.e., *understandability*). Consider, for example, an A/IS disclosing the decision to search through the kitchen, because it perceived two trapped children there. By providing a belief-based *explanation* for the decision, the system clarifies this decision and how it relates to other system elements like perceptions.

Our proposed framework has several implications. First of all, pursuing system *understandability* should be the ultimate goal, since it can improve collaboration and team performance in human-agent teams [3]. Furthermore, the framework implies that system *transparency* and *explainability* are *active* and *objective* characteristics which can be manipulated by designers to bring about the desired effects. In contrast, system *interpretability* and *understandability* are considered *passive* and *subjective* characteristics which can be measured to validate the effects of *transparency* and *explainability.*

3.4 Extended Framework

We extend our two-dimensional framework to include several often encountered XAI notions. This framework (Fig. 2) mainly illustrates the opportunities system *transparency* and *explainability* can provide to human teammates. Again, we discuss the framework in the context of a search and rescue human-agent teaming scenario where a human collaborates with a goal-driven A/IS.

The extended framework argues that when a system is *interpretable*, it is already both *controllable* and *directable.* Here, we define system *controllability* as "the extent to which human users can change or overrule functional system elements". For example, when the A/IS discloses the decision to search through the kitchen, its human teammate can overrule this decision by changing it to searching the basement instead (i.e., the system is *controllable*).

Next, we define system *directability* as "the extent to which human users can guide the actions of the system". This is different from system *controllability* in the sense that *directability* does not involve changing or overruling system elements, but rather accepting them and guiding the corresponding actions or dividing the work. For example, the human teammate could also accept the disclosed decision to search the kitchen but direct the action of the A/IS by giving the order to enter the kitchen first to assess its safety (i.e., the system is *directable*). Even though system *interpretability* already enables system *controllability* and *directability*, we argue system *understandability* will further improve

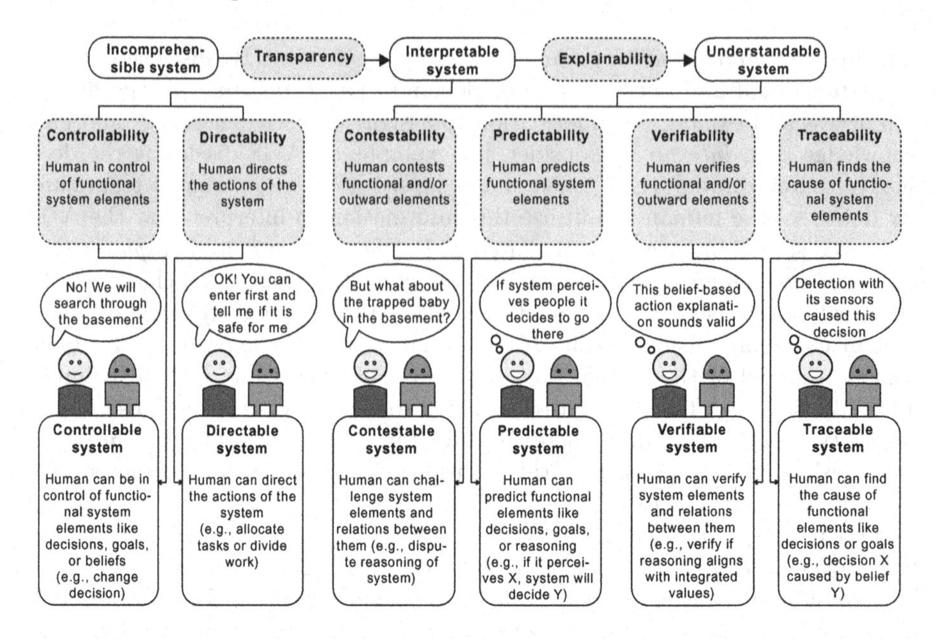

Fig. 2. Extended two-dimensional explanation framework providing distinctive definitions and relationships between key XAI concepts.

these two characteristics. For example, when the human teammate has more knowledge of the system, it can more effectively control and direct its functional elements such as actions or goals.

Furthermore, we argue that system *understandability* enables several other important notions such as system *contestability*, *predictability*, *verifiability*, and *traceability*. We define system *contestability* as "the extent to which human users can challenge or dispute system elements and the relations between them". Again, consider the example of the A/IS disclosing the decision to search through the kitchen, by clarifying it perceived two trapped people there. By doing so, the human teammate can contest this decision and dispute the underlying reason, for example by asking why they should search through the kitchen when there is a trapped baby in another room (i.e., the system is *contestable*).

We argue system *understandability* also enables system *predictability*. We define system *predictability* as "the extent to which human users can estimate future or other functional system elements". Consider the example of a system disclosing its goal to save all children first because of the norm that children are more vulnerable than adults. The human could use this explanation to predict that the agent's next actions will be focused on searching children rather than adults.

The extended framework also argues system *understandability* enables system *verifiability*. Here, we define system *verifiability* as "the extent to which human users can check that the system elements and relations between them make sense and sound valid". We do not refer to formal verification of systems using formal methods involving mathematical models of systems and analyzing them using proof-based methods. Rather, we refer to a more informal verification of the plausibility of system elements and relations between them. Again, consider the example of a system disclosing its goal to save all children first because of the norm that children are more vulnerable than adults. Based on the provided explanation the human could informally verify that the reasoning aligns with the decision and sounds valid (i.e., the system is *verifiable*).

Finally, we argue system *understandability* enables system *traceability*. We define system *traceability* as "the extent to which human users are able to find the cause of functional system elements like decisions, goals, or beliefs". Again, consider the example of the A/IS disclosing the decision to search through the kitchen, by clarifying it perceived two trapped people there. The human teammate could use the provided explanation to infer that the decision to search the kitchen was caused by the detection of two trapped people.

In summary, the extended framework argues system *interpretability* and *understandability* enable important factors such as system *controllability*, *directability*, *contestability*, *predictability*, *verifiability*, and *traceability*. These factors are crucial for and determine the success of human-agent teams [22–24,33]. Therefore, pursuing system *understandability* should be the main goal when developing AI-systems part of human-agent teams.

4 Discussion

In this work we have presented a two-dimensional explanation framework providing distinctive XAI concept definitions and relationships between them. In this section, we will discuss how our presented relationships between the concepts can be used to guide future research into these relationships. Additionally, we will describe how we believe this framework can be applied in practice.

4.1 Evaluation of Main Framework

Several assumptions arise from the proposed relationships in our presented framework. Below, we introduce these assumptions as claims and describe their corresponding requirements. Next, we discuss whether these assumptions can be evaluated, and how they offer a road map for future research.

- **Claim 1** - System *explainability* results in more knowledge/complete mental models of the system than *transparency*
  Requirement 1 - Manipulating/implementing system *transparency* and *explainability*
  Requirement 2 - Measuring user knowledge of a system
- **Claim 2** - Increased user knowledge of a system results in improved human-agent collaboration and eventually team performance
  Requirement 1 - Subjective and objective measurements of human-agent collaboration

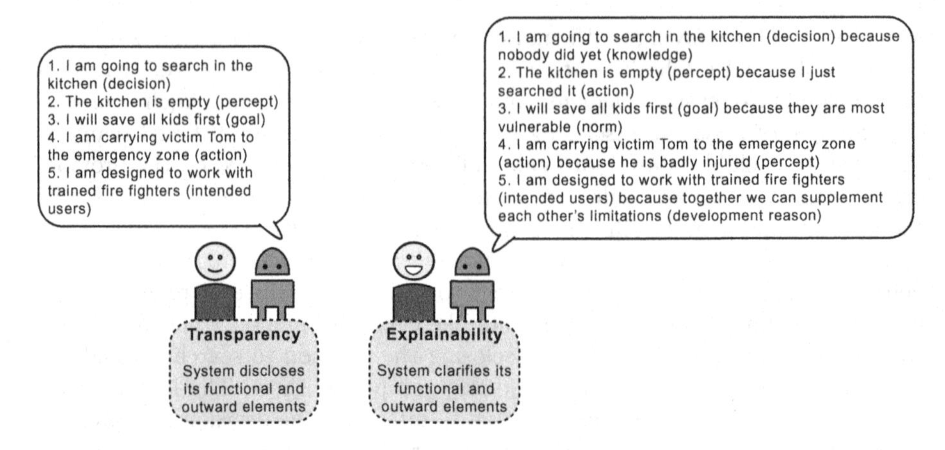

Fig. 3. Examples of system transparency and explainability in the context of a (simulated) search and rescue mission.

We will illustrate how these claims can be evaluated using the example of a simulated search and rescue mission where a human operator and self-explaining A/IS collaborate to search and rescue victims. To validate Claim 1, implementing system *transparency* and *explainability* would be required. Examples of implementing system *transparency* involve disclosure of the system's goals, decisions, and intended users. Figure 3 shows several examples of system *transparency* in the context of the search and rescue mission.

Implementing system *explainability* can be achieved in many different ways. However, a fundamental requirement is providing information about causality in the system and establishing relations between system elements. Existing approaches from the XAI literature include explanations of actions based on state information [2,19,27]; explanations of actions based on goals [5,17,18]; explanations of decisions based on demonstrating that alternative decisions would be sub-optimal [35]; and sequence-based explanations clarifying the next action(s) [5,17]. Figure 3 shows several concrete examples of system *explainability* in the context of the search and rescue operation.

Validating Claim 1 would also require the measurement of human user knowledge and understanding of the system, which can be done both subjectively and objectively. Subjective examples from the XAI literature include asking questions related to perceived understandability of the system and its model [20], and asking users to choose which of two possible system outputs is of higher quality (implicitly measures understanding) [11]. However, objectively measuring user knowledge and understanding of the system would be a more robust indicator than the subjective alternatives.

Currently, objective methods and metrics for measuring user knowledge and understanding of systems are lacking. Nevertheless, Sanneman and Shah [34] propose a relevant method based on the widely-used and empirically validated Situation Awareness Global Assessment Technique (SAGAT) [12,13]. In short, their proposed technique involves freezing simulations of representative tasks at random time points, followed by asking questions measuring user knowledge about information related to system behavior. It is crucial to first define the human informational needs related to system behavior. Accordingly, a list of questions regarding the informational needs can be specified and used to measure user knowledge of the system.

Whereas Sanneman and Shah [34] focus solely on measuring user knowledge related to AI-system behavior, the test/technique can also be extended to include information related to other relevant system elements like goals, knowledge, decisions, or even development reasons. Some example questions based on the information in Fig. 3 include "Which room will the agent search next?"; "What is the current action of the agent?"; "Why is the agent going to search in the kitchen?"; and "Why will the agent save all kids first?".

Validating Claim 2 would require the subjective and objective measurement of human-agent collaboration and team performance. Subjective measures could include user satisfaction [7] or system usability [6], whereas objective measures could include aspects like the number of victims rescued or seconds required to finish tasks. The outlined example experiment, discussed example implementations of *transparency* and *explainability*, and suggested metrics for measuring user knowledge, human-agent collaboration, and team performance can be used as a road map for future work aimed at validating the assumptions arising from our framework.

4.2 Evaluation of Extended Framework

Several assumptions arise from the proposed relationships in our extended framework as well. Below, we introduce these assumptions as claims and describe their corresponding requirements. Next, we discuss whether these assumptions can be evaluated, and how they offer a road map for future research.

- **Claim 3** - System *transparency* already enables system *controllability* and *directability*, but not system *contestability*, *predictability*, *verifiability*, and *traceability*
  Requirement 1 - Implementing system *transparency*
  Requirement 2 - Measuring system *controllability*, *directability*, *contestability*, *predictability*, *verifiability*, and *traceability*
- **Claim 4** - System *explainability* enables system *contestability*, *predictability*, *verifiability*, and *traceability*
  Requirement 1 - Implementing system *explainability*
  Requirement 2 - Measuring system *contestability*, *predictability*, *verifiability*, and *traceability*

Validating Claims 3 and 4 would require implementing system *transparency* and *explainability*, and measuring system *controllability*, *directability*, *contestability*, *predictability*, *verifiability*, and *traceability*. An example of subjectively measuring these system characteristics could be freezing the simulated experiment at random points, followed by measuring perceived system *controllability*, *directability*, *contestability*, *predictability*, *verifiability*, and *traceability*. One approach involves Likert-scale questions[3] asked to the human users. Table 2 shows example questions for each of these variables, though full questionnaires would require more research and validation of the exact scales. The outlined example experiment, discussed example implementations of *transparency* and *explainability*, and suggested metrics for measuring system *controllability*, *directability*, *contestability*, *predictability*, *verifiability*, and *traceability* can be used as a road map for future work aimed at validating the assumptions arising from our extended framework.

Table 2. Example questions for subjectively measuring the system variables in the extended framework.

Variable	Example question
Controllability	"I feel like I can change the system's decision"
Directability	"I feel like I can guide the system's behavior"
Contestability	"I feel like I can challenge the system's decision"
Predictability	"I feel like I can predict the system's next action"
Verifiability	"I feel like I can check that the system's behavior makes sense"
Traceability	"I feel like I can find the cause of the system's decision"

4.3 Application of Framework

Here we briefly address how our framework can be used in practice. Specifically, what difference can the framework make when developing systems part of human-agent teams? Consider the example of developing an autonomous and intelligent drone which should collaborate with a human operator (e.g., a firefighter) during the aftermath of an earthquake. The goal of the team is to search and rescue trapped victims as soon as possible. Our framework can be particularly helpful by mapping specific types of context and informational needs onto requiring either system *transparency* or *explainability*. For example, the drone can be developed/implemented in such a way that when the workload or time pressure is high, the drone displays *transparency* only. Similarly, contextual factors that could be mapped onto system *explainability* include low time pressure and operator workload, or when the user has an imprecise mental model of the system. In this way, the framework can contribute to developing adaptive systems able to tailor their communication of relevant information to the needs and requirements of both users and situations.

4.4 Future Work

Based on the work presented in this paper, we identify a few key ideas for future work. A possible first direction could be to conduct experiments aimed at vali-

[3] For example ranging from "Totally Disagree" to "Totally Agree" on a 7-point scale.

dating the assumptions arising from our framework. Some ideas, requirements, and examples concerning this validation have been discussed in more detail in Sect. 4.1 and Sect. 4.2.

For now, our framework focuses on sharing information regarding mental constructs like decisions or goals. A relevant suggestion for future work would be to extend the framework with a more physical domain as well by including literature/perspectives from explainable and understandable robots. For example, the role of visual and body cues could be incorporated in the framework. Furthermore, our provided framework is rather broad/general and informally defined. Therefore, another suggestion would be to formalize it and make it more concrete by providing examples in terms of different computational frameworks/architectures (e.g., transparency vs. explainability differences between agents using BDI vs. PDDL models). In addition, we currently do not consider situations where the user may be under the false impression of understanding the system, but only consider cases where their understanding actually matches the system's models/elements. We also do not consider different roles taken by human and agent, such as commander or supervisor. In future work, it would be interesting to extend the framework by including these two aspects, and see how it affects our proposed definitions.

Another future direction for this work would be to extend the framework to include context- and user-awareness required for tailoring system *transparency* and *explainability* to specific needs and requirements. The need for personalized and context-dependent system *transparency* and *explainability* is one of the main goals within XAI community and research [3]. However, the actual implementation and investigation is still somewhat in its infant stages. Currently, our proposed framework does not address context- and user-dependent system *transparency* and *explainability*, so this would be a relevant suggestion for future work. Ideas involve mapping specific types of context or user knowledge to requiring either system *transparency* or *explainability*. Furthermore, these aspects could also be mapped onto *transparency* and *explanation* modality/presentation instead of just content. Examples include mapping high workload to system *transparency*, rudimentary system knowledge to *explainability*, or visual thinkers to receiving visual *explanations* and verbal thinkers receiving textual *explanations*. Another idea involves adapting system *transparency* and *explainability* based on the interdependence relationship between human and system. For example, the system could adapt its communication based on whether joint activity is required (i.e., hard interdependence) or when joint activity is optional (i.e., soft interdependence) [21,22].

5 Conclusion

In this paper, we propose a two-dimensional explanation framework introducing clear distinctions and relationships between the key XAI notions *transparency*, *interpretability*, *explainability*, and *understandability*. This concise and comprehensive framework explicitly addresses the lack of consensus and ambiguity problem surrounding these concepts. We argue that adopting our distinctive concept definitions and mutual relations can greatly benefit XAI community, as clearly defining concepts and relationships between them is a pre-requisite for both the implementation and evaluation of these concepts. Furthermore, the framework yields a classification of AI-systems as *incomprehensible*, *interpretable*, or *understandable*, guiding the research and development to establish understandable AI (e.g., by setting requirements for *contestability*, *predictability*, *verifiability* and *traceability*).

Acknowledgements. This work is part of the research lab AI*MAN of Delft University of Technology.

References

1. Alvarado, R., Humphreys, P.: Big data, thick mediation, and representational opacity. New Lit. Hist. **48**(4), 729–749 (2017). https://doi.org/10.1353/nlh.2017.0037
2. Amir, D., Amir, O.: Highlights: summarizing agent behavior to people. In: Proceedings of the 17th International Conference on Autonomous Agents and MultiAgent Systems, pp. 1168–1176 (2018)
3. Anjomshoae, S., Najjar, A., Calvaresi, D., Främling, K.: Explainable agents and robots: results from a systematic literature review. In: 18th International Conference on Autonomous Agents and Multiagent Systems (AAMAS 2019), Montreal, Canada, 13–17 May 2019, pp. 1078–1088. International Foundation for Autonomous Agents and Multiagent Systems (2019)
4. Arrieta, A.B., et al.: Explainable artificial intelligence (XAI): concepts, taxonomies, opportunities and challenges toward responsible AI. Inf. Fusion **58**, 82–115 (2020). https://doi.org/10.1016/j.inffus.2019.12.012. http://www.sciencedirect.com/science/article/pii/S1566253519308103
5. Broekens, J., Harbers, M., Hindriks, K., van den Bosch, K., Jonker, C., Meyer, J.-J.: Do you get it? User-evaluated explainable BDI agents. In: Dix, J., Witteveen, C. (eds.) MATES 2010. LNCS (LNAI), vol. 6251, pp. 28–39. Springer, Heidelberg (2010). https://doi.org/10.1007/978-3-642-16178-0_5
6. Brooke, J.: SUS: a quick and dirty usability. In: Usability Evaluation in Industry, p. 189 (1996)
7. Chin, J.P., Diehl, V.A., Norman, K.L.: Development of an instrument measuring user satisfaction of the human-computer interface. In: Proceedings of the SIGCHI Conference on Human Factors in Computing Systems, CHI 1988, pp. 213–218. Association for Computing Machinery, New York (1988). https://doi.org/10.1145/57167.57203
8. Ciatto, G., Schumacher, M.I., Omicini, A., Calvaresi, D.: Agent-based explanations in AI: towards an abstract framework. In: Calvaresi, D., Najjar, A., Winikoff, M., Främling, K. (eds.) EXTRAAMAS 2020. LNCS (LNAI), vol. 12175, pp. 3–20. Springer, Cham (2020). https://doi.org/10.1007/978-3-030-51924-7_1
9. van Diggelen, J., et al.: Pluggable social artificial intelligence for enabling human-agent teaming. arXiv preprint arXiv:1909.04492 (2019)
10. Doran, D., Schulz, S., Besold, T.R.: What does explainable AI really mean? A new conceptualization of perspectives. CoRR abs/1710.00794 (2017). http://arxiv.org/abs/1710.00794
11. Doshi-Velez, F., Kim, B.: Towards a rigorous science of interpretable machine learning. arXiv preprint arXiv:1702.08608 (2017)

12. Endsley, M.R.: Situation awareness global assessment technique (SAGAT). In: Proceedings of the IEEE 1988 National Aerospace and Electronics Conference, vol. 3, pp. 789–795 (1988). https://doi.org/10.1109/NAECON.1988.195097
13. Endsley, M.R.: A systematic review and meta-analysis of direct objective measures of situation awareness: a comparison of SAGAT and SPAM. Hum. Factors **63**(1), 124–150 (2021). https://doi.org/10.1177/0018720819875376. pMID: 31560575
14. Goldman, A.I., et al.: Theory of mind. In: The Oxford Handbook of Philosophy of Cognitive Science, vol. 1 (2012)
15. Guidotti, R., Monreale, A., Ruggieri, S., Turini, F., Giannotti, F., Pedreschi, D.: A survey of methods for explaining black box models. ACM Comput. Surv. **51**(5) (2018). https://doi.org/10.1145/3236009
16. Gunning, D.: Explainable artificial intelligence (XAI). Defense Advanced Research Projects Agency (DARPA), nd Web **2**(2) (2017)
17. Harbers, M., van den Bosch, K., Meyer, J.: Design and evaluation of explainable BDI agents. In: 2010 IEEE/WIC/ACM International Conference on Web Intelligence and Intelligent Agent Technology, vol. 2, pp. 125–132 (2010). https://doi.org/10.1109/WI-IAT.2010.115
18. Harbers, M., Bradshaw, J.M., Johnson, M., Feltovich, P., van den Bosch, K., Meyer, J.-J.: Explanation in human-agent teamwork. In: Cranefield, S., van Riemsdijk, M.B., Vázquez-Salceda, J., Noriega, P. (eds.) COIN -2011. LNCS (LNAI), vol. 7254, pp. 21–37. Springer, Heidelberg (2012). https://doi.org/10.1007/978-3-642-35545-5_2
19. Hayes, B., Shah, J.A.: Improving robot controller transparency through autonomous policy explanation. In: 2017 12th ACM/IEEE International Conference on Human-Robot Interaction, HRI, pp. 303–312 (2017)
20. Hoffman, R.R., Mueller, S.T., Klein, G., Litman, J.: Metrics for explainable AI: challenges and prospects (2019)
21. Johnson, M., Bradshaw, J.M., Feltovich, P.J., Jonker, C.M., van Riemsdijk, B., Sierhuis, M.: The fundamental principle of coactive design: interdependence must shape autonomy. In: De Vos, M., Fornara, N., Pitt, J.V., Vouros, G. (eds.) COIN -2010. LNCS (LNAI), vol. 6541, pp. 172–191. Springer, Heidelberg (2011). https://doi.org/10.1007/978-3-642-21268-0_10
22. Johnson, M., Bradshaw, J.M., Feltovich, P.J., Jonker, C.M., van Riemsdijk, M.B., Sierhuis, M.: Coactive design: designing support for interdependence in joint activity. J. Hum.-Robot Interact. **3**(1), 43–69 (2014). https://doi.org/10.5898/JHRI.3.1.Johnson
23. Johnson, M., Vera, A.: No AI is an Island: the case for teaming intelligence. AI Mag. **40**(1), 16–28 (2019). https://doi.org/10.1609/aimag.v40i1.2842. https://ojs.aaai.org/index.php/aimagazine/article/view/2842
24. Klien, G., Woods, D.D., Bradshaw, J.M., Hoffman, R.R., Feltovich, P.J.: Ten challenges for making automation a "team player" in joint human-agent activity. IEEE Intell. Syst. **19**(6), 91–95 (2004). https://doi.org/10.1109/MIS.2004.74
25. Langley, P., Meadows, B., Sridharan, M., Choi, D.: Explainable agency for intelligent autonomous systems. In: AAAI 2017, pp. 4762–4763 (2017)
26. Lipton, Z.C.: The mythos of model interpretability: in machine learning, the concept of interpretability is both important and slippery. Queue **16**(3), 31–57 (2018)
27. Lomas, M., Chevalier, R., Cross, E.V., Garrett, R.C., Hoare, J., Kopack, M.: Explaining robot actions. In: Proceedings of the Seventh Annual ACM/IEEE International Conference on Human-Robot Interaction, HRI 2012, pp. 187–188. Association for Computing Machinery, New York (2012). https://doi.org/10.1145/2157689.2157748

28. Malle, B.F.: How the Mind Explains Behavior. Folk Explanation, Meaning and Social Interaction. MIT-Press, Cambridge (2004)
29. Malle, B.F.: Attribution theories: how people make sense of behavior. Theor. Soc. Psychol. **23**, 72–95 (2011)
30. Miller, T.: Explanation in artificial intelligence: insights from the social sciences. Artif. Intell. **267**, 1–38 (2019). https://doi.org/10.1016/j.artint.2018.07.007. https://www.sciencedirect.com/science/article/pii/S0004370218305988
31. Neerincx, M.A., van der Waa, J., Kaptein, F., van Diggelen, J.: Using perceptual and cognitive explanations for enhanced human-agent team performance. In: Harris, D. (ed.) EPCE 2018. LNCS (LNAI), vol. 10906, pp. 204–214. Springer, Cham (2018). https://doi.org/10.1007/978-3-319-91122-9_18
32. Rosenfeld, A., Richardson, A.: Explainability in human-agent systems. Auton. Agent. Multi-Agent Syst. **33**(6), 673–705 (2019). https://doi.org/10.1007/s10458-019-09408-y
33. Salas, E., Sims, D.E., Burke, C.S.: Is there a "big five" in teamwork? Small Group Res. **36**(5), 555–599 (2005). https://doi.org/10.1177/1046496405277134
34. Sanneman, L., Shah, J.A.: A situation awareness-based framework for design and evaluation of explainable AI. In: Calvaresi, D., Najjar, A., Winikoff, M., Främling, K. (eds.) EXTRAAMAS 2020. LNCS (LNAI), vol. 12175, pp. 94–110. Springer, Cham (2020). https://doi.org/10.1007/978-3-030-51924-7_6
35. Sreedharan, S., Srivastava, S., Kambhampati, S.: Hierarchical expertise level modeling for user specific contrastive explanations. In: IJCAI, pp. 4829–4836 (2018)
36. Turilli, M., Floridi, L.: The ethics of information transparency. Ethics Inf. Technol. **11**(2), 105–112 (2009). https://doi.org/10.1007/s10676-009-9187-9
37. Vilone, G., Longo, L.: Explainable artificial intelligence: a systematic review. arXiv preprint arXiv:2006.00093 (2020)
38. Walmsley, Joel: Artificial intelligence and the value of transparency. AI Soc. 1–11 (2020). https://doi.org/10.1007/s00146-020-01066-z

Towards Explainable Visionary Agents: License to Dare and Imagine

Giovanni Ciatto[2](✉), Amro Najjar[3], Jean-Paul Calbimonte[1], and Davide Calvaresi[1]

[1] University of Applied Sciences and Arts Western Switzerland HES-SO, Sierre, Switzerland
{jean-paul.calbimonte,davide.calvaresi}@hevs.ch
[2] University of Bologna, Cesena, Italy
giovanni.ciatto@unibo.it
[3] University of Luxembourg, Esch-sur-Alzette, Luxembourg
amro.najjar@uni.lu

Abstract. Since their appearance, computer programs have embodied discipline and structured approaches and methodologies. Yet, to this day, equipping machines with imaginative and creative capabilities remains one of the most challenging and fascinating goals we pursue. Intelligent software agents can behave *intelligently* in well-defined scenarios, relying on Machine Learning (ML), symbolic reasoning, and the ability of their developers for tailoring *smart behaviors* to specific application domains. However, to forecast the evolution of all possible scenarios is unfeasible. Thus, intelligent agents should autonomously/creatively adapt to the world's mutability. This paper investigates the meaning of imagination in the context of cognitive agents. In particular, it addresses techniques and approaches to let agents *autonomously imagine/simulate* their course of action and generate explanations supporting it, and formalizes thematic challenges. Accordingly, we investigate research areas including: *(i)* reasoning and automatic theorem proving to synthesize novel knowledge via inference; *(ii)* automatic planning and simulation, used to speculate over alternative courses of action; *(iii)* machine learning and data mining, exploited to induce new knowledge from experience; and *(iv)* biochemical coordination, which keeps imagination dynamic by continuously reorganizing it.

Keywords: Multi-agent systems · Imagination · BDI · Cognitive agents · XAI

1 Introduction

Imagination is among the most powerful tools humankind has ever had. Fundamentally, imagination is responsible for the *spontaneous* creation of novel ideas which do not originate directly from the human senses. Such a mental process

D. Calvaresi et al. (Eds.): EXTRAAMAS 2021, LNAI 12688, pp. 139–157, 2021.
https://doi.org/10.1007/978-3-030-82017-6_9

enabled humans to design complex concepts and artifacts, shaping the societies we live in nowadays [1].

Over the years, imagination and creativity have been considered the nemesis of discipline and structured approaches in general [2]. Indeed, at an individual level, imagination is a relatively simple process. It can be conceived as *a never-ending activity occurring within each person's mind along their entire lifetime.* Such activity copes with reorganizing a person's beliefs, perceptions, feelings, and habits repeatedly to *generate* novel believes, abilities, desires, insights about the future, and needs—which in turn may motivate novel activities.

Let us consider two simple examples commonly dealt with:

Counterfactual thinking: it is a retrospective "what if" analysis, elaborating how things could have been different (e.g., regretting a decision, *"I should have behaved differently"*) – also known as staircase wit – from which a *lesson* is supposably learned.

Speculative thinking: mentally simulating possible future scenarios according to models of *(i)* the world, and *(ii)* other agents/humans behaviours—e.g., imaging the effect a proposed example might have on the reader.

There, imagination is a key enabler for intelligent behavior.

In modern Artificial Intelligence (AI), many research efforts are devoted to the engineering of *smart mechanisms*, which enable software agents to behave *intelligently* in well-defined scenarios. Most of these mechanisms are either based on Machine Learning (ML) or on symbolic reasoning (including planning or automatic theorem proving) [3]. Nevertheless, the capability of software agents (intended as intelligent virtual entities) to behave intelligently strongly depends on their developers' capability of tailoring *smart behaviors* on the particular scenario the software agents operate into. Arguably, however, it is unfeasible for developers to forecast all possible evolutions a real-world scenario may be subject to. Accordingly, intelligent software agents should also adapt to the world by autonomously figuring out how to deal with its mutability—similarly to what a human would do. Notably, one of the significant areas where adaptability is expected to play a major role is XAI. Indeed, there is an increasing push for intelligent systems capable of explaining their own behavior [4]. However, current research efforts are mostly focused on supporting data scientists in drawing explanations in particular cases. Even when explanations are delegated to autonomous agents, their capability to generate effective explanations still relies on their developers' foresight. In other words, the problem of letting agents *autonomously* generate explanations is still open.

Human beings heavily leverage on imagination to adapt to the world. In particular, they exploit both counterfactual and speculative thinking to adapt the way they *interact* with their interlocutor. Arguably, similar mechanisms could be conceived for agents willing to attain the capability of *generating* explanations.

Accordingly, in this paper, we discuss *(i)* what imagination may mean for software agents, *(ii)* how it could be technically realized within modern agent frameworks, *(iii)* what is the role of imagination-equipped agents in modern data-driven AI, and *(iv)* how can imagination support the autonomous generation of explanations. In doing so, we restrict our scope to the case of *cognitive* agents, as their abstractions are rich enough to capture a general – yet precise – notion of imagination. In particular, we focus on the Belief-Desire-Intention (BDI) agent architectures as they represent the best viable bridge among theory and practice—being backed by effective technologies such as Jason [5].

Within the scope of this paper, we conceive *imagination* as a non-terminating background activity carried on by an agent behind the scenes, possibly while doing anything else. The imagination activity takes care of *continuously* revising an agent's internal knowledge, possibly *(i)* obliterating useless information; *(ii)* synthesizing novel information out of the current and previous experience; *(iii)* dismissing or generating desires and needs; *(iv)* critically analyzing the previous courses of actions w.r.t. their goals; *(v)* simulating possible similar/alternative behaviors to be exploited in similar situations; and *(vi)* looking for *post-hoc* motivations for their actions. Thanks to such an ability, agents would become not only able to acquire novel information but also novel capabilities (i.e., procedural knowledge), possibly acquiring the (bits of) self-awareness required to provide *explanations* about their own courses of action.

In practice, the imagination abstraction leverages mechanisms laying at the intersection of different research areas, such as: *(i)* symbolic reasoning and automatic theorem proving (which are exploited to synthesize novel knowledge via inference), *(ii)* automatic planning and simulation (which is exploited to speculate over alternative courses of action), *(iii)* machine learning and data mining (which are exploited to induce new knowledge from experience), and *(iv)* biochemical coordination (which keeps imagination dynamic by continuously reorganizing it).

The rest of the paper is organized as follows. Section 2 briefly presents the current background technologies and their state-of-the-art supporting our notion of imagination. Section 3 introduces, defines, and discusses the concept of imagination and our practical view. Section 4 elicits the challenges related to our definition of imagination and the related research areas involved. Finally, Sect. 5 concludes the paper.

2 State of the Art

The investigation of mechanisms for agents' imagination roots from cross-disciplinary components. In particular, this section provides a brief background on *(i)* human imagination mechanisms; *(ii)* cognitive agent architectures; *(iii)* imagination mechanisms including inference, data-driven learning, biochemical coordination, & simulation; and *(iv)* computational creativity.

2.1 Imagination in Humans

In the late 80s and early 90s, constructivist [6,7] and developmental [8] approaches inspired many advancements in AI. In particular, virtual agents have been equipped with "inherent" learning mechanisms, allowing them *to make sense* of their environment and exploit its affordances[1] [6,10]. Such an approach has been inspired by constructivism [11] and developmental learning, promoted by the Swiss psychologist Jean Piaget (1930s), and adapted by Drescher into a bottom-up developmental approach to interact with the surrounding environment named `Schema` mechanism [6].

In the context of autonomous agents, such `Schemas` are pieces of knowledge processed by the agents to comprehend and react to their environments. The developmental process of `Schemas` is characterized by:

Assimilation: describes how humans or agents perceive and adapt to new information, fitting them into existing cognitive schemas.
Accommodation: restructures existing `Schemas` to handle novel information.

Intuitively, the cognitive growth of an intelligent system would imply the evolution/extension of both very specific knowledge and overall dynamics coordinating the several learning-related aspects. In the last decade, constructivist approaches have been used for smart environments [12,13] and transport systems [14]. The studies contributing to these aspects provide contributions highly specialized in simplistic and very structured domains [15]. Nevertheless, such individual approaches can be hand-crafted together into architectures relying on traditional component-based software development methodologies. However, although the single components can evolve singularly, the reconciling system is constrained by the hand-crafted interconnection mechanisms. Consequently, such lack of flexibility precludes architecture-level evolution (i.e., autonomous architectural adaptation and growth of the systems) and learning [7].

The lack of generalization characterizing current solutions impedes the application of intelligent/learning systems in general-purpose scenarios, being incapable of applying themselves autonomously to arbitrary problems. Therefore, imagination cannot be a cross-system functionality.

2.2 Cognitive Agents

The philosopher Michael Bratman formalized human practical reasoning in the beliefs-desires-intentions (BDI) model as a way to explaining future-directed intention [16]. Successively, it became a model to program intelligent agents, which made its first appearance in the Rational Agency project at the Stanford Research Institute in the mid 1980s [17]. Such a model is characterized by:

Beliefs: a set of facts and rules representing an agent's epistemic memory, possibly containing its knowledge about the world, itself, and other agents.
Desires: a set of goals the agent is willing to achieve, test, or maintain.
Intentions: a set of tasks the agent is currently carrying on.
Plans: a set of *recipes* representing the agent's procedural memory, encoding the procedural know-how about tasks.

[1] The term, originally coined by Gibson, refers to what the environment offers an individual [9].

Any cognitive feature of a BDI agent may vary during its lifetime. For instance, novel beliefs appear in the agents' minds whenever they receive novel perceptions from their sensors, while stale beliefs simultaneously disappear. Similarly, novel beliefs may arise while agents interact among each other – or with humans – or as they chose to memorize some information they have deducted via reasoning. The occurrence of relevant events may provoke the desire pool's update (i.e., acquiring new goals to be achieved/tested/maintained and/or discard some goals). Agents' desires eventually lead to spawning novel intentions (activities to achieve/test/maintain goals the agent is committed to). While carrying on an intention, agents may select one or more plans among those supporting the corresponding desire's accomplishment. Plans may involve the execution of one or more actions – possibly affecting the world via actuators – or the accomplishment of further sub-goals, which may, in turn, require the execution of further plans as part of the same intention.

In the scope of this paper, it is worth highlighting that the BDI model allows agents to exhibit more complex behavior than purely reactive models, unbound to the computational overhead of other cognitive architectures [17]. Furthermore, being rooted in folk-psychology, it has been outlined as an excellent candidate to represent everyday explanations [18,19] (since it is considered as the attribution of human behavior using "everyday" terms such as beliefs, desires, intentions, emotions, and personality traits [20,21]).

The BDI model has also been identified as the most used/suitable architecture to generate explanations for goal-driven agents/robots [19,22,23]. The trend of attributing to the BDI model the suitability for XAI applications is reinforced by user studies supporting the human tendency to attribute a State of Mind (SoM) to robots and agents. In such a context, the lack of communication or misalignment due to lack of transparency can result in ill-formed SoM [24]. To avoid such a risk and the consequent drop of trust in the system, BDI agents are envisioned to employ folk-psychology to explain their SoM [19,25].

2.3 Mechanisms for Imagination

In our view, the agents' imagination process must rely on mechanisms employing different techniques, such as inference, data-driven learning, biochemistry-inspired coordination, and simulation.

Logic Inference. In computational logic [26], inference is the process of rigorously drawing conclusions out of premises. The existing inference procedures depend on the given logic formalism at hand, and they may serve different purposes (depending on their nature). Overall, there are three main sorts of inference: deductive, inductive, and abductive.

Deductive inference dictates under which conditions conclusions can be drawn out of some axioms, i.e., rules and facts considered true. In other words, deduction elicits the knowledge which is possibly implicit into the axioms.

Inductive inference aims at estimating rules out of a number of positive (and negative) examples of facts satisfying (or violating) the rule. In other words, induction attempts to generalize principles by distilling patterns from generic observations/contingencies.

Abductive inference: aims at hypothesizing which premises could provoke some evidences, given a number of rules describing how causes provoke effects. In other words, abduction attempts to speculate on the possible causes of some phenomenon (e.g., finding the most straightforward and most likely explanation for an observation), given that the general rules governing that phenomenon are known.

Logic programming (LP) technologies (e.g., Prolog) enable users – and potentially agents – to encode their knowledge into logic facts and rules, which may then be queried via logic solvers [27]. Accordingly, by endowing agents with adequate LP technologies, they can autonomously exploit inference when required [27].

Learning from Data. Data-driven AI falls into the context of the so-called machine learning (ML). Learning from data is commonly the activity performed via supervised or unsupervised learning and comprises a broad set of methods and tools such as reinforcement learning, classification, regression, time series forecasting, pattern recognition, generative models [28]. Supervised learning leverages on the existence of many input/expected-output examples and consists of looking for the best function mapping the available inputs into the corresponding expected outputs. In a sense, supervised learning is very similar to logic induction, except that it does not assume knowledge to be encoded via logic clauses, and it is better suited for *learning* from numeric data. Unsupervised learning aims at finding similarities and patterns possibly buried into numerical data without any expected outcome at hand. Thus, pieces of information are extracted from data through some optimality criterion.

By exploiting the wide availability of task-specific techniques and algorithms in ML, agents may be equipped with the capability of managing different sorts of data to serve disparate purposes [29]. For instance, by wrapping neural networks, agents may gain image and speech recognition capabilities, as well as the capability of analyzing and forecasting time-related measurements.

Finally, it is worth mentioning another relevant perspective intersecting ML and MAS: learning autonomously, continuously, and adaptively to increment skills and knowledge (a.k.a, lifelong ML or continuous learning—CL henceforth). In the context of ML, it entails updating the prediction models periodically with novel tasks and data distributions, still being able to (re)use and retain knowledge and skills over time. CL is beneficial when data or tasks' availability varies over time (i.e., no longer or not yet available), and it is imperative to consider prior knowledge [30,31].

Biochemical Coordination. Within the scope of self-organizing MAS, biochemical coordination is the study of interaction among agents mediated by biochemistry-inspired patterns. There, information is modelled as *molecules*, i.e., chunks of data characterized by a *concentration* value denoting their relevance [32]. Such molecules may *diffuse* among different locations (e.g., to represent information exchanges), *aggregate* with each others (e.g., to represent more complex data structures), and *evaporate*, (e.g., reduce their concentration as the carried information loses relevance). The concentration and nature of such molecules determine the dynamics of the systems relying on the biochemical metaphor. A number of coordination rules are commonly in place, affecting (and being affected by) the concentration of molecules within a given context, and governing information diffusion, aggregation, evaporation, or generation.

Due to their nature, such sorts of systems are inherently stochastic and fuzzy, and therefore ideal to realize resilient, robust, and self-organizing applications. In this context, pieces of information are not solely true or false, but rather more or less *concentrated.* Therefore, inconsistencies and contradictory data may simultaneously co-exist with no harm, as long as consistent truths eventually *emerge* by becoming significantly more concentrated. The combination of such features determines biochemical coordination mechanisms eligible to support imagination, as it may spawn several (possibly inconsistent) ideas, properly balancing evaporation, diffusion, and aggregation to retain only the most useful ones.

Multi-agent Based Simulation. Simulation is one of the most employed techniques to identify/reach potentially useful outcomes. Agent simulation technology has been outlined as an efficient platform helping to understand autonomous behavior and decision-making [33]. An agent-based simulation (ABS) model is a set of interacting intelligent entities that reflect, within an artificial environment, the relationships in the real world [33]. Thus, ABS is typically used for helping decision-makers cope with complex and changing environment in the domains such as UAVs (cf. [34] and the references therein), IoT, and CPS [35,36], and to model and optimize robot behavior [37].

It is worth noticing that most of the works in the ABS literature focus on inter-agent relationships and their interaction with the environment [17]. Conversely, this paper aims at tackling the intra-agent perspective, where agents should be capable of simulating multiple states of themselves and their actions within their own "mind". This internal simulation process is analogous to human "mental simulation" where humans rely on the ability to construct mental models to imagine what will happen or what could be [38–40]. Such capability has helped humans in physical reasoning [41,42], spatial reasoning, and counterfactual reasoning [43].

Similarly, agents can mimic this "mental" modeling and analyze the assumed outcomes of its own actions, identify and possibly exploring arguably reasonably paths leading to potentially creative scenarios even in robustly novel situations [40]. Such explorations might lead to totally unforeseen solutions, which without a simulation based on a trial and error approach would not have been discovered/investigate. Hence, agents may reflect upon a set of simulations representing themselves in different/alternative scenarios.

2.4 Computational Creativity

In its broadest scope, creativity is defined as the ability to generate new forms and artifacts autonomously [44]. In the literature, creativity is classified as either *biological* (the ability to generate new cells, organs, organisms, or species [44]) or *psychological* (the ability to generate new ideas and artifacts). Researches in AI have been pushing to extend the notion of creativity to virtual systems [45]. For example, a recent study, inspired from enactive AI [46,47], investigates how computational creativity can be adopted by autonomous agents [48]. Despite this progress, most of the works in this domain are either carried out at the conceptual level or solely rely on data-driven mechanisms (e.g., generative adversarial networks, a.k.a. GANs) to generate "creative" contents (e.g., music [49] or pictures [50]).

In contrast with these works (primarily ML-centered), we envision agents questioning their beliefs, knowledge, and goals continuously. In particular, agents should combine classic planning, reinforcement learning, and in-mind simulation about their future actions to simulate and possibly provide explanations about their courses of actions.

3 Imagination in Cognitive Agents

Overall, BDI agents' dynamics are moved by intentions and directed by desires. Equipped with *sensors* and *actuators*, they can respectively *perceive* and *affect* the world they live into. However, an agent's admissible pool of desires and plans is defined/constrained by human developers. Indeed, developers tend to dictate agents' initial desires and plans to keep their dynamics predictable and controllable. However, this prevents the full exploitation of agents' autonomy, adaptability, and, ultimately, intelligence.

Arguably, to let agents access a higher degree of intelligence, they should be endowed with the capability of spontaneously generating new desires, acquiring novel beliefs, and learning novel plans. Briefly speaking, we consider imagination as the activity devoted to supporting such capability. Thus, we define imagination as an agent's intention aimed at *maintaining* its *innate* desires of being *creative*, *curious*, and *effective*. More precisely, in our framework, agents are assumed to be endowed with (at least) one *maintenance* desire since their creation. Such desire pushes them to (attempt to) be creative, curious, and effective whenever they can. To be creative, an agent should keep looking for novel information, as well as novel ways to do what it needs to do (i.e., it must keep trying to enrich its belief and plan bases). To be curious, an agent should keep exploring the world and search for novel stimuli or just doing things to learn something new (i.e., it must keep trying to enrich its desires). To be effective and prove the way it deliberates and acts, an agent should keep improving its epistemic and procedural knowledge (i.e., improve its belief and plan bases).

To accomplish such an innate desire, BDI agents must spawn an intention that will be part of them for their whole lifetime. The basic functioning of this intention is relatively straightforward: *to keep revising the agents' beliefs, desires, and plans to generate novel epistemic/procedural knowledge or improve the current one.* We call this intention "imagination".

To accomplish its purpose, the imagination intention may leverage and combine several basic mechanisms coming from different branches of AI. Imagination can exploit mechanisms deriving from the classes of activities listed below (independently from technical details). For example,

knowledge acquisition is the process of converting raw data (i.e., percepts or beliefs) into general and reusable knowledge (e.g., in the form of logic rules or sub-symbolic predictors)

knowledge synthesis is the process of inferring or distilling novel knowledge out of pre-existing ones

speculation is the process of exploring alternative truths, situations, or courses of actions based on previous experiences

knowledge revision is the process of criticizing the pre-existing knowledge, possibly evicting stale or wrong information

The remainder of this section analyzes how mechanisms from the many branches of AI may be exploited to support such activities. Figure 1 provides a summarizing characterization.

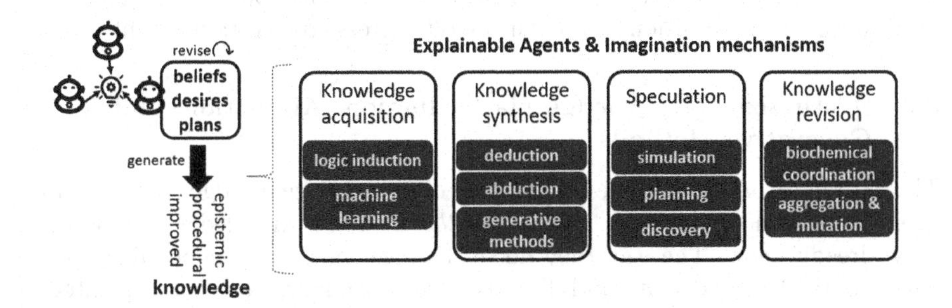

Fig. 1. Imagination in cognitive agents: AI mechanisms.

3.1 Acquiring Knowledge via Learning and Induction

BDI agents acquire information either by perceiving the environment or by communicating with other agents. In both cases, information comes in the form of raw data and can be *stored* either symbolically or sub-symbolically. Each datum represents a particular event from the external world. While the single event may be potentially useless *per se*, the frequent occurrence of similar events may generate value in the long run. Indeed, agents – similarly to humans – may distill valuable *knowledge* out of statistically relevant anterior experience (i.e., data).

Differently from data, however, knowledge is an *aggregated* and *reusable* form of information. It must be reusable because that is what makes it valuable enough to memorize it. It must be aggregated because agents' cognitive resources (such as computational power and memory) are inherently limited, in practice, and such limitations affect how and to what extent data can be actually reused. However, the way data is actually aggregated to make it reusable depends remarkably on its nature.

In case data is symbolically represented, it is interpreted as logic *facts* and stored into symbolic knowledge bases that the agent may efficiently update and query. When this is the case, logic *induction* may be used to distill *rules* out of facts. While facts are contingent, rules are synthetic, and they may be reused in several similar situations. Furthermore, symbolic rules are human-intelligible. Thus, they can be used by agents as a basis to construct explanations for their reasoning or behavior.

Conversely, when data is represented sub-symbolically, machine learning can be exploited to draw knowledge out of it. When this is the case, data is interpreted as tensors of numbers used to *train* a predictor (e.g., a neural network). This usually makes knowledge both aggregated and reusable, despite not directly intelligible (and explainable) for humans. Accordingly, induction can be exploited by agents willing or requiring to manipulate symbolic information, either because they need to take discrete decisions or because they care about the intelligibility of their decisional process. Conversely, machine learning can be exploited by agents needing to manipulate sub-symbolic information—possibly because they need to take fuzzy decisions, and can tolerate errors to a certain extent.

3.2 Synthesizing Knowledge via Deduction, Abduction, and Generative Methods

The external world is not the only source of valuable knowledge. Indeed, intelligent agents should also be able to *synthesize* novel knowledge out of what they already know. The way they do so, however, may vary depending on the nature of the knowledge at hand. For instance, when knowledge is represented in symbolic form, *deductive* or *abductive* reasoning procedures may be exploited to infer novel information out of it. Conversely, when knowledge is sub-symbolically represented, *generative* methods may be exploited instead.

In particular, deductive reasoning may be exploited to make *implitic* knowledge explicit. In fact, deduction can derive specific facts out of rules. In other words, it is dual w.r.t. induction. Thanks to deduction, agents may for instance select useful knowledge for the contingent situation they are immersed into, out of general rules. Similarly, abductive reasoning may be exploited by agents willing to draw *hypotheses* about the causes which lead to a particular situation. In other words, abduction let agents synthesize likely facts which justify the facts they already know to be true, according to the rules they know to hold in a particular context. Accordingly, abduction is one of the mechanisms supporting speculative thinking.

Conversely, generative methods – such as GAN – may be exploited by agents needing to produce human-comprehensible representations (e.g. audio, video, etc.) of categories they already know how to recognize and manipulate—such as faces [51], shapes [52], handwriting [53], speech, etc. These representations might serve as key enablers for explainable synthesized knowledge. In turn, generative methods may be key enablers for *(i)* computational creativity, as they let agents produce original representations; *(ii)* counterfactual thinking, as they support the generation of *variants* of any given concept; and *(iii)* effective human-machine interactions, as they let agents enrich their interactions with humans with randomly-generated examples or analogies.

3.3 Speculating via Simulation and Planning

Mentally simulating scenarios is a fundamental human capability [39]. Once people have enough information about the characterization and dynamics of the surrounding world, simulating the effects of their actions becomes a common practice (to a certain extent). Often, it is only after having *mentally* simulated the most likely outcomes of their course of actions that an individual chooses how to act. Then, by comparing the *actual* outcomes with the expected ones, humans may learn how to improve their behavior w.r.t. their goals. Furthermore, even when a direct experience is lacking, simulating the possible effects of a given action is still better than acting randomly.

In the AI literature, *planning* is the activity performed by agents willing to deliberate what to do in a particular context. Planning algorithms commonly leverage on rich descriptions of *(i)* the environment, *(ii)* agents' actions, *(iii)* their effects, and *(iv)* some description of the target goal the agent is willing to achieve. Through such descriptions, planning algorithms (attempt to) compute viable workflows of actions that should lead agents towards the target goal. However, even when only a few agents and small deterministic environments are involved, planning is computationally costly. Therefore, when complex non-deterministic environments are in place (where several agents interact in non-trivial ways), planning may quickly become unfeasible.

Scientific researchers often tackle the complexity of systems by *simulating* simplified parametric models executed multiple times, with randomly generated parameters. Doing so allows drawing statistical conclusions based on the data generated by such *in-silico* experiments. Accordingly, agents may follow a similar approach, in their minds, to decide what to do or what to expect. In other words, agents may leverage simulation to realize *speculative* thinking.

MAS have been exploited for the purpose of simulation since their very beginning—cf. ABS in Sect. 2.3. However, currently, most ABS research efforts are devoted to exploiting MAS *in* simulations rather than the opposite. Conversely, the idea of letting an agent simulate itself and its environment is quite new—other than very challenging. According to such a perspective, we envision equipping each agent with an ABS sub-system capable of simulating an entire MAS. Using *inner* simulations, agents could then try out different actions and sequence of actions that are otherwise too costly or even dangerous to try in the real environment, other than retrying the same scenario over and over again with different rules or parameters. Similarly to humans, agents may then exploit such capability to autonomously discover plans, rules, or even policies for situations that they have either already experienced, or not.

3.4 Adaptively Revising Knowledge via Biochemical Coordination

Simulating the world to discover novel plans, rules, or – more generally – background knowledge may eventually lead to the creation of chunks of *potentially* useless information. By doing so, efficiency issues in both information storage and retrieval may arise. However, a more conservative strategy may prevent agents from discovering novel and *potentially* useful information. Accordingly, some general strategies should be exploited to allow each agent to decide what knowledge to retain and discard dynamically.

Here we welcome the idea that no "one size fits all" solution exists to select knowledge based on expected utility. Indeed, any predefined strategy may be affected by the biases of who designed it or be tailored to a particular scenario while being sub-optimal in other ones. For this reason, we argue that an *adaptive* strategy based on a biochemical metaphor would be preferable.

From such a perspective, we assume the many mechanisms proposed so far – inference, machine learning, planning, simulation – to produce information in the form of molecules. Agents' minds can then be conceived as containers of molecules of different sorts—e.g., beliefs, neurons, plans, etc. Such molecules' concentration may increase over time as the corresponding information may be produced multiple times or on a per-usage basis. For instance, the same rule may be induced from different data in different instants or be frequently used. Similarly, different runs of a simulation may lead to the frequent exploitation of similar courses of action. This, in turn, may lead an agent to increase the concentration of one or more plans. At the same time, we assume the information is subject to *evaporation* on a uniform basis. In other words, all sorts of information evaporate at the same pace. As a global effect, only relevant information would be able to survive, in the long run – where by relevant we mean either frequently generated or frequently used, without requiring to *a-priori* define what is actually relevant.

Finally, aggregation and mutation mechanisms may fit the picture by supporting random and periodic modifications or combinations of pre-existing information—e.g., by merging similar plans/rules into more articulated ones, or by slightly altering some parameter of a wrapped neural network to change its behavior. Such random modifications may then be adaptively confirmed or discarded, depending on whether they result to be relevant or not for the agent. In the former case, their concentration would increase, whereas in the latter case, it would decrease.

4 Open Challenges

As we discussed in the previous sections, imagination within the context of agent intentions entails the provision of continuous knowledge acquisition, synthesis, revision and exploration. While we identified both the existing approaches (cf. Sect. 2) as well as the specific research areas (cf. Sect. 3) to investigate, we acknowledge the existence of key challenges to tackle in the process. These are summarized as follows:

C1) *Knowledge heterogeneity.* Acquisition of knowledge is essential to feed a creative process, whether it is intending to combine information to derive new insights, for synthesizing explainable knowledge, for confronting different views, or even for exploration of uncharted territory. Nevertheless, agents will be exposed to the challenge of extreme variety among the different knowledge sources that they run across. Solving semantic and data-representation heterogeneity issues arising from this diversity will be a necessary step. An example would be using knowledge graph matching and fusion techniques [54]. Moreover, given the need for integrating symbolic and sub-symbolic sources, tools and techniques for a coherent integration between both will have to be studied [55].

C2) *Goal generation.* A fundamental step in a creative cycle is establishing clear goals, even if these may be updated in the future. While a goal may define an overall scope for the development of creative activities, in some cases, the goals may not be entirely known *a priori*. In such conditions, goal generation [56] must be part of the creative process that needs to be incorporated into the agent model [57].

C3) *Knowledge alignment.* Even when knowledge heterogeneity has been addressed, it will still be required to align different understandings of observed phenomena relevant to the creative plan's scope. For example, if knowledge sources' provenance is dissimilar, simply aligning terminologies and concepts is not enough [58]. At this point, it is crucial to study models that allow handling contradictions, assumptions, explainable outcomes, and interpretations as part of the knowledge alignment task [59,60].

C4) *Information uncertainty.* Creative agents must take into account not only the potential inaccuracy of their information sources but also the eventual uncertainty of their own artificial imagination. The exploration and navigation over radically new ideas and approaches entail high risk, meaning that oftentimes they may lead to dead-ends. Agents may need to incorporate risk management strategies [61] allowing them not to constrain themselves only to *safe* knowledge but leave enough space for behavioral models that adapt to different levels of uncertainty. This also applies to uncertainty in XAI outcomes and their consequences on inter-agent agreements.

C5) *Reasoning complexity.* The generation of new knowledge may require reasoning over potentially large and/or complex knowledge graphs [62]. Depending on the complexity of these graphs' underlying logics, reasoning tasks may become increasingly expensive in terms of computation. Moreover, the agents' autonomous nature will necessitate further exploration of decentralized reasoning techniques, including partial knowledge and probabilistic approaches. An additional challenge to tackle is the combination of explicable results of data-driven AI predictions. Multi-agent speculative reasoning may need to be combined with machine learning outcomes to address this challenge.

C6) *Hypotheses evaluation.* Agents will be able to propose hypotheses that may need to be validated or refuted [63]. This ability should be accompanied by a robust framework for managing assumptions, claims, justifications, explanations, and proofs [64,65]. As explained in the previous point, reasoning and sub-symbolic outcomes have to be evaluated with respect to the hypotheses. Agents may eventually have different or plainly contradictory points of view, for which reconciling mechanisms may need to put in place. While in some cases competitive approaches may be preferred (e.g., working towards the same goal but under different imaginative hypotheses and assumptions), in others, it might be necessary to align and establish a cooperation scheme.

C7) *Explicable knowledge revision.* When the results of explicable machine learning and, in general, generated sub-symbolic knowledge are produced, agents need to navigate through them and understand their implication over existing information [66]. This may lead to invalidating previous beliefs or to changing the uncertain status or certain facts. The challenge of explaining these decisions, and providing justification of the imaginative paths taken by a community of agents, shall be addressed to understand the path that leads to creative activities. The provenance of knowledge and the changes may lead to even reconsidering information that was deemed false or invalid in a previous iteration.

C8) *Exploration.* Agent imagination requires substantial space for the exploration of new knowledge and experimentation through novel approaches. Although exploratory agents have been studied in the past [67], it remains a challenge to establish a formal framework for discovery in large knowledge spaces. Approaches like link traversal of knowledge graphs may serve as a starting point, although they may need to be extended to a cooperative scenario where different agents run exploratory tasks under coordination mechanisms.

C9) *Accountability.* Imaginative processes in multi-agent systems entail the exploration and creation of new knowledge, as well as the validation of previous and new ideas. The consequences of these actions may lead to decisions and actions for which there should be clear responsibilities. In that context, the provenance information emanating from exploratory processes and knowledge revision decisions will need to be associated with trust mechanisms allowing to ensure proper attribution to an agent or a person embodied by an agent. Furthermore, accountability [68] in terms of ethical and even legal terms should be studied, not only from a purely technical perspective (e.g., accountability networks, knowledge graph ledgers) but also from a psycho-social point of view (i.e., human-agent accountability).

5 Conclusions

This paper provided a ground for discussing the meaning of imagination in the setting of cognitive agents, selected possible tools and approaches, and elicited the envisioned contextual challenges. In particular, the investigated research are *(i)* reasoning and automatic theorem proving, *(ii)* automatic planning and simulation, *(iii)* machine learning and data mining, and *(iv)* biochemical coordination. Finally, the intuitions proposed directions collapsed in the definition of challenges in the areas of knowledge heterogeneity, goal generation/definition, knowledge alignment, information uncertainty, reasoning complexity, hypothesis evaluation, explicable knowledge revision, exploration, and accountability.

To address these challenges, in the future we plan to explore a number of practical research directions aimed at creating the technological playground for supporting our notion of imagination. For instance, the problem of letting agent programming technologies support several logics and inference procedures, is far from being solved [69]. A similar statement holds for the simulation of large-scale MAS composed by cognitive agents. For this reason, our first efforts shall be devoted to *(i)* the development (resp. extension) of novel (resp. existing) agent programming framework to support inductive, and abductive reasoning – for instance, via the 2P-KT technology [70] –, *(ii)* the development of simulation frameworks for cognitive agents, supporting virtualization of both space and time – for instance via the Alchemist simulator [71] –, *(iii)* blending (either existing or novel) agent programming frameworks and mainstream ML frameworks—such as TensorFlow, PyTorch, etc. Conversely, concerning the design of and biochemical coordination at the single agent level, we argue that further research is needed. Along this path, our first step will consist of a formalization, aimed at further clarifying possibilities and challenges.

Acknowledgements. This work has been partially supported by the CHIST-ERA grant CHIST-ERA-19-XAI-005, and by *(i)* the Swiss National Science Foundation (G.A. 20CH21_195530), *(ii)* the Italian Ministry for Universities and Research, *(iii)* the Luxembourg National Research Fund (G.A. INTER/CHIST/19/14589586), *(iv)* the Scientific and Research Council of Turkey (TÜBİTAK, G.A. 120N680).

References

1. Stout, D.: Stone toolmaking and the evolution of human culture and cognition. Philos. Trans. Roy. Soc. B Biol. Sci. **366**(1567), 1050–1059 (2011)
2. Cave, S.: The problem with intelligence: its value-laden history and the future of AI. In: Proceedings of the AAAI/ACM Conference on AI, Ethics, and Society, pp. 29–35 (2020)
3. Honavar, V.: Symbolic artificial intelligence and numeric artificial neural networks: towards a resolution of the dichotomy. In: Sun, R., Bookman, L.A. (eds.) Computational Architectures Integrating Neural and Symbolic Processes. SECS, vol. 292, pp. 351–388. Springer, Boston (1995). https://doi.org/10.1007/978-0-585-29599-2_11

4. Ciatto, G., Schumacher, M.I., Omicini, A., Calvaresi, D.: Agent-based explanations in AI: towards an abstract framework. In: Calvaresi, D., Najjar, A., Winikoff, M., Främling, K. (eds.) EXTRAAMAS 2020. LNCS (LNAI), vol. 12175, pp. 3–20. Springer, Cham (2020). https://doi.org/10.1007/978-3-030-51924-7_1
5. Bordini, R.H., Hübner, J.F.: BDI agent programming in AgentSpeak using *Jason*. In: Toni, F., Torroni, P. (eds.) CLIMA 2005. LNCS (LNAI), vol. 3900, pp. 143–164. Springer, Heidelberg (2006). https://doi.org/10.1007/11750734_9
6. Drescher, G.L.: Made-Up Minds: A Constructivist Approach to Artificial Intelligence. MIT Press, Cambridge (1991)
7. Thórisson, K.R.: From constructionist to constructivist AI. In: AAAI Fall Symposium: Biologically Inspired Cognitive Architectures (2009)
8. Blank, D., Kumar, D., Meeden, L., Marshall, J.B.: Bringing up robot: fundamental mechanisms for creating a self-motivated, self-organizing architecture. Cybern. Syst. Int. J. **36**(2), 125–150 (2005)
9. Gibson, J.J.: The Senses Considered as Perceptual Systems (1966)
10. Brooks, R.A.: Intelligence without representation. Artif. Intell. **47**(1–3), 139–159 (1991)
11. Piaget, J.: La naissance de l'intelligence chez l'enfant. Delachaux et niestlé, Paris (1936)
12. Najjar, A., Reignier, P.: Constructivist ambient intelligent agent for smart environments. In: 2013 IEEE International Conference on Pervasive Computing and Communications Workshops (PERCOM Workshops), pp. 356–359. IEEE (2013)
13. Alyafi, A.A., et al.: From usable to incentive building energy management systems. Modélisation et utilisation du contexte (Model. Using Context) **2**(1), 1–30 (2018)
14. Guériau, M., Armetta, F., Hassas, S., Billot, R., El Faouzi, N.-E.: A constructivist approach for a self-adaptive decision-making system: application to road traffic control. In: 2016 IEEE 28th International Conference on Tools with Artificial Intelligence (ICTAI), pp. 670–677. IEEE (2016)
15. Georgeon, O.L., Morgan, J.H., Ritter, F.E.: An algorithm for self-motivated hierarchical sequence learning. In: Proceedings of the International Conference on Cognitive Modeling, Philadelphia, PA, ICCM-164, pp. 73–78. Citeseer (2010)
16. Bratman, M., et al.: Intention, Plans, and Practical Reason, vol. 10. Harvard University Press, Cambridge (1987)
17. Adam, C., Gaudou, B.: BDI agents in social simulations: a survey. Knowl. Eng. Rev. **31**(3), 207–238 (2016)
18. Norling, E.: Folk psychology for human modelling: extending the BDI paradigm. In: Proceedings of the Third International Joint Conference on Autonomous Agents and Multiagent Systems, vol. 1, pp. 202–209 (2004)
19. Broekens, J., Harbers, M., Hindriks, K., van den Bosch, K., Jonker, C., Meyer, J.-J.: Do you get it? User-evaluated explainable BDI agents. In: Dix, J., Witteveen, C. (eds.) MATES 2010. LNCS (LNAI), vol. 6251, pp. 28–39. Springer, Heidelberg (2010). https://doi.org/10.1007/978-3-642-16178-0_5
20. Churchland, P.M.: Folk psychology and the explanation of human behavior. Philos. Perspect. **3**, 225–241 (1989)
21. Malle, B.F.: How people explain behavior: a new theoretical framework. Pers. Soc. Psychol. Rev. **3**(1), 23–48 (1999)
22. Neerincx, M.A., van der Waa, J., Kaptein, F., van Diggelen, J.: Using perceptual and cognitive explanations for enhanced human-agent team performance. In: Harris, D. (ed.) EPCE 2018. LNCS (LNAI), vol. 10906, pp. 204–214. Springer, Cham (2018). https://doi.org/10.1007/978-3-319-91122-9_18

23. Anjomshoae, S., Najjar, A., Calvaresi, D., Främling, K.: Explainable agents and robots: results from a systematic literature review. In: 18th International Conference on Autonomous Agents and Multiagent Systems (AAMAS 2019), Montreal, Canada, 13–17 May 2019, pp. 1078–1088. International Foundation for Autonomous Agents and Multiagent Systems (2019)
24. Hellström, T., Bensch, S.: Understandable robots-what, why, and how. Paladyn J. Behav. Robot. **9**(1), 110–123 (2018)
25. Kaptein, F., Broekens, J., Hindriks, K., Neerincx, M.: Personalised self-explanation by robots: the role of goals versus beliefs in robot-action explanation for children and adults. In: 2017 26th IEEE International Symposium on Robot and Human Interactive Communication (RO-MAN), pp. 676–682. IEEE (2017)
26. Paulson, L.C.: Computational logic: its origins and applications. Proc. Roy. Soc. A Math. Phys. Eng. Sci. **474**(2210), 20170872 (2018)
27. Calegari, R., Ciatto, G., Denti, E., Omicini, A.: Logic-based technologies for intelligent systems: state of the art and perspectives. Information **11**(3), 1–29 (2020). Special Issue "10th Anniversary of Information-Emerging Research Challenges"
28. Russell, S., Norvig, P.: Artificial intelligence: a modern approach (2002)
29. Ciatto, G., Calegari, R., Omicini, A., Calvaresi, D.: Towards XMAS: eXplainability through multi-agent systems. In: Savaglio, C., Fortino, G., Ciatto, G., Omicini, A. (eds.) AI&IoT 2019 - Artificial Intelligence and Internet of Things 2019, Volume 2502 of CEUR Workshop Proceedings, pp. 40–53. Sun SITE Central Europe, RWTH Aachen University, November 2019
30. Liu, B.: Lifelong machine learning: a paradigm for continuous learning. Front. Comput. Sci. **11**(3), 359–361 (2017)
31. Cui, Y., Ahmad, S., Hawkins, J.: Continuous online sequence learning with an unsupervised neural network model. Neural Comput. **28**(11), 2474–2504 (2016)
32. Fernandez-Marquez, J.L., Di Marzo Serugendo, G., Montagna, S., Viroli, M., Arcos, J.L.: Description and composition of bio-inspired design patterns: a complete overview. Nat. Comput. **12**(1), 43–67 (2013). https://doi.org/10.1007/s11047-012-9324-y
33. Wooldridge, M.J., Jennings, N.R.: Intelligent agents: theory and practice. Knowl. Eng. Rev. **10**(2), 115–152 (1995)
34. Mualla, Y., Bai, W., Galland, S., Nicolle, C.: Comparison of agent-based simulation frameworks for unmanned aerial transportation applications. Procedia Comput. Sci. **130**, 791–796 (2018)
35. Calvaresi, D., Marinoni, M., Sturm, A., Schumacher, M., Buttazzo, G.: The challenge of real-time multi-agent systems for enabling IoT and CPS. In: Proceedings of the International Conference on Web Intelligence, pp. 356–364 (2017)
36. Calvaresi, D., Albanese, G., Calbimonte, J.-P., Schumacher, M.: SEAMLESS: simulation and analysis for multi-agent system in time-constrained environments. In: Demazeau, Y., Holvoet, T., Corchado, J.M., Costantini, S. (eds.) PAAMS 2020. LNCS (LNAI), vol. 12092, pp. 392–397. Springer, Cham (2020). https://doi.org/10.1007/978-3-030-49778-1_30
37. Urieli, D., MacAlpine, P., Kalyanakrishnan, S., Bentor, Y., Stone, P.: On optimizing interdependent skills: a case study in simulated 3D humanoid robot soccer. In: AAMAS, vol. 11, p. 769 (2011)
38. Johnson-Laird, P.N.: Inference with mental models. In: The Oxford Handbook of Thinking and Reasoning, pp. 134–145 (2012)
39. Gentner, D., Stevens, A.L.: Mental models Lawrence Erlbaum associates, Hillsdale, New Jersey (1983)

40. Hamrick, J.B.: Analogues of mental simulation and imagination in deep learning. Curr. Opin. Behav. Sci. **29**, 8–16 (2019)
41. Hegarty, M.: Mechanical reasoning by mental simulation. Trends Cogn. Sci. **8**(6), 280–285 (2004)
42. Battaglia, P.W., Hamrick, J.B., Tenenbaum, J.B.: Simulation as an engine of physical scene understanding. Proc. Natl. Acad. Sci. **110**(45), 18327–18332 (2013)
43. Harris, P.L.: The Work of the Imagination. Blackwell Publishing, Hoboken (2000)
44. Boden, M.A., et al.: The Creative Mind: Myths and Mechanisms. Psychology Press (2004)
45. Boden, M.A.: Creativity and ALife. Artif. Life **21**(3), 354–365 (2015)
46. Froese, T., Ziemke, T.: Enactive artificial intelligence: investigating the systemic organization of life and mind. Artif. Intell. **173**(3–4), 466–500 (2009)
47. De Loor, P., Manac'h, K., Tisseau, J.: Enaction-based artificial intelligence: toward co-evolution with humans in the loop. Minds Mach. **19**(3), 319–343 (2009). https://doi.org/10.1007/s11023-009-9165-3
48. Guckelsberger, C., Salge, C., Colton, S.: Addressing the "why?" in computational creativity: a non-anthropocentric, minimal model of intentional creative agency (2017)
49. Mogren, O.: C-RNN-GAN: continuous recurrent neural networks with adversarial training. arXiv preprint arXiv:1611.09904 (2016)
50. Park, T., Liu, M.-Y., Wang, T.-C., Zhu, J.-Y.: GauGAN: semantic image synthesis with spatially adaptive normalization. In: ACM SIGGRAPH 2019 Real-Time Live! p. 1 (2019)
51. Genovese, A., Piuri, V., Scotti, F.: Towards explainable face aging with generative adversarial networks. In: 2019 IEEE International Conference on Image Processing (ICIP), pp. 3806–3810. IEEE (2019)
52. Biffi, C., et al.: Explainable anatomical shape analysis through deep hierarchical generative models. IEEE Trans. Med. Imaging **39**(6), 2088–2099 (2020)
53. Zhu, Y., Suri, S., Kulkarni, P., Chen, Y., Duan, J., Kuo, C.-C.J.: An interpretable generative model for handwritten digits synthesis. In: 2019 IEEE International Conference on Image Processing (ICIP), pp. 1910–1914. IEEE (2019)
54. Azmy, M., Shi, P., Lin, J., Ilyas, I.F.: Matching entities across different knowledge graphs with graph embeddings. arXiv preprint arXiv:1903.06607 (2019)
55. d'Avila Garcez, A.S., Broda, K.B., Gabbay, D.M.: Neural-Symbolic Learning Systems: Foundations and Applications. Springer, London (2012)
56. Broersen, J., Dastani, M., Hulstijn, J., van der Torre, L.: Goal generation in the BOID architecture. Cogn. Sci. Q. **2**(3–4), 428–447 (2002)
57. Ren, Z., Dong, K., Zhou, Y., Liu, Q., Peng, J.: Exploration via hindsight goal generation. arXiv preprint arXiv:1906.04279 (2019)
58. Euzenat, J.: Interaction-based ontology alignment repair with expansion and relaxation. In: IJCAI 2017–26th International Joint Conference on Artificial Intelligence, pp. 185–191. AAAI Press (2017)
59. Chocron, P., Pareti, P.: Vocabulary alignment for collaborative agents: a study with real-world multilingual how-to instructions. In: IJCAI, pp. 159–165 (2018)
60. Jiménez-Ruiz, E., Payne, T.R., Solimando, A., Tamma, V.: Limiting logical violations in ontology alignment through negotiation. In: Proceedings of the Fifteenth International Conference on Principles of Knowledge Representation and Reasoning, pp. 217–226 (2016)
61. Lorenz, M., Gehrke, J.D., Langer, H., Timm, I.J., Hammer, J.: Situation-aware risk management in autonomous agents. In: Proceedings of the 14th ACM International Conference on Information and Knowledge Management, pp. 363–364 (2005)

62. Bellomarini, L., Laurenza, E., Sallinger, E., Sherkhonov, E.: Reasoning under uncertainty in knowledge graphs. In: Gutiérrez-Basulto, V., Kliegr, T., Soylu, A., Giese, M., Roman, D. (eds.) RuleML+RR 2020. LNCS, vol. 12173, pp. 131–139. Springer, Cham (2020). https://doi.org/10.1007/978-3-030-57977-7_9
63. Seeme, F.B., Green, D.G.: Pluralistic ignorance: emergence and hypotheses testing in a multi-agent system. In: 2016 International Joint Conference on Neural Networks (IJCNN), pp. 5269–5274. IEEE (2016)
64. Letia, I.A., Groza, A.: Arguing with justifications between collaborating agents. In: McBurney, P., Parsons, S., Rahwan, I. (eds.) ArgMAS 2011. LNCS (LNAI), vol. 7543, pp. 102–116. Springer, Heidelberg (2012). https://doi.org/10.1007/978-3-642-33152-7_7
65. Olofsson, J., Hendeby, G., Lauknes, T.R., Johansen, T.A.: Multi-agent informed path planning using the probability hypothesis density. Auton. Robots **44**(6), 913–925 (2020). https://doi.org/10.1007/s10514-020-09904-1
66. Confalonieri, R., Weyde, T., Besold, T.R., del Prado Martín, F.M.: Using ontologies to enhance human understandability of global post-hoc explanations of black-box models. Artif. Intell. **296**, 103471 (2021)
67. Pathak, D., Agrawal, P., Efros, A.A., Darrell, T.: Curiosity-driven exploration by self-supervised prediction. In: International Conference on Machine Learning, pp. 2778–2787. PMLR (2017)
68. Baldoni, M., Baroglio, C., Boissier, O., May, K.M., Micalizio, R., Tedeschi, S.: Accountability and responsibility in agent organizations. In: Miller, T., Oren, N., Sakurai, Y., Noda, I., Savarimuthu, B.T.R., Cao Son, T. (eds.) PRIMA 2018. LNCS (LNAI), vol. 11224, pp. 261–278. Springer, Cham (2018). https://doi.org/10.1007/978-3-030-03098-8_16
69. Calegari, R., Ciatto, G., Mascardi, V., Omicini, A.: Logic-based technologies for multi-agent systems: a systematic literature review. Auton. Agents Multi-Agent Syst. **35**(1), 1:1–1:67 (2021). https://doi.org/10.1007/s10458-020-09478-3. Collection "Current Trends in Research on Software Agents and Agent-Based Software Development"
70. Ciatto, G., Calegari, R., Siboni, E., Denti, E., Omicini, A.: 2P-Kt: logic programming with objects & functions in Kotlin. In: Calegari, R., Ciatto, G., Denti, E., Omicini, A., Sartor, G. (eds.) WOA 2020–Proceedings of the 21th Workshop "From Objects to Agents", Volume 2706 of CEUR Workshop Proceedings, Aachen, Germany, October 2020, pp. 219–236. Sun SITE Central Europe, RWTH Aachen University, Bologna, Italy, 14–16 September 2020 (2020)
71. Pianini, D., Montagna, S., Viroli, M.: Chemical-oriented simulation of computational systems with ALCHEMIST. J. Simul. **7**, 202–215 (2013). https://doi.org/10.1057/jos.2012.27

Towards an XAI-Assisted Third-Party Evaluation of AI Systems: Illustration on Decision Trees

Yongxin Zhou[1](✉), Matthieu Boussard[2](✉), and Agnes Delaborde[3]

[1] Université Grenoble Alpes, LIG, 38000 Grenoble, France
yongxin.zhou@univ-grenoble-alpes.fr
[2] Craft AI, 140 bis rue de Rennes, 75006 Paris, France
matthieu.boussard@craft.ai
[3] LNE - Laboratoire national de métrologie et d'essais, 29 Avenue Roger Hennequin, 78197 Trappes Cedex, France
agnes.delaborde@lne.fr

Abstract. We explored the potential contribution of eXplainable Artificial Intelligence (XAI) for the evaluation of Artificial Intelligence (AI), in a context where such an evaluation is performed by independent third-party evaluators, for example in the objective of certification. The experimental approach of this paper is based on *"explainable by design"* decision trees that produce predictions on health data and bank data. Results presented in this paper show that the explanations could be used by the evaluators to identify the parameters used in decision making and their levels of importance. The explanations would thus make it possible to orient the constitution of the evaluation corpus, to explore the rules followed for decision-making and to identify potentially critical relationships between different parameters. In addition, the explanations make it possible to inspect the presence of bias in the database and in the algorithm. These first results lay the groundwork for further additional research in order to generalize the conclusions of this paper to different XAI methods.

Keywords: Artificial intelligence · Explainable AI · Evaluation of AI · Experimental methods

1 Introduction

There are strong needs from public authorities for a coherent and strict verification of Artificial Intelligence (AI) [10] systems. So, those systems must be tested according to reference testing methods that are not aimed at improving performance, but at proving that the output behaviors of the system are safe and compliant with regulation. In the European regulatory framework, external evaluation – that is to say, performed by an independent third-party entity rather than by the developers themselves – is required for critical systems, such

D. Calvaresi et al. (Eds.): EXTRAAMAS 2021, LNAI 12688, pp. 158–172, 2021.
https://doi.org/10.1007/978-3-030-82017-6_10

as medical devices. For non-critical systems, designers of AI solutions must prove through self-declaration that their solutions are law-compliant.

The inspection of AI by a third party currently tends to rely on audits (inspection based on declarations and direct observation) and on tests on "representative samples". The determination of the test database is a challenge: all the factors influencing the system's decision-making must be known to the evaluator, who can then choose the relevant configurations of the input data, and set the distribution of these configurations within the test database. In addition, even the information provided by the designer may be insufficient. It is therefore necessary to be able to rely on elements of information which, without necessarily revealing the designer's industrial secrets, can enlighten the evaluator on the most optimal way to proceed with the tests.

The 2020 technical report from the European Commission on the robustness and explainability of AI [8] states that the integration of AI components in products and services, and their use in sensitive contexts, both require the intervention from regulatory organizations to avoid potential harm to EU citizens. Explainable Artificial Intelligence (XAI) is thus recognized as a potential lever for the deployment of trustworthy AIs. We took the hypothesis that XAI could constitute a valuable tool for evaluators, since it could provide information to facilitate the evaluation and make it more transparent. The exploration focuses on the use case of automatic prediction.

This study focuses on the use case of "explainable by design" decision trees that produce predictions on health data and bank data. The paper is structured as follows: Sect. 2 presents several limitations associated to the traditionally used methods for the evaluation of AI, and how the added-value of XAI has rarely been explored in this context. Section 3 focuses on the experimental method designed to estimate the usability of such an XAI model for AI evaluation. Section 4 presents the results of the experiments. Conclusions and outlook are given thereafter in Sect. 5.

2 State of the Art

2.1 Issues in the Evaluation of Artificial Intelligence

The evaluation of AI has two main objectives: its functional assessment, and its regulatory compliance. From a functional point of view, the evaluation provides a quantified and objective estimate of performance through benchmarking, for example to identify the causes of under-performance or to compare the system with the technological state of the art. From an economic, regulatory and societal point of view the evaluation allows assessing the degree of compliance of the system with safety, ethical and acceptability requirements.

The process of evaluation of AI may require either testing the system on representative samples, or an audit of the software, or both. The audit relies on the inspection of the elements constituting the AI algorithms, for example as recommended by the French High Autority for Health in the context of AI-driven medical devices [3] (including, in particular, a description of samples used for initial model learning or relearning, and a description of the models used).

The evaluation is performed either in "black box" conditions, or in "white box": the white box evaluation consists in inspecting the source code of the system, an example being cash register systems [12] where national finance laws require the inspection of the critical functions of such systems. These systems are rule-based, which enables direct inspection. But such direct inspection of AI algorithms, due to their complexity and probabilistic nature, is not an adapted strategy for performance assessment or compliance verification. While some recent work in formal method evaluation shows promising advances [6], the immediate prospects are still limited. In addition to that the evaluator may not even have access to technical information, for example when a client requests a comparison of different products on the market. The independent evaluator thus relies mainly on "black box" evaluation, by submitting input test data to the system, and by observing the resulting outputs, without necessarily knowing or controlling the mechanisms that led to these results.

Testing AI systems through data sets traditionally consists of three main phases: the first stage is the selection of test data; then, the design of a reference data set (the ground truth associated to the test data, often provided by a human expert) and obtaining the hypothesis data set (outputs from the system). The final phase consists in performing the comparison between the reference and the hypothesis, through metrics (F-measure, accuracy, etc.) to estimate the performance of the evaluated system. The selection of pertinent test data represents a challenge for the external and independent evaluation of AI systems. For instance, the most challenging issues concern the estimation of the domain of functioning of the system: do the data appropriately cover it? Can we spot that the set of data is out of the scope of the system (which may then explain low performance)? Or has the system been trained on the same set of test data? The last situation often happens if the data is hard to collect in the domain of application (for example, with natural language applications on low-resource languages), and the evaluation may need to be performed with open-access data.

Another challenge of AI evaluation lies in the selection of the evaluation metrics. Indeed, the performance metric is sometimes binary and the indicators are easy to observe. In the case of an autonomous vehicle, evaluating its stopping function before a red light implies a performance metric based on the success or failure of the stop. But such a simplistic metric is not adapted for decisions involving several parameters. Evaluating the global stopping feature of an autonomous vehicle thus needs to include external factors, such as the level of criticality of the stop when, for example, an individual is in front of the vehicle (opposed to an inanimate object). A score fully representative of the performance must then take into account all the factors of influence of the decision.

Independent evaluation of AI thus requires the access to information of the internal mechanisms of the system, so as to identify the factors of influence of the decision-making, the domain of functioning, etc. The information conveyed by an AI algorithm able to explain its decisions could help address several of the traditional issues of AI evaluation.

2.2 XAI for Third-Party Evaluation

Literature traditionally identifies five types of target audiences for XAI, who present different interests in an explanation of the decision-making [1,2,7]: domain experts (e.g. doctors, bank advisor), individuals impacted by the decisions of the model, regulatory entities/agencies, developers who design the AI solutions, and managers who deploy AI solutions.

We note that there are two common objectives for all users: the need to understand the model and the compliance with a set of specifications (e.g. regulatory). Our study focuses on an alternate user type, the "inspectors" (insurance agent, auditor, public authorities, third-party independent evaluator, etc.), who share commonalities with both the regulatory entities and data scientists. This public presents specific expectancies that have not really been broached by the community up to now. As indicated in [8], "the concepts of robustness and explainability of AI systems have emerged as key elements for a future regulation of this technology". XAI should help the evaluator to ensure the responsible use of AI techniques. The link between the explainability of a system and its auditability is also highlighted in [2]. However, apart from these general considerations on the importance of explainability for such an audience, the literature does not provide arguments to justify the provision of explanations for the external evaluation.

3 Methodology

3.1 Context

In this work, we took as a use case models that are interpretable by design, which means that the algorithm making the prediction also provides explanations. In the context of this work, the explanations are represented in the form of a decision tree. Half of the samples are used for training and the other half are used as test data. In addition, the algorithm's prediction is quantified using evaluation metrics: precision, recall, f-measure.

We note that the context and scope of this work do not allow for a fully representative study. Indeed, to achieve the general research objective and to verify the research hypothesis, it would require having different AI solutions, as well as different XAI techniques in order to carry out a comparative study. The main purpose of the experimental method described here is to lay the foundations of the method, and to define all the research and methodological questions which should be resolved in subsequent work.

3.2 Experimental Data

We conducted our experiments on two data sets: a publicly available dataset on health domain, and a private data set in the banking domain.

Health Data - The health data set used is entitled "Risk factors associated with low infant birth weight" [11,14], also named "LBW" (low birth weight). The original data set contains 189 samples, where each sample represents a newborn baby. Each sample is described through 10 attributes, which provide information about the baby (related to weight) and the mother (among which age, ethnicity, health history). The classification for each sample is represented by the binary attribute "low", which indicates whether a baby's birth weight is less than 2500 g (low = 1) or greater (low = 0). Table 1 describes the attributes.

Table 1. LBW health database: the 10 attributes and their description, adapted from [14].

Attribute	Description
low	Indicator of birth weight less 2.5 kg (1 = low; 0 = normal)
age	Mother's age in years (min: 14, max: 45)
lwt	Mother's weight in pounds at last menstrual period (min: 80 lbs, max: 250 lbs)
race	Mother's race (1 = white, 2 = black, 3 = other)
smoke	Smoking status during pregnancy (1 = smoker, 0 = non-smoker)
ptl	Number of previous premature labours (min: 0, max: 3)
ht	History of hypertension (1 = yes, 0 = no)
ui	Presence of uterine irritability (1 = presence, 0 = absence)
ftv	Number of physician visits during the first trimester (min: 0, max: 6)
bwt	Birth weight in grams (min: 709 g, max: 4990 g)

One can note that the low number of samples (189 samples), makes it unsuitable for a relevant medical study, and that the distribution of the "race" factor is not homogeneously distributed: 96 samples for "white" (51%), 67 for "other" (35%), and only 26 for "black" (14%).

Bank Data - The experiment was also conducted on a private data set issued by a bank specialized in consumer loans (works, household appliances, automobile, etc.). Bank customers mainly use a website to apply for a loan, with a low approval rate (14.75%). This data base is not publicly available for business property reasons, but was made accessible for the needs of our experiment. The scientific relevance of presenting results on non-publicly available data may be questioned; data sharing is highly constrained by economic pressures for competitiveness, and initiatives for data federation are still at the research stage (one can cite, for example, the European program GAIA-X [4]). The present article provides however a thorough description of this data, and allows extending the observations performed on the public LBW corpus to a real-life application.

The data set contains 17,011 samples, in CSV format. Each sample presents a request made by a customer on the bank's website, and contains 33 attributes that describe the type of credit requested, the customer's income and personal situation, etc. The attributes mentioned in this study are presented in

Table 2. The decision taken (attribute "Result_credit") will be either "Refused" or "Approved". In this case, the goal of the AI application is to provide advice to the user at the end of their website request, to help them increase the chances of being approved.

Table 2. Bank database: attributes and their description (excerpt).

Attributes	Descriptions
Date	Date of the start of the loan request
Product	Type of loan product (usually indicate what the lean is for)
Amount	Amount of money asked for the loan (in euros)
Salary	User's salary (in euros/month)
Other_revenue	User's other type of revenue (in euros per month)
Revenue	Salary + Other_revenue
Gender	Gender of the user
Work_contract	Kind of professional contract of the user
Cluster_localisation	Geographical cluster
Cluster_number	Number part of the cluster code
Cluster_letter	Letter part of the cluster code
Result_credit	Is the loan request approved?
Timezone	Timezone of the user

3.3 Experiments

Our research hypothesis is that the explanations can help refining the constitution of a relevant test database, and identifying potential biases of an AI algorithm, such as model bias and discrimination bias [9]. In particular, we expected that explanation provided by decision trees may facilitate test data preparation: obtaining data that fully matches the domain of functioning of the system can be impossible (for logistic reasons such as cost of collection or labelling, for example). XAI may help estimating the tolerance of the algorithm to variations in the testing data.

The work is exploratory in nature, on a research subject still under-explored by the scientific community; the experimental protocol was therefore built incrementally during this work based upon desk research and the observations made during the trials. We explored several data bases (including the health and bank databases presented in this paper) to observe the predictions of the AI algorithm and the corresponding explanations. Then, we carried out our initial tests on the health data use case, because it contains few samples and is therefore suitable for carrying out manipulations and observations. We then chose the bank data use case to perform all the tests and finalize the experimental protocol.

We designed and tested a set of manipulations that can be systematically applied to enhance the evaluator's testing protocol, among which:

- Changes to the configuration of the algorithm (depth and minimum number of samples per leaf). The objective is to identify potential necessary features for explanation-assisted evaluation;
- Various data groupings have been carried out in particular for the observation of bias (via ethical and moral parameters) and implicit errors (via parameters deemed as important for decision-making);
- Semantic groupings of data in order to estimate their proximity of weight in decision-making. For example, the "Salary", "Other_revenue" and "Revenue" parameters are estimated to belong to the same semantic field;
- Variations in the size of the learning and test corpus, in order to estimate the influence caused by the size of the test data corpus (for example in the case of supervised learning systems);
- Removal of parameters (for example parameters with an ethical impact) in order to estimate their influence on decision-making;
- Transformation of the parameters' type (for example, changing from a continuous to a discrete dimension for the "age" parameter) in order to estimate the impact on the explanations.

4 Results

This section presents the most relevant observations performed on both use cases (health and bank), which highlights the potential usability of XAI for third-party evaluation.

4.1 Health Data

Data Groupings - The goal of these experiments is to explore how the explanations can provide additional information to the evaluator. The evaluator may, by expertise, estimate that certain attribute values are not relevant for the evaluation, or that they may cause a bias of the algorithm which should be verified. For example, taking ethnic origin into account in automatic decision making can lead to ethical biases, such as misclassification for certain ethnicities, or any type of discrimination towards vulnerable groups or minorities.

The first test consisted in grouping the data according to the "race" attribute. Three models have been generated, each running only on the samples of a same value ("black", "white", "other"). Figure 1 shows the results for each model, including the results obtained when all samples are processed.

The scores show that the model running only on "black" samples has lower performance than all other models, either in terms of average performance or in terms of distribution (scores can reach zero). We also note that the difference in precision scores between "black" and "white" is particularly marked. These differences are also observed with the "other" model. From these tests, we can say

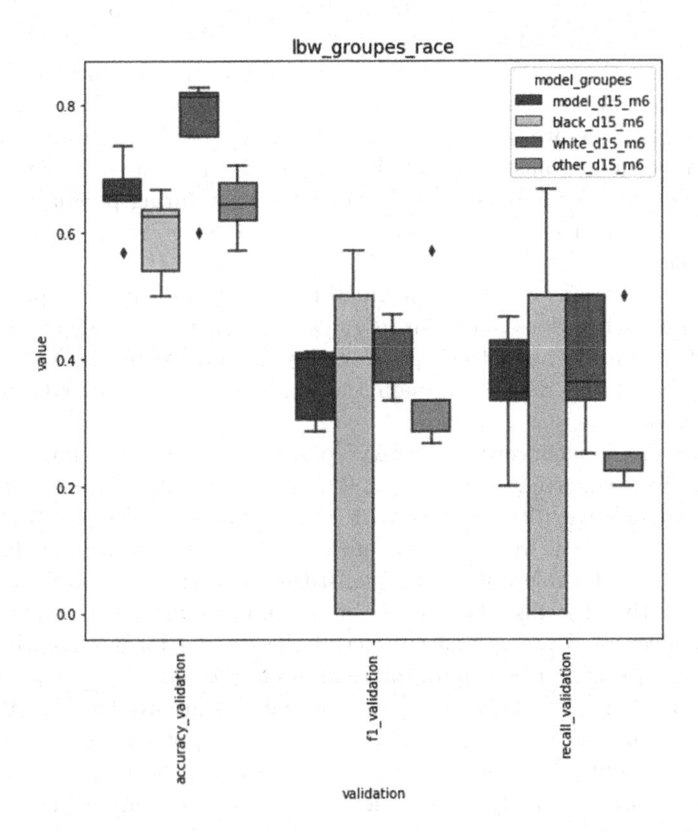

Fig. 1. Accuracy, F-measure and recall for several groupings of the "race" attribute (all samples: *model_d15_m6*, "black": *black_d15_m6*, "white": *white_d15_m6*, "other": *other_d15_m6)*. (health data)

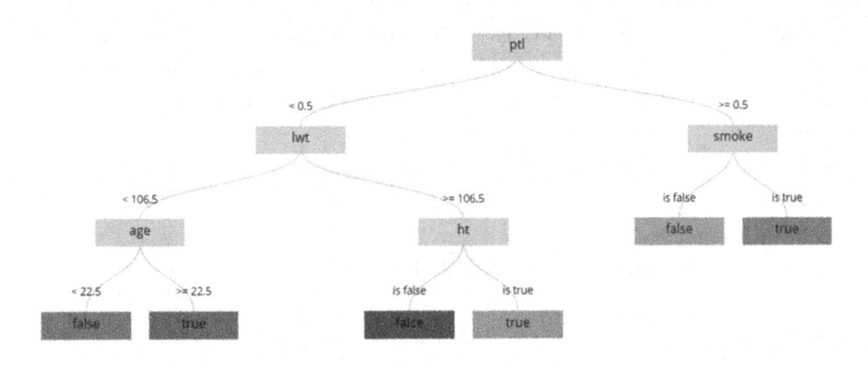

Fig. 2. Decision tree of the model, computed on all samples.

that the algorithm tends to show a biased functioning, which presents differences in performance according to ethnicity.

We pursued the exploration of this potential ethical bias through the analysis of the explanations. The explanations are presented as a decision tree, represented in Fig. 2. Here we can see that the parameters used in decision making are "ptl" (number of previous premature labours), "lwt" (mother's weight in pounds at last menstrual period), "smoke" (smoking status during pregnancy), "age", and "ht" (history of hypertension). The attribute "race" is not a parameter in the decision.

These explanations inform the evaluators that even if there is a performance bias in the model, it does not seem to guide decision making. The explanations provided by XAI may allow the evaluator to verify how "disputable" parameters are processed in the model, through the analysis of their effective weight in decision making.

Another analysis concerned the "ftv" parameter (number of physician visits during the first trimester, ranging from 0 to 6). The results presented in Table 3 show that models with ftv superior to 1, their F1 and recall scores are all zero, which may indicate the importance of such samples in the decision making. This information may be relevant for the constitution of the test data set, since it would imply that the distribution of this attribute must be carefully set during data collection. The explanation trees indicate that for models where ftv is superior to 1, decision trees are dramatically reduced (simple selection of either true or false). One can notice that only few samples are available for these values. These explanations could tend to justify the importance of resorting to a reasonable amount of samples of each value, to place the model in conditions where it can run "optimally" – or at least, not biased by the testing conditions themselves.

Table 3. Performance scores and decision trees for several groupings of the "ftv" attribute. (health data)

Samples	Samples	Precision	F1	Recall	Decision tree
All samples	95	0.568	0.305	0.346	(Complex decision tree, not reproduced here)
ftv = 0	50	0.720	0.461	0.353	(Complex decision tree, not reproduced here)
ftv = 1	24	0.792	0	0	if "ptl" <0.5, then "false"; else "true"
ftv = 2	15	0.867	0	0	"false"
ftv = 3	4	0.250	0	0	"true"
ftv = 4	2	1	0	0	"false"
ftv = 5	No sample for "ftv = 5"				
ftv = 6	*Only one sample for "ftv = 6"*				"false"

Transform the Type of Attributes - The evaluator may not have test data that is of the same exact type as the data the system has been trained on.

For example, an algorithm may have been developed to process age values in continuous format (20, 21 years old, etc.), but the evaluation corpus will only present age categories ("young adult", "Senior", etc.). We will therefore explore how the system reacts to variations in the types of data, and whether decision trees can provide us with additional information to guide an optimal presentation of the data.

From these ideas, we transformed the age values from continuous format to four categories, and then to two categories while respecting the decision frontier decided in the initial model (22.5 years old). Results are presented in Table 4.

Table 4. Performance scores for different types of the "age" attribute for different value format. (health data)

Age format	Categories and number of samples	Precision	F1	Recall
Type: continuous	189 samples	0.568	0.305	0.346
Type: enum	- [14–18] (Youth): 35 samples -]18–25] (YoungAdult): 100 samples -]25–35] (MiddleAge): 51 samples -]35–60] (Senior): 3 samples	0.695	0.473	0.5
Type: enum	- [14–22.5) (Youth): 94 samples - [22.5–45] (NotYouth): 95 samples	0.695	0.491	0.538

We observe that initially "age" was an important parameter in decision making (see Fig. 2), and our manipulation made it disappear, replaced by the "ftv" (number of physician visits during the first trimester), the new decision tree is presented in Fig. 3. Our "categorical" corpus chosen for the evaluation therefore had an impact on the mechanisms of the algorithm, which we were able to visualize thanks to the explanations.

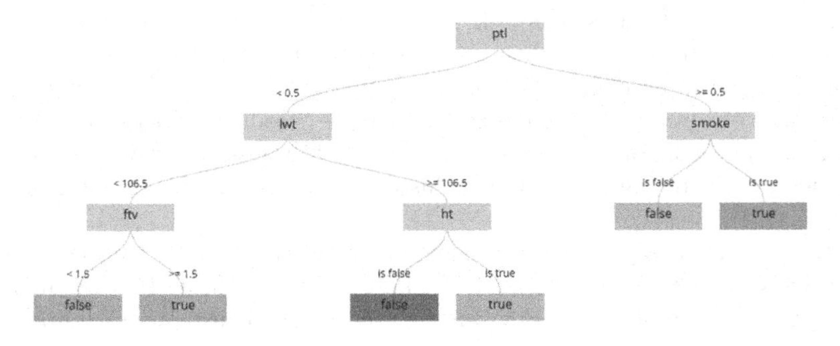

Fig. 3. Decision tree with the attribute "age" in 4 categories: Youth, YoungAdult, MiddleAge, Senior (health data).

Next, we would like to verify whether our division, chosen only on the basis of *a priori* expert knowledge, influenced the explanations. We made a transformation of the age attribute into two categories while respecting the decision

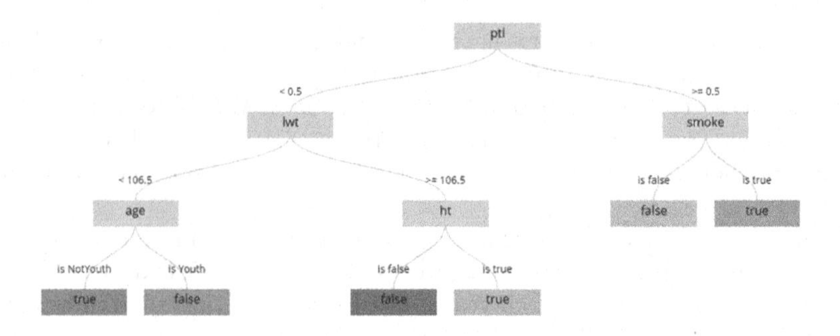

Fig. 4. Decision tree with the attribute "age" in 2 categories: Youth, NotYouth (health data).

frontier in the initial model (22.5 years old). We notice that the decision tree (see Fig. 4) is the same as the initial one, and the performance scores are a little higher. This allows us to identify a tolerance threshold in the variations that the evaluation dataset may present: in the case presented, in order not to bias the operation of the algorithm too much (ask it to operate on a type of data outside its operating domain), it is possible to perform the tests on a database divided according to a slice of age 22.5 years.

We note that this manipulation can only be carried out if the developer of the algorithm provides the evaluator with the database that they used during the development, because this database had to guide the configuration of the parameters of his model. Carrying out this manipulation on a database selected by the evaluator alone would not be relevant.

4.2 Bank Data

Exploration of Parameters - In addition to graphical visualization of the explanations (decision trees), the explanations also provide the occurrence counts of the parameters used in the decision-making process. A first observation of the parameters (see Table 5) indicates that the prediction performed on the banking database ignores 6 parameters: "Date", "Profession", "Work_contract", "District", "Cluster_number" and "Cluster_letter" (they are absent from the table).

The exploration of the parameters count highlights that "Timezone" is used in the decision-making. Relying on the timezone could indicate that the moment (summer time vs standard time) when the user makes their request is important. The context of the banking sector at a given time is relevant information, as it may be more or less advantageous for a bank to allocate a loan depending on the timing. In our view, this is only a descriptive attribute (at certain times loans have or have not been granted), which may lend itself to data mining by a data scientist or a financial analyst, but it does not appear to be a meaningful basis for future decision making. In addition, we noted earlier that the "Date"

Table 5. Number of occurrences of the 26 parameters in the tree. (bank data)

Parameter	#	Parameter	#	Parameter	#
Since_profession	12	Gender	9	Flag_homephone	3
Revenue_minus_installment	11	Amount	8	Current_housing	2
Salary	11	New_1st_holder	7	Flag_cellphone	2
Installment	11	Other_revenue	6	Education	2
Revenue	10	Number_dependents	6	Civility	2
Age	10	Flag_employmentphone	5	Cluster_localisation	1
Since_housing	10	Two_holders	4	Product	1
Duration	9	Timezone	3	Client_status	1
Revenue_on_installment	9	Origin_code	3		

parameter was not used in decision making, which could lead us to believe that the use of "Timezone" can be an implicit error in the model.

From the exploration of parameters, we also note that decision-making is based on the gender of the loan applicant. Non-discrimination based on gender is a European obligation, mentioned for example in the Directive 2004/113/EC for the equal treatment of men and women in the access to goods and services [5]. We note however that in the present use case, the decision is not for loan granting, but meant to produce advice to the user. Further enquiries should then be conducted to determine the legality of basing this automatic decision-making on gender. In the context of the study, however, this highlights how the explanations can drive the exploration performed by a third-party evaluator.

In these cases, the explanations allow identifying the essential parameters of the algorithm, which may be the starting point of further enquiries. In addition, an exploration of the explanations may allow the evaluator to identify a possible cause of under-performance of the algorithm, or the invalidity of a decision-making process in the context of external auditing.

Remove Some Parameters - Some attributes may be absent from the test database, or under-represented, for example because these elements are hard to model at the moment of database collection. In this phase of the experimentation, we want to verify whether removing certain attributes from the database would have an impact on decision making, through the explanations provided. Results are shown in Table 6.

Three parameters, used in the decision making process, seem to have a link with the economic autonomy of the applicant: "Salary", "Other_revenue" and "Revenue". In order to see if the absence of any of this information has a significant impact on the model, we performed tests by removing one of the three parameters each time. We only observe slight variations in performance. Such an observation could have been performed without explanation from the decision tree. However, as shown in the last column, we notice that the variable "Work_contract" which is normally not used, seems to become a decision

Table 6. Accuracy, Recall and F1 scores of the tests by eliminating certain parameters (first column). The last column lists the parameters that are not used by the algorithm. (bank data)

Removed	Acc.	F1	Rec.	Unused variables
–	0.853	0.421	0.345	Date, Profession, District, Work_contract, Cluster_number, Cluster_letter
Revenue	0.857	0.428	0.345	Date, Profession, District, Work_contract, Cluster_number, Cluster_letter, Revenue
Other_revenue	0.850	0.390	0.309	Date, Profession, District, Work_contract, Cluster_number, Other_revenue
Salary	0.855	0.420	0.339	Date, Profession, District, Cluster_number, Cluster_letter, Salary, Product, Cluster_localisation
Gender	0.853	0.424	0.349	Date, Profession, District, Work_contract, Cluster_letter, Product, Gender
Timezone	0.854	0.422	0.345	Date, Profession, District, Work_contract, Cluster_number, Cluster_letter, Timezone

factor when "Salary" is not available for decision making (at the same time, the variables "Product" and "Cluster_localisation" are no longer used in decision making). Similarly, another variable "Cluster_letter" which is normally not used, seems to become a decision factor when "Other_revenue" is not available for decision making. This manipulation tends to show that explanations may provide a more thorough understanding of the impact of the parameters on decision making, that can validate the choices made for the test corpus.

In addition, we also carried out tests by eliminating the two parameters "Timezone" and "Gender" previously mentioned. In the case of the removal of "Timezone", we notice that the difference in performance is insignificant, in comparison with the performance when no parameter was removed (first row of the table). On the other hand, this deletion did not lead to a modification of the parameters used by the algorithm (see last column). This test therefore allows us to verify that a parameter, identified as being able to generate or explain an error in decision-making, seems to have little influence on the performance and operation of the algorithm.

In the case of "Gender", the performance also remains globally stable. However, one point of attention concerns the fact that there seems to be a link with the parameter "Product" (what the loan will be used for): if "Gender" is deleted, then "Product" has no more weight in the decision making. We understand that the database has descriptive value: there is naturally a distribution in terms of gender and use of the loan within the population. However, this manipulation allows us to draw the attention of the evaluator toward a potential gender bias of the algorithm, who would derive from the database that "for a given gender, it is generally this type of credit that is granted". An algorithm trained on such information would then be based on a detrimental causal relationship, which could be explored further by the evaluator.

5 Conclusions and Outlook

Public authorities express a pressing need for methods and tools for a thorough inspection and validation of AI, including requirements on traceability, non-discrimination, or transparency, which could potentially be verified thanks to XAI tools. Indeed, as by [13], "Machine learning models can only be debugged and audited when they can be interpreted". Our research work aims to explore the contribution of XAI to third-party evaluation, in particular by providing additional information for test data selection and bias inspection. The study presents an experimental protocol based on an explainable by design prediction system, exemplified on two use cases (health and banking).

The various observations presented in this paper show that the explanations could make it possible to:

- Indicate the parameters used in decision-making, which makes it possible to identify, for example, parameters having an ethical impact (e.g. the ethnic origin of the health use case) or relating to an implicit error of the model (e.g. the use of the timezone in the banking use case);
- Indicate the level of importance of the parameters in decision-making, in order to estimate whether the parameters deemed "critical", either for performance or inspection of bias, really have a weight or not (e.g. ethnic origin in health, or gender in banking);
- Guide the constitution of the evaluation corpus, by providing information on the domain of operation of the algorithm (e.g. the groupings of data acceptable by the algorithm for age in health data);
- Explore the rules followed for decision making, and thus identify potentially critical relationships between different parameters (e.g. the relationship between gender and type of loan in the bank use case);

As there is a clear distinction between models that can be interpreted by design, also referred to as "transparent", and the inherently non-transparent algorithms such as neural networks that can be explained using external or "post hoc" XAI techniques [1,8], it will be necessary to complement this study with analysis of other techniques of XAI, for other types of AI algorithms. This work nonetheless encourages further exploration of XAI for the community of third-party evaluators, for it may help producing testing procedures that match better the algorithm characteristics (parameters, operating area, etc.). This work also shows that the conclusions drawn from the explanations need to be analyzed in conjunction with several additional observations, including expert knowledge in the domain of application (healthcare, banking, etc.) and with the descriptive information provided, for example, through an audit of the machine learning algorithm.

References

1. Arrieta, A.B., et al.: Explainable artificial intelligence (XAI): concepts, taxonomies, opportunities and challenges toward responsible AI. Inf. Fusion **58**, 82–115 (2020)
2. Beaudouin, V., et al.: Flexible and context-specific AI explainability: a multidisciplinary approach. Available at SSRN 3559477 (2020)
3. (CNEDiMTS): LPPR: Dossier submission to the medical device and health technology evaluation committee (2020)
4. Collective: GAIA-X: driver of digital innovation in Europe. Featuring the next generation of data infrastructure (2020)
5. Council of the European Union: COUNCIL DIRECTIVE 2004/113/EC of 13 December 2004 implementing the principle of equal treatment between men and women in the access to and supply of goods and services (2004)
6. Girard-Satabin, J., Charpiat, G., Chihani, Z., Schoenauer, M.: CAMUS: a framework to build formal specifications for deep perception systems using simulators. arXiv preprint arXiv:1911.10735 (2019)
7. Gunning, D.: Explainable artificial intelligence (XAI). Defense Advanced Research Projects Agency (DARPA), nd Web 2(2) (2017)
8. Hamon, R., Junklewitz, H., Sanchez Martin, J.I.: Robustness and Explainability of Artificial Intelligence. Publications Office of the European Union (2020)
9. Hellstróm, T., Dignum, V., Bensch, S.: Bias in machine learning - what is it good for? (2020)
10. High-Level Expert Group on AI: The assessment list for trustworthy artificial intelligence (ALTAI) for self assessment. European Commission (2020). https://doi.org/10.2759/002360
11. Hosmer, D.W., Jr., Lemeshow, S., Sturdivant, R.X.: Applied Logistic Regression, vol. 398. Wiley, New York (2013)
12. LNE/CITI/CH: Standard for the certification of cash register systems. lne.fr website, pp. 20–40 (2019)
13. Molnar, C.: Interpretable Machine Learning (2019)
14. Venables, W., Ripley, B.: Modern Applied Statistics with S. Springer, New York (2002). https://doi.org/10.1007/978-0-387-21706-2

What Does It Cost to Deploy an XAI System: A Case Study in Legacy Systems

Sviatlana Höhn[1](✉) and Niko Faradouris[2]

[1] University of Luxembourg, 2 Av. de'l Université, Esch-sur-Alzette, Luxembourg
sviatlana.hoehn@uni.lu

[2] smartShift Technologies GmbH, Augustaanlage 59, Mannheim, Germany
nfaradouris@smartshifttech.com
http://satoss.uni.lu/sviatlana,
https://www.smartshifttech.com

Abstract. Enterprise Resource Planning (ERP) software is used by businesses and extended via customisation. Automated custom code analysis and migration is a critical issue at ERP release upgrade times. Despite research advances, automated code analysis and transformation require a huge amount of manual work related to parser adaptation, rule extension and post-processing. These operations become unmanageable if the frequency of updates increases from yearly to monthly intervals. This article describes how the process of custom code analysis to custom code transformation can be automated in an explainable way. We develop an aggregate taxonomy for explainability and analyse the requirements based on roles. We explain in which steps on the new code migration process machine learning is used. Further, we analyse additional effort needed to make the new way of code migration explainable to different stakeholders.

Keywords: Explainable automated source-code transformation · Multi-modal conversational interfaces · Explainability taxonomy

1 Introduction

Enterprise resource planning (ERP) systems, such as SAP Business One[1] and Oracle E-Business Suite[2], have been implemented by many large and medium size companies [5]. The ERP software market grew globally by 10% and reached $ 35 billion in 2018 [18]. The costs of initial implementation range for medium to large scale businesses from $ 150.000 to over $10 million.

Although ERP system vendors recommend to limit code customisation and adapt the business processes to the system, studies report a significant number

[1] https://www.sap.com/products/business-one.html.
[2] https://www.oracle.com/de/applications/ebusiness/.

Supported by FNR Luxembourg INTER-SLANT grant 13320890.

D. Calvaresi et al. (Eds.): EXTRAAMAS 2021, LNAI 12688, pp. 173–186, 2021.
https://doi.org/10.1007/978-3-030-82017-6_11

of customisations in existing ERP systems. For instance, 74% of the companies studied in [9] have adjusted their systems to a degree between moderate and high. The possibility to adapt an ERP system to the business needs and processes of an organisation (i.e. create custom programs) is one of the key requirements for the choice of a specific ERP software [33].

An ERP upgrade implies major changes caused by an implementation of a new version of an already installed ERP system [5]. ERP upgrades can add up to 25–33% of the initial implementation costs for one upgrade [33], most of them caused by labour costs. In average, the costs can increase by $ 1,5 million (between $37.500 and $3,3 million). According to [36], usually a major ERP system upgrade is needed every three years. Cloud-based SAP solutions, however, move to a quarterly release cycle. Current change in the market leader's platform to the SAP S/4HANA ERP software and HANA database forces SAP customers to spend resources on migration of their custom code base [31]. ABAP (Advanced Business Application Programming), the proprietary programming language of SAP, is normally used for custom programs.

Several companies developed rule-based solutions to support SAP customers in their custom code migration tasks within SAP upgrade projects. However, all state-of-the-art solutions share two major problems:

1. Frequent manual updates of the rule-based solution would be required in order to keep it compatible with quarterly updates of the cloud-based SAP ERP solutions, but they are time-consuming and labour-expensive.
2. The reports generated by the current set of tools usually consist several hundreds of spreadsheets. Usually experts review these reports using standard Business Intelligence software and decide which custom programs must be maintained. This is expensive and exhausting.

This is why, technological innovation that transfers academic advances in Machine Learning (ML) and Robotic Process automation (RPA) to the field of software transformation in legacy ERP systems is urgently needed. To tackle this challenge, we developed a plan for knowledge transfer for this specific problem in a close collaboration with one of the technology leaders in the domain of automated ERP custom-code upgrade, smartShift Technologies[3]. Although SAP itself offers a standard solution for custom code upgrade, it only covers unused code and some database compatibility checks. The smartShift tools also analyse syntax and functional errors in the custom code. This is why we chose this solution as our baseline. The business case discussed here is the service of automated custom code migration offered by smartShift.

One of the important questions within this business case is the question of decision explainability. First, the service providers needs to ensure the quality of analysis and migration with a certain level of accuracy. Second, the service provider can be made liable for problems in the work of the ERP system after migration caused by the wrong decisions made by the AI system. Third, errors

[3] https://smartshifttech.com.

in the process of code transformation can delay other dependent tasks in the upgrade project and cause additional costs, which should be avoided.

The process of code migration is monitored on both sides, the service provider and the customer. This is why the explanations need to be reasonable for multiple stakeholders and multiple roles. With this motivation, we focus on the question: **Is additional effort needed in order to make a transformation from an existing rule-based to an ML-based system in an explainable way?**

Because the notion of explainability is a complex phenomenon by itself, we first create an aggregate taxonomy of explainability as explained in Sect. 2. Section 3 briefly describes the steps in the innovation process. Further in Sect. 4, we use the aggregate taxonomy to evaluate feasibility and cost of a potential transformation of a rule-based system to an ML-based system used in the ERP domain. Section 5 discusses the results and future research directions.

2 Dimensions of Explainability

Scholar literature on XAI developed several approaches for managing explainability. The works deal with such issues as interpretability [21,26], fairness [40] and transparency [7]. Each of these terms is recognised as very complex, so that efforts have been made to break them down to a well-defined set of features or parameters [2,39].

It has been recognised that explainability of ML systems has to target system-related aspects, such as data, models and algorithms [2,39], and user-related features, such as the user's knowledge of the domain and the interface [12,37]. Some authors propose to include explainability as a non-functional requirement in the process of software design by focusing on the real users' need for explanations of the system's behaviour [6]. This, in turn, is very difficult given that the notions of *interpretable*, *intelligible*, *explainable* and *understandable* AI are vague and overlapping. For example, [39] discusses multiple definitions of these terms from literature and, finally, uses *interpretability* as a synonym to all termsthat describe the process and the result of human sense-making by using, training and modifying an ML system. In contrast [13] uses the term *transparency* for "decisions affecting us explained to us in terms, formats, and languages we can understand" [13, p. 29].

What is more, case studies such as [12] show that in industrial applications of XAI, different users have different needs (data scientists want to see the data while ML engineers want to see the trained model).

In order to make the explainability requirement manageable, we created an aggregate of existing taxonomies for XAI [2,13,39]. The interface dimension has not been mentioned in the earlier taxonomies, however, it is an important aspect. As pointed out in [2], interactive explanations are rarely studied; a few examples not mentioned in [2] include [1,16].

The aggregated taxonomy is presented in Table 1. It includes six dimensions: user, object, scope, directness, interaction and interface. The taxonomy can be used as follows: first we need to understand who will be the user of the explanations. Then, based on the description of the role or the persona, we need to determine the object of explanations (can be more than one).

Table 1. Aggregated dimensions of explainability based on [2,12,13,39] and own work

Dimension	Values	Reference
User	e.g. based on roles or personas	[2,12]
Object	Data: bias, causality, Model: optimiser, training, prediction Evaluation: accuracy, fairness of the decisions, safety and reliability of the system	[2,13,39]
Scope	Local vs. global	[2]
Directness	Directly interpretable, post-hoc explanation after training, surrogate approximation model	[2,13]
Interaction	Static vs. interactive	[2]
Interface	Visualisation, voice, text, virtual or augmented reality, multi-modal	Own work

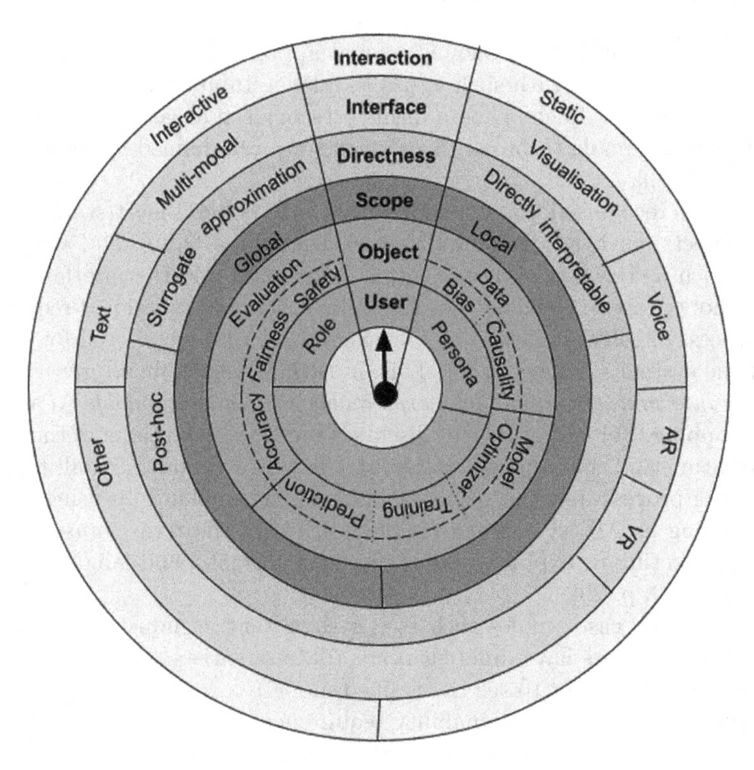

Fig. 1. Cut colour paper circles, write the dimensions on them and assemble in the center, then rotate and bring the needed dimensions together on a line to construct a vector describing your case

For every object, explanations can be of different scope (local or global). The models used for each object can be either directly interpretable (e.g. a small and simple set of rules) or can be explained post-hoc or using a surrogate, approximation model. The explanations for each of those cases can be either static or interactive. For both static and interactive explanations, we can use different interfaces, e.g. conversational, visual and multi-modal. Figure 1 shows how this taxonomy can be visually used to choose the right features.

3 Steps in the Innovation Process

This section explains the changes needed in the custom-code upgrade process in order to go from the rule-based existing system to an ML-based system with a multi-modal conversational interface. In theory it is possible to have multiple interfaces for the same XAI system (e.g. an app, a website and an Alexa skill. However, we chose a multi-modal conversational interface for our business case, as explained below.

3.1 Custom Code Analysis

Custom code analysis includes detection of unused custom code, syntactic, semantic and functional analysis. It addresses two key requirements for a successful ERP upgrade project:

1. System clean-up and
2. Replacement of custom programs whose functionality is already covered by the standard functionality of the target version (obsolete programs).

System clean-up includes identification of duplicates, outdated, or unused functions in the custom code. Between 40% and 70% of the customisations in an average SAP system are not in use [10]. Replacement of obsolete custom programs with standard functionality implies a detailed analysis of the features in the target ERP version [5]. We challenge scholar state of the art by applying natural language processing (NLP) methods, models and tools to legacy programming languages. We analyse the ABAP custom code on syntactic, semantic and functional levels.

In the first step we develop a general method based on NLP state-of-the-art to identify and track custom programs that are obsolete, no longer in use or incompatible with application functionality, data model and interfaces. The automation of the custom code analysis happens in four stages:

1. **Detection of unused custom code.** It includes custom programs that have not been activated in the ERP system for more than N months (in practice, $6 \leq N \leq 18$), including underutilized dependencies (modules that are used only partially and contain blocks of "dead code") [41].

2. **Syntactic analysis**. It includes checks with respect to the target ERP related to the syntactic correctness of the custom programs, the correctness of the data model, and the API interface compatibility with the target platform. This step can be formulated as a machine-translation problem in which the translation happens from code with syntactic errors to correct code [28].
3. **Semantic analysis.** This phase detects code clones. Following [32] we distinguish between structural clones (exact copies of code, copies with renaming and/or modification) and semantic clones (programs that do the same without being structurally similar). Code clones are intentionally introduced by programmers because they help to address changes quickly, save procedure calls and are encouraged by template-based programming. However, code clones increase maintenance costs, cause bug propagation and have negative impact on design and system understanding [32]. Scientists distinguish four types of semantic clones with different grades of syntactic similarities. Type 1 of code clones are syntactically very similar, while Type 4 code clones have no syntactic similarities. The problem can be technically solved using neural models, for example convolutional [42] or rule-based approaches, for example graph-based [35].
4. **Functional analysis.** This step detects obsolete programs. Detection of obsolete programs is one instance of the Type 4 semantic clone detection. Neural models have been shown to be most successful for this task [42]. Custom code is usually unstructured bulk of programs while SAP uses so-called tile architecture in which all programs are classified by business area. Therefore, we can also specify the detection of obsolete custom programs as a text classification task with the classes from SAP core. Deep Learning-based text classification approaches are in this case also the most successful [25].

These four stages of the custom code analysis deliver facts about the actual state of the software code in a particular ERP system with respect to a particular target release version. Thus, these analysis steps need to be repeated for each source-target pair each time.

Usually, the results of such an analysis are presented to a business analyst in the form of hundreds of spreadsheets (the number depends on the system's size and the number of identified issues in the custom code). The business analyst makes decisions related to the code transformation using standard business intelligence software. We decided to automate this process using multi-modal conversational interfaces. We store the facts learned about the current state of the system in a knowledge base. The next section explains the implementation of the multi-model conversational interface.

3.2 Multi-modal Conversational Interface

Conversational interfaces, also called chatbots, support explainability of complex technical systems [16] and facilitate complex problem solving [11]. Multi-modal conversational interfaces have been explored in the bio-informatics domain [8] and fashion retail applications [20]. The authors in [8] formulate seven design

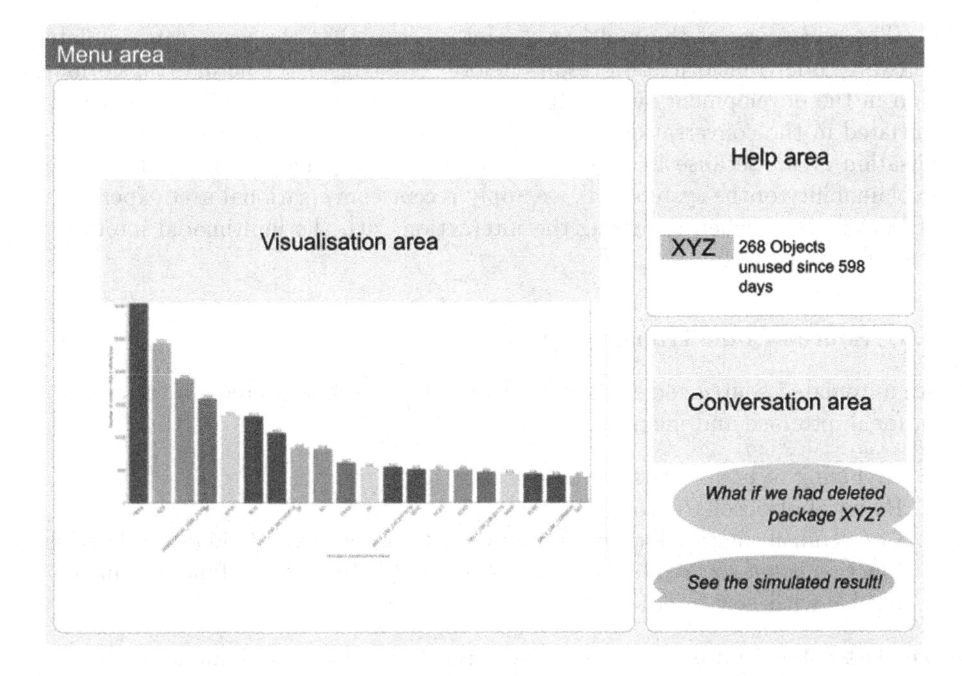

Fig. 2. Preliminary design of the multi-modal conversational interface

principles for multi-modal conversational interfaces which we use for the design of our system. Figure 2 shows the current modalities.

The visualisation area presents the contextual view on the data from the knowledge base as requested by the user via the conversation area. The conversation area is designed and developed as a standard chatbot design pipeline consisting of skills, natural language understanding (NLU) module and natural language generation (NLG) module

1. Skills are defined manually using the SAP CAI interface[4]. Each skill contains the domain knowledge of the analysed custom code, e.g. obsolete code, unused code etc. Therefore, the skills are the same for each system to be analysed and explored with the help of the new interface.
2. Because Intent-based language understanding is commonly used in NLU platforms for task-based interactions. Intents represent the identified meaning of a user's utterance. NLU models are usually trained on a large number of examples of user inputs. Because no conversational training data for the custom code upgrade domain are available, we use other types of documents such as analysis reports and written customer communication.
3. For the NLG module, we define the chatbot's interaction profile that is consistent with the corporate environment. We use the persona methodology to define the chatbot's personality aligned with the company's brand [23].

[4] https://cai.tools.sap, former recast.ai.

The advantage of the multi-modal interface is that the users can simulate software code transformation results before they trigger the actual transformation in the development system. The so-called "what-if"-analysis can be easily initiated in the conversational area of the interface, and visualised in the visualisation area. Because bad user experience also may negatively influence the explainability of the system [34], we apply recent conversational user experience (CUX) findings when designing the interaction with the multimodal interface, see for instance [14].

3.3 Source-Code Transformation

An automated source-code transformation can be then triggered via the conversational interface and includes:

1. Archiving of unused custom programs;
2. Removal of obsolete programs;
3. Code transformation for custom programs that are currently in use and whose function is not covered by the core programs. Modification of functions incompatible with new interfaces (i.e. cloud interfaces).

The latter item, automated source-code transformation, is also known as source code translation or language migration. Academic literature reports various types of source-code translation: translation from a programming language to pseudo-code and back [30]; from one programming language to a different programming language [17]; from code with errors to code without errors (code repair) [24]; from proprietary languages to non-proprietary languages (e.g. ABAP to Java) [27]; from source code to natural language (source code summarisation) [15]; and from a natural language to API code templates [29]. In this project we perform transformations from ABAP to ABAP (proprietary language) while correcting errors in the code.

4 Explainability Requirements and Costs

As explained in Sect. 3, ML is used in several phases of the analysis and transformation process, and also for the conversational part of the multi-modal interface. In this section we analyse, what kind of explainability we need at which step, and what are their costs. In order to obtain the requirements for explainability, we use the aggregate taxonomy presented in Sect. 2. We summarise in Table 2 how the explainability requirements change if the proposed innovation replaces the existing rule-based system.

Table 2. Changes in explainability requirements caused by transformation from the rule-based solution to the ML-based solution described in Sect.3, differences marked in bold

Dimension	Before	After
User	ABAP developer, project manager, business analyst	ABAP developer, project manager, business analyst, **ML/NLP engineer**
Object	Data: causality Model: classification Evaluation: accuracy, safety	Data: **bias**, causality Model: **optimiser**, **training**, prediction Evaluation: accuracy, safety
Scope	Local: single rules	Local: **single prediction of an error** **Global: behaviour of the code analyser and its parts, behaviour of the NLU**
Directness	Directly interpretable	**Post-hoc, surrogate**
Interaction	Static	**Interactive**
Interface	Spreadsheets	**Multi-modal, conversational**

4.1 Explainability Requirements

As suggested in Sect. 3.1, the code transformation task can be defined as a machine-translation problem. NLP methods have been successfully used for similar tasks for software code (see Sect. 3.3 for references). Neural machine-learning models deliver the most accurate results on tasks similar to machine translation, including software code migration. However, these models are the least interpretable. Although attempts are made to make neural models explainable [3,38], the explanations are mostly limited to visualisation of functions and variables. Such explanations are only accessible to ML experts. For instance, if visualisation of a function in the model training process has a particular curve, what does it mean for the expert? How would an expert explain this curve to a non-expert (simplified explanations)? This question must find an answer in the design of the multi-modal interface: the knowledge stored in the knowledge base needs to be formulated in natural-language utterances and/or visual elements in a way accessible to the target user group.

Because different roles may need different explanations, we analyse here the explainability requirements by role:

1. ML/NLP engineer: is interested in validating the models and the data for a particular release version and a particular set of custom programs. This role would need static views on data (both bias and causality) and models (optimizer and training) as well as evaluation of the accuracy. Static visualisations are an appropriate way used in most ML tools to generate surrogate-based and post-hoc explanations. This role would need explanations on the global level, but may need to have access to single predictions for evaluation.

2. ABAP developer: wants to know why a particular program has been classified as incorrect or obsolete. The explanation needs to be static, local and post-hoc at the level of model.
3. Project manager: needs to ensure a smooth execution of the project and is interested in the global picture of the code analysis in order to plan human resources for post-processing and quality assessment. Explanations for this role need to be interactive, global, post-hoc, the object of the explanations will be mostly model and evaluation.
4. Business analyst: is interested in the global picture and in simulations. This role would need interactive, global explanations of the surrogate model.

A set of explainability tools similar to AIX360 described in [2] can be built and offered to the users via multi-modal interface.

Because the user of the analysis results will mainly interact with the new multi-modal interface, static explanations can be generated in the process of interaction with the interface. The explanations can have the form of natural-language utterances generated by the chatbot, or visual shown in the visualisation area, or a synchronised version of both areas where a visualisation is generated and an utterance explains what the visualisation is supposed to show. The conversational area of the interface does not have to be restricted on text input. As CUX research suggests, a wise combination of textual and visual elements improves the CUX [14], and consequently, it improves transparency of the system to the target user group. Therefore, we use the 12 heuristics suggested in [14] for the interface and interaction design.

In addition, humans usually use less precise formulations in their requests that computer systems store in the models. As Hagras (2018) [13] and Alonso et al. ([1]) point out, humans are able to communicate effectively with imprecisely defined labels such as *slow*, *slightly* and *infrequent*, while computer systems need a specific numerical value for such labels. The numerical values for these labels would be different for different people, but the conversational interface needs to map them to precise numbers. Both works suggest using fuzzy logic to deal with such issues. One possible consequence of this could be additional effort in the system's explanations of the sort "Did you mean...?"

4.2 Explainability Costs

The number of tools that support explainability tasks is growing and the range of the problems tackled becomes wider and wider. We can see the explainability requirements for which no tools or methods currently exist as infinitely large. For those requirements that are currently supported by tools, we set the cost at 1. For directly interpretable parts of the systems, we set the costs as 0.

As shown in Table 2, the new system needs to be made explainable for different roles. Each role may need different parts of the system to be explained in different ways. Therefore, costs of explainability can be different for different roles. In Table 3 we specify the explainability costs for each of the parts of the new technology.

Table 3. Explainability costs for each step in the innovation process

Step	Explainability type	Cost
Unused code detection	Directly interpretable	0
Syntactic analysis	Post-hoc or surrogate	1
Semantic analysis		
- Structural clones	Directly interpretable	0
- Semantic clones	Post-hoc or surrogate	1
Functional analysis	Post-hoc or surrogate	1
NLU	Post-hoc	1
NLG	Directly interpretable	0
Simulation	Surrogate	1

As we can see from Table 3, it is possible to find an explainable model for each part so that costs are never infinitely large. Specific methods for explainable neural models have been developed, for example [4,19,22,38]. However, they may be insufficient for users who are not ML engineers. The multi-modal conversational interface can help to solve this issue in the form of feature generalisation and zoom-in/zoom-out visualisations. Also, simplified explanations as described above would be required.

5 Conclusions

This research was motivated by the question, whether an XAI system would cause addition costs as compared to non-XAI version. We analysed a use case from the SAP custom code transformation domain. To evaluate the costs we proposed an aggregated taxonomy for XAI based on six dimensions: user, object, scope, directness, interaction and interface.

We found that the new approach will cause additional effort at all six dimensions. However, due to recent progress in the implementation of explainable neural models such as [4,19,22,38], the costs are not infinitely large. Nevertheless, the creation of a multi-modal interactive interface for explanations would require integration of the explainable models in the process from the beginning.

The proposed approach brings the risk that the explainable methods described in the academic literature are not suitable for deployment in a deployed system. However, tool sets for XAI such as AIX360 show that it is technically possible at least in some cases.

References

1. Alonso, J.M., et al.: Interactive natural language technology for explainable artificial intelligence. In: Heintz, F., Milano, M., O'Sullivan, B. (eds.) TAILOR 2020. LNCS (LNAI), vol. 12641, pp. 63–70. Springer, Cham (2021). https://doi.org/10.1007/978-3-030-73959-1_5

2. Arya, V., et al.: One explanation does not fit all: a toolkit and taxonomy of AI explainability techniques. arXiv preprint arXiv:1909.03012 (2019)
3. Bach, S., Binder, A., Montavon, G., Klauschen, F., Müller, K.R., Samek, W.: On pixel-wise explanations for non-linear classifier decisions by layer-wise relevance propagation. PloS One **10**(7), e0130140 (2015)
4. Baldassarre, F., Azizpour, H.: Explainability techniques for graph convolutional networks. arXiv preprint arXiv:1905.13686 (2019)
5. Barth, C., Koch, S.: Critical success factors in ERP upgrade projects. In: Industrial Management and Data Systems (2019)
6. Chazette, L., Schneider, K.: Explainability as a non-functional requirement: challenges and recommendations. Requirements Eng. **25**(4), 493–514 (2020). https://doi.org/10.1007/s00766-020-00333-1
7. Chiticariu, L., Li, Y., Reiss, F.: Transparent machine learning for information extraction: state-of-the-art and the future. EMNLP (tutorial) (2015)
8. Crovari, P., Pidó, S., Garzotto, F., Ceri, S.: Show, Don't tell. Reflections on the design of multi-modal conversational interfaces. CONVERSATIONS 2020. LNCS, vol. 12604, pp. 64–77. Springer, Cham (2021). https://doi.org/10.1007/978-3-030-68288-0_5
9. Davenport, T.H., Harris, J.G., Cantrell, S.: Enterprise systems and ongoing process change. Bus. Process Manage. J. **10**(1), 16–26 (2004). https://doi.org/10.1108/14637150410518301
10. Eder, S., Junker, M., Jürgens, E., Hauptmann, B., Vaas, R., Prommer, K.H.: How much does unused code matter for maintenance? In: 2012 34th International Conference on Software Engineering (ICSE), pp. 1102–1111. IEEE (2012)
11. Fast, E., Chen, B., Mendelsohn, J., Bassen, J., Bernstein, M.S.: Iris: a conversational agent for complex tasks. In: Proceedings of the 2018 CHI Conference on Human Factors in Computing Systems, pp. 1–12 (2018)
12. Ferreira, J.J., Monteiro, M.D.S.: Do ML experts discuss explainability for AI systems? a discussion case in the industry for a domain-specific solution. arXiv preprint arXiv:2002.12450 (2020)
13. Hagras, H.: Toward human-understandable, explainable AI. Computer **51**(9), 28–36 (2018)
14. Höhn, S., Bongard-Blanchy, K.: Heuristic evaluation of COVID-19 chatbots. In: Følstad, A., et al. (eds.) CONVERSATIONS 2020. LNCS, vol. 12604, pp. 131–144. Springer, Cham (2021). https://doi.org/10.1007/978-3-030-68288-0_9
15. Iyer, S., Konstas, I., Cheung, A., Zettlemoyer, L.: Summarizing source code using a neural attention model. In: Proceedings of the 54th Annual Meeting of the Association for Computational Linguistics (Volume 1: Long Papers), pp. 2073–2083 (2016)
16. Jentzsch, S.F., Höhn, S., Hochgeschwender, N.: Conversational interfaces for explainable AI: a human-centred approach. In: Calvaresi, D., Najjar, A., Schumacher, M., Främling, K. (eds.) EXTRAAMAS 2019. LNCS (LNAI), vol. 11763, pp. 77–92. Springer, Cham (2019). https://doi.org/10.1007/978-3-030-30391-4_5
17. Karaivanov, S., Raychev, V., Vechev, M.: Phrase-based statistical translation of programming languages. In: Proceedings of the 2014 ACM International Symposium on New Ideas, New Paradigms, and Reflections on Programming and Software, pp. 173–184 (2014)
18. Kostoulas, J., Anderson, R., Pang, C.: Market share analysis: Erp software, worldwide, 2018. Gartner (2019)

19. Kumar, P., Singh, A., Kumar, P., Kumar, C.: An explainable machine learning approach for definition extraction. In: Bhattacharjee, A., Borgohain, S.K., Soni, B., Verma, G., Gao, X.-Z. (eds.) MIND 2020. CCIS, vol. 1241, pp. 145–155. Springer, Singapore (2020). https://doi.org/10.1007/978-981-15-6318-8_13
20. Liao, L., Zhou, Y., Ma, Y., Hong, R., Chua, T.S.: Knowledge-aware multimodal fashion chatbot. In: Proceedings of the 26th ACM International Conference on Multimedia, pp. 1265–1266 (2018)
21. Lipton, Z.C.: The mythos of model interpretability: in machine learning, the concept of interpretability is both important and slippery. Queue **16**(3), 31–57 (2018)
22. Mahoney, C.J., Zhang, J., Huber-Fliflet, N., Gronvall, P., Zhao, H.: A framework for explainable text classification in legal document review. In: 2019 IEEE International Conference on Big Data (Big Data), pp. 1858–1867. IEEE (2019)
23. Mano Ferreira, C., Hoehn, S.: Crafting conversational agents' personality in a user-centric context. In: Proceedings of the 31st Benelux Conference on Artificial Intelligence (BNAIC 2019) and the 28th Belgian Dutch Conference on Machine Learning (Benelearn 2019), pp. 1–2 (2019)
24. Mesbah, A., Rice, A., Johnston, E., Glorioso, N., Aftandilian, E.: Deepdelta: learning to repair compilation errors. In: Proceedings of the 2019 27th ACM Joint Meeting on European Software Engineering Conference and Symposium on the Foundations of Software Engineering, pp. 925–936 (2019)
25. Minaee, S., Kalchbrenner, N., Cambria, E., Nikzad, N., Chenaghlu, M., Gao, J.: Deep learning based text classification: a comprehensive review. arXiv preprint arXiv:2004.03705 (2020)
26. Murdoch, W.J., Singh, C., Kumbier, K., Abbasi-Asl, R., Yu, B.: Definitions, methods, and applications in interpretable machine learning. Proc. Nat. Acad. Sci. **116**(44), 22071–22080 (2019)
27. Nandivada, V.K., Nanda, M.G., Dhoolia, P., Saha, D., Nandy, A., Ghosh, A.: A framework for analyzing programs written in proprietary languages. In: Proceedings of the ACM International Conference Companion on Object Oriented Programming Systems Languages and Applications Companion, pp. 289–300 (2011)
28. Nguyen, A.T., Nguyen, T.T., Nguyen, T.N.: Lexical statistical machine translation for language migration. In: Proceedings of the 2013 9th Joint Meeting on Foundations of Software Engineering, pp. 651–654 (2013)
29. Nguyen, A.T., Rigby, P.C., Nguyen, T., Palani, D., Karanfil, M., Nguyen, T.N.: Statistical translation of english texts to API code templates. In: 2018 IEEE International Conference on Software Maintenance and Evolution (ICSME), pp. 194–205. IEEE (2018)
30. Oda, Y., Fudaba, H., Neubig, G., Hata, H., Sakti, S., Toda, T., Nakamura, S.: Learning to generate pseudo-code from source code using statistical machine translation (t). In: 2015 30th IEEE/ACM International Conference on Automated Software Engineering (ASE), pp. 574–584. IEEE (2015)
31. Olson, D.L., Staley, J.: Case study of open-source enterprise resource planning implementation in a small business. Enterprise Inf. Syst. **6**(1), 79–94 (2012)
32. Rattan, D., Bhatia, R., Singh, M.: Software clone detection: a systematic review. Inf. Softw. Technol. **55**(7), 1165–1199 (2013)
33. Rothenberger, M.A., Srite, M.: An investigation of customization in ERP system implementations. IEEE Trans. Eng. Manage. **56**(4), 663–676 (2009)
34. Shariat, J., Saucier, C.S.: Tragic Design: The Impact of Bad Product Design and How to Fix It. O'Reilly Media Inc., Sebastopol (2017)

35. Svacina, J., Simmons, J., Cerny, T.: Semantic code clone detection for enterprise applications. In: Proceedings of the 35th Annual ACM Symposium on Applied Computing, pp. 129–131 (2020)
36. Technologies, S.: Removing unused code matters (2017). https://smartshifttech.com/removing-unused-code-matters/
37. Tomsett, R., Braines, D., Harborne, D., Preece, A., Chakraborty, S.: Interpretable to whom? a role-based model for analyzing interpretable machine learning systems. arXiv preprint arXiv:1806.07552 (2018)
38. Vaughan, J., Sudjianto, A., Brahimi, E., Chen, J., Nair, V.N.: Explainable neural networks based on additive index models. arXiv preprint arXiv:1806.01933 (2018)
39. Ventocilla, E., et al.: Towards a taxonomy for interpretable and interactive machine learning. In: XAI Workshop on Explainable Artificial Intelligence, pp. 151–157 (2018)
40. Verma, S., Rubin, J.: Fairness definitions explained. In: 2018 IEEE/ACM International Workshop On Software Fairness (fairware), pp. 1–7. IEEE (2018)
41. Wang, P., Yang, J., Tan, L., Kroeger, R., Morgenthaler, J.D.: Generating precise dependencies for large software. In: 2013 4th International Workshop on Managing Technical Debt (MTD), pp. 47–50. IEEE (2013)
42. Yu, H., Lam, W., Chen, L., Li, G., Xie, T., Wang, Q.: Neural detection of semantic code clones via tree-based convolution. In: 2019 IEEE/ACM 27th International Conference on Program Comprehension (ICPC), pp. 70–80. IEEE (2019)

XAI Applications

Explainable AI (XAI) Models Applied to the Multi-agent Environment of Financial Markets

Jean Jacques Ohana[1,5], Steve Ohana[2,5], Eric Benhamou[2,3,5](✉), David Saltiel[2,4,5], and Beatrice Guez[2,5]

[1] Homa Capital, Paris, France
[2] Homa Capital, Jerusalem, Israel
eric.benhamou@lamsade.dauphine.fr
[3] AI for Alpha, Neuilly-sur-Seine, France
[4] Lamsade, Dauphine, Paris, France
[5] LISIC ULCO, Calais, France

Abstract. Financial markets are a real life multi-agent system that is well known to be hard to explain and interpret. We consider a gradient boosting decision trees (GBDT) approach to predict large S&P 500 price drops from a set of 150 technical, fundamental and macroeconomic features. We report an improved accuracy of GBDT over other machine learning (ML) methods on the S&P 500 futures prices. We show that retaining fewer and carefully selected features provides improvements across all ML approaches. Shapley values have recently been introduced from game theory to the field of ML. They allow for a robust identification of the most important variables predicting stock market crises, and of a local explanation of the crisis probability at each date, through a consistent features attribution. We apply this methodology to analyse in detail the March 2020 financial meltdown, for which the model offered a timely out of sample prediction. This analysis unveils in particular the contrarian predictive role of the tech equity sector before and after the crash.

Keywords: Explainable AI · GBDT · Multi-agent environment · Financial markets meltdown

1 Introduction

Financial markets are a real life multi agent systems that is well known to be hard to forecast and to explain. The prediction of equity crashes, although particularly challenging due to their infrequent nature and the non-stationary features of financial markets, has been the focus of several important works in the past decades. For instance, [38] have proposed a deterministic log-periodic model with finite-time explosion to represent equity prices bubbles and crashes. In more recent works, [14] and [35] have introduced machine-learning approaches to the

D. Calvaresi et al. (Eds.): EXTRAAMAS 2021, LNAI 12688, pp. 189–207, 2021.
https://doi.org/10.1007/978-3-030-82017-6_12

prediction of global equity crises, emphasizing the importance of cross-market contagion effects. Our goal in this work is to use an AI model that is accurate and explainable to solve the multi-agent system environment of financial markets. Hence, we introduce a gradient boosting decision tree (GBDT) approach to predict large falls in the S&P500 equity index, using a large set of technical, fundamental and macroeconomic features as predictors. Besides illustrating the value of carefully selecting the features and the superior accuracy of GBDT over other ML approaches in some types of small/imbalanced data sets classification problems, our main contribution lies in the explanation of the model predictions at any date. Indeed, from a practitioner viewpoint, understanding why a model provides certain predictions is at least as important as its accuracy. Although the complexity of AI models is often presented as a barrier to a practitioner understanding of their local predictions, the use of SHAP (SHapley Additive exPlanation) values, introduced for the first time by [24] in machine-learning applications, makes AI models more explainable and transparent. Shapley values are the contributions of each individual feature to the overall crash logit probability. They represent the only set of attributions presenting certain properties of consistency and additivity, as defined by Lundberg. In particular, the most commonly used variable importance measurement methodologies fail to pass the consistency test, which makes it difficult to compare their outputs across different models. Shapley values enlighten both the global understanding and the local explanations of machine learning prediction models, as they may be computed at each point in time.

In our context, this approach first allows us to determine which features most efficiently predict equity crashes (and in which global direction). We infer from a features importance analysis that the S&P500 crash probability is driven by a mix of pro-cyclical and counter-cyclical features. Pro-cyclical features consist either of positive economic/equity market developments that remove the prospect of large equity price drops, or alternatively of negative economic/equity market shocks that portend deadly equity downward spirals. Among these pro-cyclical features, we find the 120-day S&P500 Price/Earnings ratio percent change, the global risk aversion level, the 20-day S&P500 sales percent change, the six-months to one-year US 2 Yrs and 10 Yrs rates evolution, economic surprises indices, and the medium-term industrial metals, European equity indices and emerging currencies price trends. Conversely, counter-cyclical features can either be positive economic/equity market anticipations predating large equity price corrections, or negative shocks involving a reduced risk of equity downside moves. Important contrarian indicators are the put/call ratio, the six-months S&P500 sales percent change, the 100-day Nasdaq 100 Sharpe ratio and the U.S. 100-day 10 Yrs real interest change. Sharpe ratios are well known to be predictive in asset management. Indeed it can be shown that Sharpe ratio makes a lot of sense for manager to measure their performance. The distribution of Sharpe ratio can be computed explicitly [4]. Sharpe ratio is not an accident and is a good indicator of manager performance [7]. It can also be related to other performance measures like Omega ratio [6] and other performance ratios [5].

A second crucial contribution of Shapley values is to help uncover how different features locally contribute to the logit probability at each point in time. We apply this methodology to analyse in detail the unfolding of the events surrounding the March 2020 equity meltdown, for which the model offered a timely prediction out of sample. On January 1, 2020, the crash probability was fairly low, standing at 9.4%. On February 3, we observe a first neat increase in the crash probability (to 27%), driven by the 100-day Nasdaq Sharpe Ratio contrarian indicator. At the onset of the Covid crash, on March 2, 2020, most pro-cyclical indicators concurred to steeply increase the crash probability (to 61%), as given by Fig. 5, in a way that proved prescient. Interestingly, the Nasdaq 100 index had already started its correction by this date, prompting the tech sector contrarian indicators to switch back in favor of a decreased crash probability. On April 1st, the crash probability plummeted back to 29%, as the Nasdaq 100 appeared oversold while the Put/Call ratio reflected extremely cautious market anticipations. Overall, the analysis unveils the role of the tech sector as a powerful contrarian predictor before and after the March 2020 crash.

1.1 Related Works

Our work can be related to the ever growing field of machine learning applications to financial markets forecasting. Indeed, robust forecasting methods have recently garnered a lot of interest, both from finance scholars and practitioners. This interest can be traced back as early as the late 2000's where machine learning started to pick up. Instead of listing the large amount of works, we will refer readers to various works that reviewed the existing literature in chronological order.

In 2009, [3] surveyed already more than 100 related published articles using neural and neuro-fuzzy techniques derived and applied to forecasting stock markets, or discussing classifications of financial market data and forecasting methods. In 2010, [22] gave a survey on the application of artificial neural networks in forecasting financial market prices, including exchange rates, stock prices, and financial crisis prediction as well as option pricing. And the stream of machine learning was not only based on neural network but also genetic and evolutionary algorithms as reviewed in [2].

More recently, [39] reviewed the application of cutting-edge NLP techniques for financial forecasting, using text from financial news or tweets. [34] covered the wider topic of machine learning, including deep learning, applications to financial portfolio allocation and optimization systems. [30] focused on the use of support vector machine and artificial neural networks to forecast prices and regimes based on fundamental and technical analysis. Later on, [37] discussed some of the challenges and research opportunities, including issues for algorithmic trading, back testing and live testing on single stocks and more generally prediction in financial market. Finally, [36] reviewed deep learning as well as other machine learning methods to forecast financial time series. As the hype has been recently mostly on deep learning, it comes as no surprise that most reviewed works relate to this field. One of the only works, to our knowledge, that refers to gradient boosted decision tree applications is [21].

Recently, [35], used machine learning to provide early signal to predict financial crises. The article emphasizes that regional crashes may spread to the whole market, increase the probability of re-occurrence of crises in the near term and show universal and characteristic behavior that machine learning can capture. Likewise, [18] proved that machine learning techniques are able to extract and identify dominant predictive signals, that includes variations on momentum, liquidity, and volatility. They show that machine learning methods are able to provide predictive gains thanks to capturing nonlinear interactions missed by other methods. The aim of this article is to present a machine learning planning algorithm that captures universal and reproducible behaviours to timely invest in and divest from equity markets. On another theme, [8,9,11] or [10] showed that deep reinforcement learning are a good alternative to traditional portfolio methods. However, if one wants specifically to target crisis detection, a good alternative is rather to tackle this planning exercise as a supervised learning problem.

Interestingly, Gradient boosting decision trees (GBDT) are almost non-existent in the financial market forecasting literature. As is well-known, GBDT are prone to over-fitting in regression applications. However, they are the method of choice for classification problems as reported by the ML platform Kaggle. In finance, the only space where GBDT have become popular is the credit scoring and retail banking literature. For instance, [13] or [28] reported that GBDT are the best ML method for this specific task as they can cope with limited amount of data and very imbalanced classes.

When classifying stock markets into two regimes (a ‘normal’ one and a ‘crisis’ one), we are precisely facing very imbalanced classes and a binary classification challenge. In addition, when working with daily observations, we are faced with a ML problem with a limited number of data. These two points can seriously hinder the performance of deep learning algorithms that are well known to be data greedy. Hence, our work investigates whether GBDT can provide a suitable method to identify stock market regimes. In addition, as a byproduct, GBDT provide explicit decision rules (as opposed to deep learning), making it an ideal candidate to investigate regime qualification for stock markets. In this work, we apply our methodology to the US S&P 500 futures prices. In unreported works, we have shown that our approach may easily and successfully be transposed to other leading stock indices like the Nasdaq, the Eurostoxx, the FTSE, the Nikkei or the MSCI Emerging futures prices.

Concerning explainable AI (XAI), there has been plenty of research on transparent and interpretable machine learning models, with comprehensive surveys like [1] or [16]. [33] discusses at length XAI motivations, and methods. Furthermore, [23] regroup XAI into three main categories for understanding, diagnosing and refining. It also presents applicable examples relating to the prevailing state-of-the-art with upcoming future possibilities. Indeed, explainable systems for machine learning have has applied in multiple fields like plant stress phenotyping [17], heat recycling, fault detection [25], capsule Gastroenterology [26] and loan attribution [27]. Furthermore, [27] use a combination of LIME and Shapley for understanding and explaining models outputs. Our application is about financial markets which is a more complex multi-agent environment where global explainability is more needed, hence the usage of Shapley to provide intuition and insights to explain and understand model outputs.

1.2 Contribution

Our contributions are threefold:

- We specify a valid GBDT methodology to identify stock market regimes, based on a combination of more than 150 features including financial, macro, risk aversion, price and technical indicators.
- We compare this methodology with other machine learning (ML) methods and report an improved accuracy of GBDT over other ML methods on the S&P 500 futures prices.
- Last but not least, we use Shapley values to provide a global understanding and local explanations of the model at each date, which allows us to analyze in detail the model predictions before and after the March 2020 equity crash.

1.3 Why GBDT?

The motivations for Gradient boosting decision trees (GBDT) are multiple:

- GBDT are the most suitable ML methods for small data sets classification problems. In particular, they are known to perform better than their state-of-the-art cousins, Deep Learning methods, for small data sets. As a matter of fact, GBDT methods have been Kagglers' preferred ones and have won multiple challenges.
- GBDT methods are less sensitive to data re-scaling, compared to logistic regression or penalized methods.
- They can cope with imbalanced data sets.
- They allow for very fast training when using the leaf-wise tree growth (compared to level-wise tree growth).

2 Methodology

In a normal regime, equity markets are rising as investors get rewarded for their risk-taking. This has been referred to as the 'equity risk premium' in the financial economics literature [29]. However, there are subsequent downturns when financial markets switch to panic mode and start falling sharply. Hence, we can simply assume that there are two equity market regimes:

- a *normal* regime where an asset manager should be positively exposed to benefit from the upward bias in equity markets.

- and a *crisis* regime, where an asset manager should either reduce its equity exposure or even possibly short-sell when permitted.

We define a crisis regime as an occurrence of index return below the historical 5% percentile, computed on the training data set. The 5% is not taken randomly but has been validated historically to provide meaningful levels, indicative of real panic and more importantly forecastable. For instance, in the S&P 500 market, typical levels are returns of -6 to -5% over a 15-day horizon. To predict whether the coming 15-day return will be below the 5% percentile (hence being classified as in crisis regime), we use more than 150 features described later on. Simply speaking, these 150 features are variables ranging from risk aversion measures to financial metrics indicators like 12-month-forward sales estimates, earning per share, Price/Earnings ratio, economic surprise indices (like the aggregated Citigroup index that compiles major figures like ISM numbers, non farm payrolls, unemployment rates, etc.).

We only consider two regimes with a specific focus on left-tail events on the returns distribution because we found it easier to characterize extreme returns than to predict outright returns using our set of financial features. In the ML language, our regime detection problem is a pure supervised learning exercise, with a two-regimes classification. Hence the probability of being in the normal regime and the one of being in the crisis regime sum to one.

Daily price data are denoted by P_t. The return over a period of d trading days is simply given by the corresponding percentage change over the period: $R_t^d = P_t/P_{t-d} - 1$. The crisis regime is determined by the subset of events where returns are lower or equal to the historical 5% percentile denoted by C. Returns that are below this threshold are labeled "1" while the label value for the normal regime is set to "0". Using traditional binary classification formalism, we denote the training data $X = \{x_i\}_{i=1}^N$ with $x_i \in \mathbb{R}^D$ and their corresponding labels $Y = \{y_i\}_{i=1}^N$ with $y_i \in 0, 1$. The goal is to find the best *classification* function $f^*(x)$ according to the temporal sum of some specific loss function $\mathcal{L}(y_i, f(x_i))$ as follows:

$$f^* = \arg\min_f \sum_{i=1}^N \mathcal{L}(y_i, f(x_i))$$

Gradient boosting assumes the function f to take an additive form:

$$f(x) = \sum_{m=1}^T f_m(x) \tag{1}$$

where T is the number of iterations. The set of weak learners $f_m(x)$ is designed in an incremental fashion. At the m-th stage, the newly added function, f_m is chosen to optimize the aggregated loss while keeping the previously found weak learners $\{f_j\}_{j=1}^{m-1}$ fixed. Each function f_m belongs to a set of parameterized base learners that are modeled as decision trees. Hence, in GBDT, there is an obvious design trade-off between taking a large number of boosted rounds and very simple based decision trees or a limited number of base learners but of larger size. From our experience, it is better to take small decision trees to avoid over-fitting and an important number of boosted rounds. In this work, we use 500 boosted rounds. The intuition between this choice is to prefer a large crowd of experts that difficultly memorize data and should hence avoid over-fitting compared to a small number of strong experts that are represented by large decision trees. Indeed, if these trees go wrong, their failure is not averaged out, as opposed to the first alternative. Typical implementations of GBDT are XGBoost, as presented in [15], LightGBM as presented [19], or Catboost as presented [31]. We tested both XGBoost and LightGBM and found threefold speed for LighGBM compared to XGBoost for similar learning performances. Hence, in the rest of the paper, we will focus on LightGBM.

For our experiments, we use daily observations of the S&P 500 merged back-adjusted (rolled) futures prices using Homa internal market data. Our daily observations are from 01 Jan 2003 to 15 Jan 2021. We split our data into three subsets: a training sample from 01 Jan 2003 to 31 Dec 2018, a validation sample used to find best hyper-parameters from 01 Jan 2019 to 31 Dec 2019 and a test sample from 01 Jan 2020 to 15 Jan 2021.

2.1 GBDT Hyperparameters

The GBDT model contains a high number of hyper-parameters to be specified. From our experience, the following hyper-parameters are very relevant for imbalanced data sets and need to be fine-tuned using evolutionary optimisations as presented in [12]: min sum hessian in leaf, min gain to split, feature fraction, bagging fraction and lambda l2. The max depth parameter plays a central role in the use of GBDT. On the S&P 500 futures, we found that very small trees with a max depth of one performs better over time than larger trees. The 5 parameters mentioned above are determined as the best hyper parameters on the validation set.

2.2 Features Used

The model is fed by more than 150 features to derive a daily 'crash' probability. These data can be grouped into 6 families:

- **Risk aversion metrics** such as equities', currencies' or commodities' implied volatilities, credit spreads and VIC forward curves.
- **Price indicators** such as returns, sharpe ratio of major stock markets, distance from long term moving average and equity-bond correlation.
- **Financial metrics** such as sales growth or Price/Earnings ratios forecast 12 month forward.
- **Macroeconomic indicators** such as economic surprises indices by region and globally as given by Citigroup surprise index.
- **Technical indicators** such as market breath or put-call ratio.
- **Rates** such as 10 Yrs and 2 Yrs U.S. rates, or break-even inflation information.

2.3 Process of Features Selection

Using all raw features would add too much noise in our model and would lead to biased decisions. We thus need to select or extract only the most meaningful features. As we can see in Fig. 1, we do so by removing the features in 2 steps:

- Based on gradient boosting trees, we rank the features by importance or contribution.
- We then pay attention to the severity of multicollinearity in an ordinary least squares regression analysis by computing the variance inflation factor (VIF) to remove co-linear features. Considering a linear model $Y = \beta_0 + \beta_1 X_1 + \beta_2 X_2 + .. + \beta_n X_n + \epsilon$, the VIF is equal to $\frac{1}{1-R_j^2}$, R_j^2 being the multiple R^2 for the regression of X_j on other covariates. The VIF reflects the presence of collinear factors that increase the variance in the coefficient estimates.

At the end of this 2-part process, we only keep 33% of the initial features.

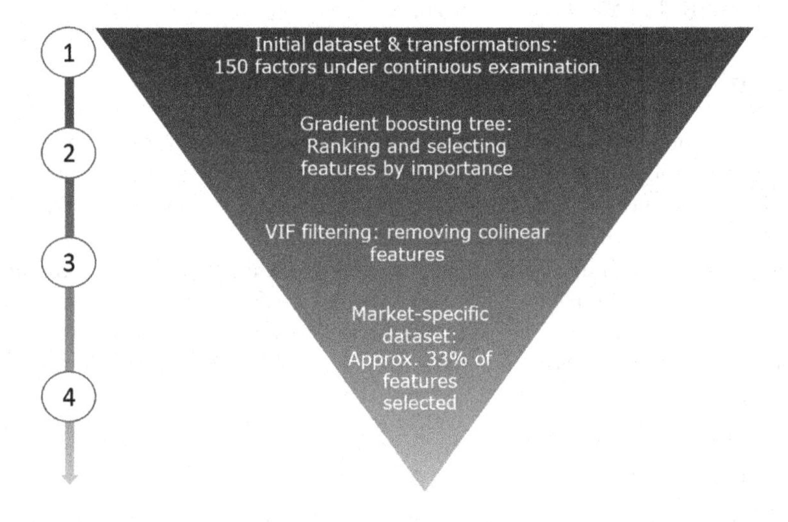

Fig. 1. Probabilities of crash

In the next section, we will investigate whether removing correlated features improves the out-of-sample precision of the model.

3 Results

3.1 Model Presentation

Although our work is mostly focused on the GBDT model, we compare it against common ML models, namely:

- a support vector model with a radial basis function kernel, a γ parameter of 2 and a C parameter of 1 (RBF SVM). We use the sklearn implementation. The two hyper parameters γ and C are found on the validation set.
- a Random Forest (RF) model, whose max depth is set to 1 and boosted rounds are set to 500. We purposely tune the RF model similarly to our GBDT model in order to benefit from the above mentioned error averaging feature. We found this parameter combination to perform well for annual validation data sets ranging from year 2015 onward on the S&P 500 market. We note that a max depth of 1 does not allow for interaction effects between features.
- a first deep learning model, referred to in our experiment as Deep FC (for fully connected layers), which is naively built with three fully connected layers (64, 32 and one for the final layer) with a drop out in of 5% between and Relu activation, whose implementation details rely on tensorflow keras 2.0.
- a second more advanced deep learning model consisting of two layers referred to in our experiment as deep LSTM: a 64 nodes LSTM layer followed by a 5% dropout followed by a 32 nodes dense layer followed by a dense layer with a single node and a sigmoïd activation.

For both deep learning models, we use a standard Adam optimizer whose benefit is to combine adaptive gradient descent with root mean square propagation [20].

We train each model using either the full set of features or only the filtered ones, as described in Fig. 1. Hence, for each model, we add a suffix 'raw' or 'FS' to specify if the model is trained on the full set of features or after features selections. We provide the performance of these models according to different metrics, namely accuracy, precision, recall, f1-score, average precision, AUC and AUC-pr in Table 1. The GBDT with features selection is superior according to all metrics and outperforms in particular the deep learning model based on LSTM, confirming the consensus reached in the ML community as regards classification problems in small and imbalanced data sets.

Table 1. Model comparison

Model	Accuracy	Precision	Recall	F1-score	Avg precision	AUC	AUC-pr
GBDT FS	**0.89**	**0.55**	**0.55**	**0.55**	**0.35**	**0.83**	**0.58**
Deep LSTM FS	0.87	0.06	0.02	0.05	0.13	0.74	0.56
RBF SVM FS	0.87	0.03	0.07	0.06	0.13	0.50	0.56
Random Forest FS	0.87	0.03	0.07	0.04	0.13	0.54	0.56
Deep FC FS	0.87	0.01	0.02	0.04	0.13	0.50	0.56
Deep LSTM Raw	0.84	0.37	0.33	0.35	0.21	0.63	0.39
RBF SVM Raw	0.87	0.02	0.01	0.05	0.13	0.50	0.36
Random Forest Raw	0.86	0.30	0.09	0.14	0.14	0.53	0.25
GBDT Raw	0.86	0.20	0.03	0.05	0.13	0.51	0.18
Deep FC Raw	0.85	0.07	0.05	0.02	0.13	0.49	0.06

3.2 AUC Performance

Figure 2 provides the ROC Curve for the two best performing models, the GBDT and the Deep learning LSTM model with features selection. ROC curves enables to visualize and analyse the relationship between precision and recall and to investigate whether the model makes more type I or type II errors when identifying market regimes. The receiver operating characteristic (ROC) curve plots the true positive rate (sensitivity) on the vertical axis against the false positive rate (1 - specificity, fall-out) on the horizontal axis for all possible threshold values. The two curves are well above the *blind guess* benchmark that is represented by the dotted red line. This effectively demonstrates that these two models have some predictability power, although being far from a perfect score that would be represented by a half square. Furthermore, the area under the GBDT curve with features selection is 0.83, to be compared with 0.74, the one of the second best model (deep LSTM), also with Features selection. Its curve, in blue, is mostly over the one of the second best model (deep LSTM), in red, which indicates that in most situations, GBDT model performs better than the deep LSTM model.

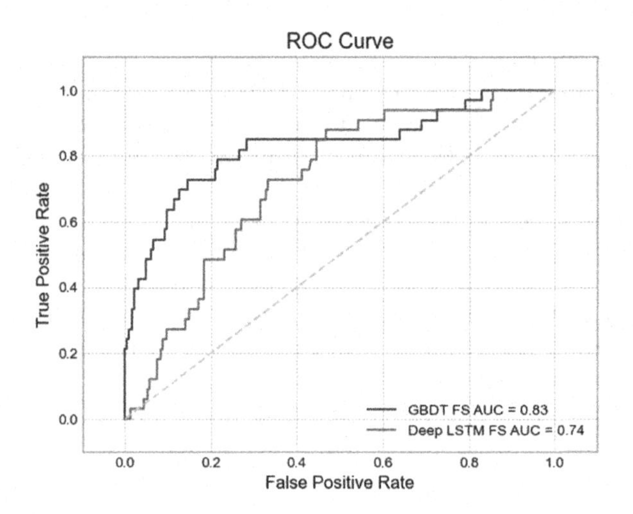

Fig. 2. ROC curve of the two best models (Color figure online)

4 Understanding the Model

4.1 Shapley Values

Building on the work of [24], we use Shapley values to represent the contribution of each feature to the crisis probability. SHAP (SHapley Additive exPlanation) values explain the output of a function f as a sum of the effects of each feature. It assigns an importance value to each feature that represents the effect on the model prediction of including that feature. To compute this effect, a model $f_{S\cup\{i\}}$

is trained with that feature present, and another model f_S is trained with the feature withheld. This method hence requires retraining the model on all feature subsets $S \subseteq M \backslash \{i\}$, where M is the set of all features. Then, predictions from the two models are compared through the difference $f_{S\cup\{i\}}(x_{S\cup\{i\}}) - f_S(x_S)$, where x_S represents the values of the input features in the set S. Since the effect of withholding a feature depends on other features in the model, the preceding differences are computed on all possible differences $f_{S\cup\{i\}}(x_{S\cup\{i\}}) - f_S(x_S)$, for all possible subsets $S \subseteq M \backslash \{i\}$. Shapley values are then constructed as a weighted average of all these differences, as follows:

Definition 1. ***Shapley Value.*** *The Shapley value Φ_i attributed to feature i is defined as:*

$$\Phi_i = \sum_{S \subseteq M \backslash \{i\}} \frac{|S|!\,(|M| - |S| - 1)!}{|M|!} \left(f_{S\cup\{i\}}(x_{S\cup\{i\}}) - f_S(x_S)\right)$$

where $|A|$ refers to the cardinal of the set A, M is the complete set of features, S is the subset of features used, and x_S represents the values of the input features in the set S. Proofs from game theory shows that Shapley values are the only possible consistent approach such that the sum of the feature attributions is equal to the output of the function we are to explain. Another approach for providing explanation is LIME, which is a model-agnostic local approach. Compared to LIME as presented in [32], Shap has the advantage of consistency, and focus at global interpretability versus local for LIME.

The exact computation of SHAP values is challenging. In practice, assuming features' independence, we approximate $f_S(x_S)$ by the Shapley 'sampling value', i.e. the conditional mean of the global model prediction $\hat{f}(X)$ (calibrated on the complete set of features), marginalizing over the values x_C of features that are not included in set S:

$$f_S(x_S) = \mathbb{E}[\hat{f}(X)|x_S] \approx \int \hat{f}(x_S, x_C)p(x_C)dx_C$$

In our application, with max depth of 1, interaction effects between features are discarded, which allows to compute Shapley values trivially from Eq. (1).

4.2 Shapley Interpretation

We can rank the Shapley values by order of magnitude importance, defined as the average absolute Shapley value over the training set of the model: this is essentially the average impact on model output when a feature becomes "hidden" from the model. Furthermore, the correlation of a feature to its Shapley value provides insight into the effect of this feature on the probability of a stock market crash. Figure 3 represents the joint behavior of Shapley and features values to better grasp their non-linear dependencies.

Concerning Fig. 3, we observe that the most significant feature is the 250 days percent change in S&P 500 Price Earnings ratio (using the forward 1 Yr Earnings of the index as provided by Bloomberg). This reflects the presence of persistent cycles during which market participants' bullish anticipations regarding future earnings growths and market valuations translate into reduced downside risk for equity prices.

By the same token, a positive (resp. negative) 250-day change in the US 2 Yrs yield characterizes a regime of growth (resp. recession) in equities. A positive change in the Bloomberg Base Metals index is associated to a reduced crash probability. The same reasoning applies to the FX Emerging Basket, the S&P Sales evolution, the Euro Stoxx distance to its 200-day moving average and the EU Economic Surprise Index. Similarly, a higher (resp. lower) Risk Aversion implies a higher (resp. lower) crash probability and the same relationship is observed for the realized 10-day S&P 500 volatility.

Interestingly, the model identifies the Put/Call ratio as a powerful contrarian indicator. This is a well-known feature highly examined by traditional financial experts. Indeed, a persistently low level of the Put/Call ratio (as reflected by a low 20-day moving average) reflects overoptimistic expectations and therefore an under-hedged market. Last but not least, the Nasdaq 100 is identified as a contrarian indicator: the higher the Nasdaq 20-day percent change and the higher the Nasdaq 100-day and 250-day Sharpe Ratios, the higher the crash probability. More generally, foreign markets are used pro-cyclically (Euro Stoxx, BCOM Industrials, FX emerging) whereas most domestic price indicators are used counter-cyclically (Nasdaq 100, S&P 500). This is an example where we can see some strong added value from the machine learning approach over a human approach, as the former combines contrarian and trend following signals while the latter is generally biased towards one type of signals.

4.3 Joint Features and Shapley Values Distribution

Because some of the features have a strongly non-linear relation to the crisis probability, we also display in Fig. 3 the joint behavior of features and Shapley values at each point in time. The y-axis reports the Shapley values, i.e. the feature contributions to the model output in log-odds (we recall that the GDBT model has a logistic loss) while the color of the dot represents the value of that feature at each point in time. This representation uncovers the non-linearities in the relationship between the Shapley values and the features.

For instance, a large 250-day increase in the P/E ratio (in red color) has a negative impact on the crash probability, everything else equal. The same type of dependency is observed for the change in US 10 Yrs and 2 Yrs yields: the higher (resp. lower) the change in yield, the lower (resp. higher) the crash probability.

However, the dependency of the Shapley value to the 120-day BCOM Industrial Metals Sharpe ratio is non-linear. Elevated Sharpe ratios portend a lower crash probability while the relation vanishes for low Sharpe ratios. An ambiguous dependency is also observed for the 100-day Emerging FX Sharpe ratio, which explains its muted correlation to the Shapley value. This observation is all the more striking as this feature is identified as important in terms of its global absolute impact on the model output. By the same token, the 20-day percent change in S&P 500 Sales does not display a linear relationship to the crash probability. First of all, mostly elevated values of the change in sales are used by the model. Second, large increases in S&P 500 sales are most of the time associated with a drop in the crash probability, but not in every instance.

The impact of the distance of the Euro Stoxx 50 to its 200-day moving average is mostly unambiguous. Elevated levels in the feature's distribution generally (but not systematically) involve a decrease in the crash probability, whereas low levels of this feature portend an increased likelihood of crisis. The 20-day Moving Average of the Put/Call Ratio intervenes as a linear contrarian predictor of the crash probability.

The 20-day percent change in the Nasdaq 100 price is confirmed as a contrarian indicator. As illustrated in Fig. 3, the impact of the 20-day Nasdaq 100 returns is non-linear, as negative returns may predict strongly reduced crash probabilities, while positive returns result in a more moderate increase in the crash probability. Conversely, the 20-day Euro Stoxx returns have a pro-cyclical linear impact on the crash probability, as confirmed by Fig. 3. As previously stated, the GBDT model uses non-US markets in a pro-cyclical way and U.S. markets in a contrarian manner.

4.4 Local Explanation of the Covid March 2020 Meltdown

Not only can Shapley values provide a global interpretation of features' impacts, as described in Sect. 4.2 and in Fig. 3, but they can also convey local explanations at every single date.

The Fig. 4 provides the Shapley values for the model on February, 2020. At this date, the model was still positive on the S&P 500 as the crash probability was fairly low, standing at 9.4%. The 6% 120-day increase in the P/E ratio, the low risk aversion level, reflecting ample liquidity conditions, and the positive EU Economic Surprise index all concurred to produce a low crash probability. However, the decline in the US LIBOR rate, which conveyed gloomy projections on the U.S. economy, and the elevated Put/Call ratio, reflecting excessive speculative behavior, both contributed positively to the crash probability. On February 3, we observe a first steep increase in the crash probability, driven by the 100-day Nasdaq Sharpe Ratio contrarian indicator. At the onset of the Covid crash, on March 2, 2020, the crash probability dramatically increased on the back of deteriorating industrial metals dynamics, falling Euro Stoxx and FTSE prices, negative EU economic surprises and decreasing S&P 500 P/E, which caused the model to identify a downturn in the equities' cycle. This prediction eventually proved prescient. Interestingly, the Nasdaq 100 index had already started its correction by this date, prompting the tech sector contrarian indicators to switch back in favor of a decreased crash probability.

We can do the same exercise for April 1, 2020. The crash probability plummeted as contrarian indicators started to balance pro-cyclical indicators: the Nasdaq 100 appeared oversold while the Put/Call ratio reflected extremely cautious market anticipations. During several months, the crash probability stabilized between 20% and 30% until the start of July, which showed a noticeable

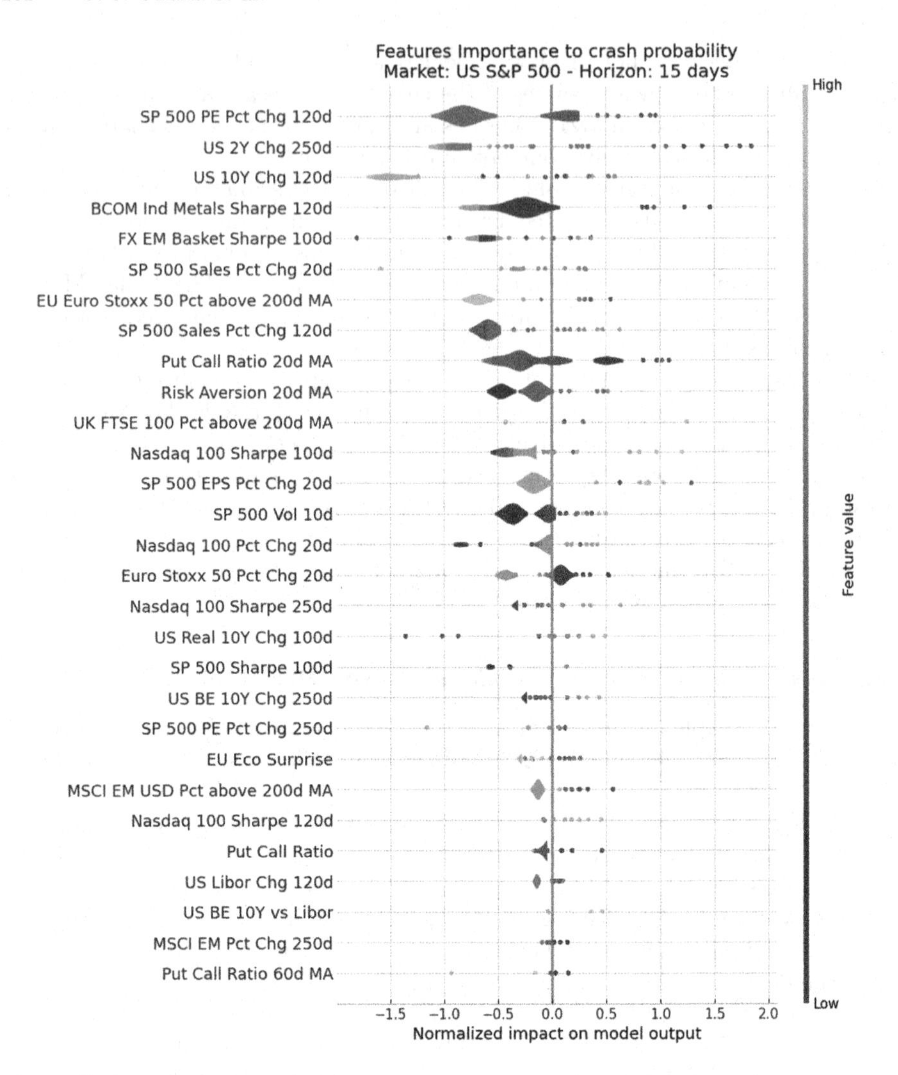

Fig. 3. Marginal contribution of features with full distribution

decline of the probability to 11.2%. The P/E cycle started improving and the momentum signals on base metals and other equities started switching side. Although the crash probability fluctuated, it remained contained throughout the rest of the year.

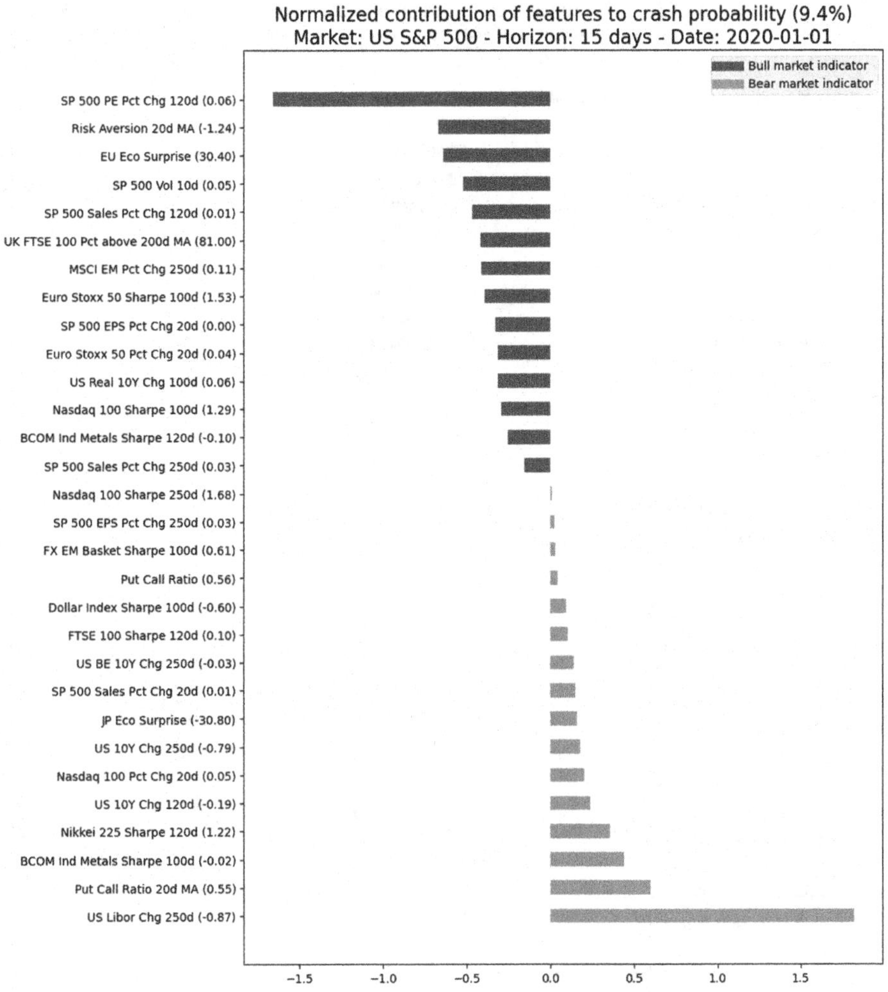

Fig. 4. Shapley values for 2020-01-01

Last but not least, if we look at Shapley value at the beginning of December 2020, we can draw further conclusions. At the turn of the year, most signals were positive on the back of improving industrial metals, recovering European equity markets dynamics, and improving liquidity conditions (reflected by a falling dollar index and a low Risk Aversion). Although this positive picture is balanced by falling LIBOR rates and various small contributors, the features' vote sharply leans in favor of the bullish side.

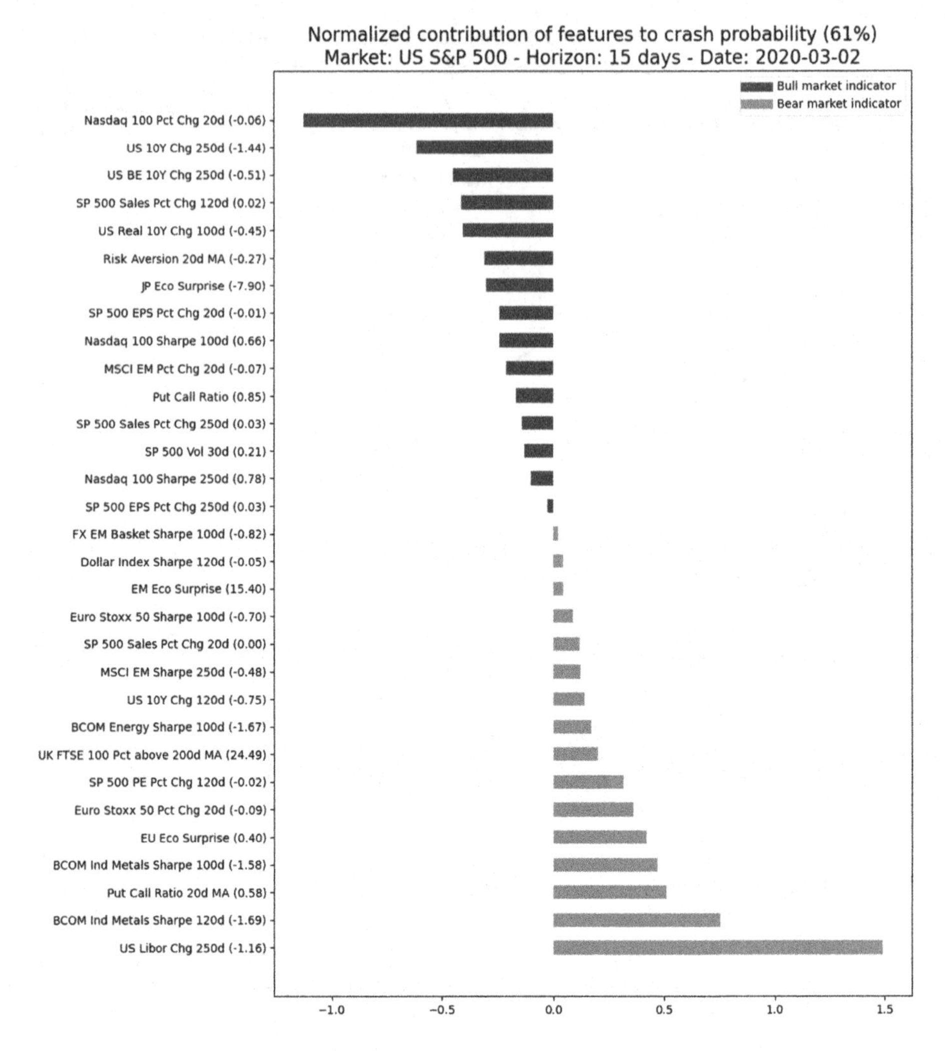

Fig. 5. Shapley values for 2020-03-02

5 Conclusion

In this paper, we have shown how the GBDT method may classify financial markets into normal and crisis regimes, using 150 technical and fundamental features. When applied to the S&P 500, the method yields a high out-of-sample AUC score, which suggests that the machine is able to efficiently learn from previous crises. Our approach also displays an improved accuracy compared to other ML methods, confirming the relevance of GBDT in solving highly imbalanced classification problems with a limited number of observations. Furthermore, the proposed approach could be applied to other fields where crisis detection is key and where data are non stationary like climate catastrophe, weather patterns earthquakes, shocks in power grids and any other economical crash and more generally any extreme events in times series.

AI models complexity are often a barrier to a practitioner understanding of their local predictions. Yet, from the practitioner viewpoint, understanding why a model provides a certain prediction as at least as important as the accuracy of this prediction. Shapley values allow for a global understanding of the model behavior and for a local explanation of each feature's contribution to the crash probability at each observation date. This framework shed light on the unfolding of the model predictions during the events that surrounded the March 2020 equity meltdown. In particular, we unveiled the role of the tech equity sector as a powerful contrarian predictor during this episode.

A few caveats are in order to conclude this paper. First, the model is short-term in nature and should be employed with an agile flexible mindset, rather than to guide strategic investment decisions. Second, one must be careful not becoming overconfident about the forecasting ability of the model, even on a short-term horizon as the model only captures a *probability* of crash risk, generating false positive and false negative signals. More importantly, as financial markets exhibit a strongly non stationary behavior, it is subject to large out of sample prediction errors should new patterns emerge.

References

1. Adadi, A., Berrada, M.: Peeking inside the black-box: a survey on explainable artificial intelligence (XAI). IEEE Access **6**, 52138–52160 (2018)
2. Aguilar-Rivera, R., Valenzuela-Rendón, M., Rodríguez-Ortiz, J.: Genetic algorithms and Darwinian approaches in financial applications: a survey. Expert Syst. Appl. **42**(21), 7684–7697 (2015)
3. Atsalakis, G.S., Valavanis, K.P.: Surveying stock market forecasting techniques - part II: soft computing methods. Expert Syst. Appl. **36**(3, Part 2), 5932–5941 (2009)
4. Benhamou, E.: Connecting Sharpe ratio and Student t-statistic, and beyond. ArXiv (2019)
5. Benhamou, E., Guez, B.: Incremental Sharpe and other performance ratios. J. Stat. Econom. Methods **2018** (2018)
6. Benhamou, E., Guez, B., Paris, N.: Omega and Sharpe ratio. ArXiv (2019)
7. Benhamou, E., Saltiel, D., Guez, B., Paris, N.: Testing Sharpe ratio: luck or skill? ArXiv (2019)
8. Benhamou, E., Saltiel, D., Ohana, J.J., Atif, J.: Detecting and adapting to crisis pattern with context based deep reinforcement learning. In: International Conference on Pattern Recognition (ICPR). IEEE Computer Society (2021)
9. Benhamou, E., Saltiel, D., Ohana, J.J., Atif, J., Laraki, R.: Deep reinforcement learning (DRL) for portfolio allocation. In: Dong, Y., Ifrim, G., Mladenić, D., Saunders, C., Van Hoecke, S. (eds.) ECML PKDD 2020. LNCS (LNAI), vol. 12461, pp. 527–531. Springer, Cham (2021). https://doi.org/10.1007/978-3-030-67670-4_32
10. Benhamou, E., Saltiel, D., Ungari, S., Mukhopadhyay, A.: Bridging the gap between Markowitz planning and deep reinforcement learning. In: Proceedings of the 30th International Conference on Automated Planning and Scheduling (ICAPS): PRL. AAAI Press (2020)

11. Benhamou, E., Saltiel, D., Ungari, S., Mukhopadhyay, A.: Time your hedge with deep reinforcement learning. In: Proceedings of the 30th International Conference on Automated Planning and Scheduling (ICAPS): FinPlan. AAAI Press (2020)
12. Benhamou, E., Saltiel, D., Vérel, S., Teytaud, F.: BCMA-ES: a Bayesian approach to CMA-ES. CoRR abs/1904.01401 (2019)
13. Brown, I., Mues, C.: An experimental comparison of classification algorithms for imbalanced credit scoring data sets. Expert Syst. Appl. **39**(3), 3446–3453 (2012)
14. Chatzis, S., Siakoulis, A.P.V., Stavroulakis, E., Vlachogiannakis, N.: Forecasting stock market crisis events using deep and statistical machine learning techniques. Expert Syst. Appl. **112**, 353–371 (2018)
15. Chen, T., Guestrin, C.: XGBoost: a scalable tree boosting system. CoRR abs/1603.02754 (2016)
16. Choo, J., Liu, S.: Visual analytics for explainable deep learning. CoRR abs/1804.02527 (2018)
17. Ghosal, S., Blystone, D., Singh, A.K., Ganapathysubramanian, B., Singh, A., Sarkar, S.: An explainable deep machine vision framework for plant stress phenotyping. Proc. Natl. Acad. Sci. **115**(18), 4613–4618 (2018)
18. Gu, S., Kelly, B., Xiu, D.: Empirical asset pricing via machine learning. Rev. Financ. Stud. **33**(5), 2223–2273 (2020)
19. Ke, G., et al.: LightGBM: a highly efficient gradient boosting decision tree. In: Guyon, I., et al. (eds.) Advances in Neural Information Processing Systems, vol. 30, pp. 3146–3154. Curran Associates, Inc. (2017)
20. Kingma, D., Ba, J.: Adam: a method for stochastic optimization (2014)
21. Krauss, C., Do, X.A., Huck, N.: Deep neural networks, gradient-boosted trees, random forests: statistical arbitrage on the S&P 500. Eur. J. Oper. Res. **259**(2), 689–702 (2017)
22. Li, Y., Ma, W.: Applications of artificial neural networks in financial economics: a survey. In: 2010 International Symposium on Computational Intelligence and Design, vol. 1, pp. 211–214 (2010)
23. Liu, S., Wang, X., Liu, M., Zhu, J.: Towards better analysis of machine learning models: a visual analytics perspective. CoRR abs/1702.01226 (2017)
24. Lundberg, S., Lee, S.I.: A unified approach to interpreting model predictions (2017)
25. Madhikermi, M., Malhi, A.K., Främling, K.: Explainable artificial intelligence based heat recycler fault detection in air handling unit. In: Calvaresi, D., Najjar, A., Schumacher, M., Främling, K. (eds.) EXTRAAMAS 2019. LNCS (LNAI), vol. 11763, pp. 110–125. Springer, Cham (2019). https://doi.org/10.1007/978-3-030-30391-4_7
26. Malhi, A., Kampik, T., Pannu, H.S., Madhikermi, M., Främling, K.: Explaining machine learning-based classifications of in-vivo gastral images. In: 2019 Digital Image Computing: Techniques and Applications (DICTA), p. 7, December 2019
27. Malhi, A., Knapic, S., Främling, K.: Explainable agents for less bias in human-agent decision making. In: Calvaresi, D., Najjar, A., Winikoff, M., Främling, K. (eds.) EXTRAAMAS 2020. LNCS (LNAI), vol. 12175, pp. 129–146. Springer, Cham (2020). https://doi.org/10.1007/978-3-030-51924-7_8
28. Marceau, L., Qiu, L., Vandewiele, N., Charton, E.: A comparison of deep learning performances with others machine learning algorithms on credit scoring unbalanced data. CoRR abs/1907.12363 (2019)
29. Mehra, R., Prescott, E.: The equity premium: a puzzle. J. Monet. Econ. **15**(2), 145–161 (1985)

30. Nti, I.K., Adekoya, A.F., Weyori, B.A.: A systematic review of fundamental and technical analysis of stock market predictions. Artif. Intell. Rev. **53**(4), 3007–3057 (2019). https://doi.org/10.1007/s10462-019-09754-z
31. Prokhorenkova, L., Gusev, G., Vorobev, A., Dorogush, A.V., Gulin, A.: CatBoost: unbiased boosting with categorical features. In: Bengio, S., Wallach, H., Larochelle, H., Grauman, K., Cesa-Bianchi, N., Garnett, R. (eds.) Advances in Neural Information Processing Systems, vol. 31, pp. 6638–6648. Curran Associates, Inc. (2018)
32. Ribeiro, M.T., Singh, S., Guestrin, C.: "why should i trust you?": Explaining the predictions of any classifier. In: Proceedings of the 22nd ACM SIGKDD International Conference on Knowledge Discovery and Data Mining, pp. 1135–1144. Association for Computing Machinery (2016)
33. Rosenfeld, A., Richardson, A.: Explainability in human-agent systems. CoRR abs/1904.08123 (2019)
34. Rundo, F., Trenta, F., di Stallo, A.L., Battiato, S.: Machine learning for quantitative finance applications: a survey. Appl. Sci. **9**(24), 5574 (2019)
35. Samitas, A., Kampouris, E., Kenourgios, D.: Machine learning as an early warning system to predict financial crisis. Int. Rev. Financ. Anal. **71**, 101507 (2020)
36. Sezer, O.B., Gudelek, M.U., Ozbayoglu, A.M.: Financial time series forecasting with deep learning: a systematic literature review: 2005–2019. arXiv preprint arXiv:1911.13288 (2019)
37. Shah, D., Isah, H., Zulkernine, F.: Stock market analysis: a review and taxonomy of prediction techniques. Int. J. Financ. Stud. **7**(2), 26 (2019)
38. Sornette, D., Johansen, A.: Significance of log-periodic precursors to financial crashes. Quant. Finance **1**, 452–471 (2001)
39. Xing, F.Z., Cambria, E., Welsch, R.E.: Natural language based financial forecasting: a survey. Artif. Intell. Rev. **50**(1), 49–73 (2017). https://doi.org/10.1007/s10462-017-9588-9

Toward XAI & Human Synergies to Explain the History of Art: The Smart Photobooth Project

Egberdien van der Peijl[1], Amro Najjar[1], Yazan Mualla[2(✉)], Thiago Jorge Bourscheid[1], Yolanda Spinola-Elias[3], Daniel Karpati[1], and Sana Nouzri[1]

[1] AI-Robolab/ICR, Computer Science and Communications, University of Luxembourg, 4365 Esch-sur-Alzette, Luxembourg
egberdien.vanderpeijl@ext.uni.lu, thiago.bourscheid.001@student.uni.lum, {sana.nouzri, Amro.Najjar}@uni.lu

[2] CIAD, Univ. Bourgogne Franche-Comté, UTBM, 90010 Belfort, France
yazan.mualla@utbm.fr

[3] Dpto. Dibujo, University of Seville, C/Larana n°3, 41003 Seville, Spain
yspinola@us.es

Abstract. The advent of Artificial Intelligence (AI) has brought about significant changes in our daily lives with applications including industry, smart cities, agriculture, and telemedicine. Despite the successes of AI in other "less-technical" domains, human-AI synergies are required to ensure user engagement and provide interactive expert knowledge. This is notably the case of applications related to art since the appreciation and the comprehension of art is considered to be an exclusively human capacity. This paper discusses the potential human-AI synergies aiming at explaining the history of art and artistic style transfer. This work is done in the context of the "Smart Photobooth" a project which runs within the AI & Art pavilion. The latter is a satellite event of Esch2022 European Capital of Culture whose main aim is to reflect on AI and the future of art. The project is mainly an outreach and knowledge dissemination project, it uses a smart photo-booth, capable of automatically transforming the user's picture into a well-known artistic style (*e.g.*, impressionism), as an interactive approach to introduce the principles of the history of art to the open public and provide them with a simple explanation of different art painting styles. Whereas some of the cutting-edge AI algorithms can provide insights on what constitutes an artistic style on the visual level, the information provided by human experts is essential to explain the historical and political context in which the style emerged. To bridge this gap, this paper explores Human-AI synergies in which the explanation generated by the eXplainable AI (XAI) mechanism is coupled with insights from the human expert to provide explanations for school students as well as a wider audience. Open issues and challenges are also identified and discussed.

Keywords: AI & Art · XAI · Agents · Neural style transfer · Cultural heritage

D. Calvaresi et al. (Eds.): EXTRAAMAS 2021, LNAI 12688, pp. 208–222, 2021.
https://doi.org/10.1007/978-3-030-82017-6_13

1 Introduction

In the last decade, AI has become omnipresent with applications spanning from autonomous vehicles, agriculture, and industry. In recent years, this wave has also spread to new domains such as digital history and cultural heritage. In this particular context, making the cultural heritage more accessible and more engaging by bringing it closer to more audiences is considered to be one of the main contributions of the latest generation of AI systems [15]. For instance, MonuMAI is a smartphone app allowing for artistic knowledge dissemination. MonuMAI classifies photos (*e.g.*, taken for a monument's facade) and classifies it into different architectonic style and provide visual hints explaining why the photo belongs to the style. To do that, MonumMAI relies on deep learning classifiers supported by expert knowledge [36]. Other works in the literature aim to increase the accessibility of cultural heritage for target user groups. This can be either for users with special needs (*e.g.*, elderly people with mobility constraints [33,34]), or to enhance the level of audience engagement by providing comprehensible and interactive content. In this paper, we present the Smart Photobooth project, a project within the context of Esch2022 European capital of culture.

The Smart Photobooth is an outreach project aiming to disseminate knowledge about the history of art and artistic styles as well as the latest machine learning mechanisms and their applications in the domain of art generation and classification. The smart booth relies on Neural Style Transfer (NST) [30] to transfer an input image (typically the portrait of the user) to one of the most famous artistic styles (*e.g.*, cubism). NST relies on machine learning Generative Adversarial Networks (GAN) to achieve the transfer [30].

After the user gets their transformed portrait, they receive both *(i)* a short tutorial explaining the style and its position in the history of art based on well-known paintings from the style and *(ii)* an AI explanation highlighting the style features appearing in the transferred user portrait.

Recently, eXplainable Artificial Intelligence (XAI) has been identified as a powerful approach to support these outreach efforts by helping to interpret the otherwise incomprehensible inner-workings of sophisticated machine learning mechanisms such as the advanced GANs powering the NST process [12,15]. In particular, XAI has been suggested as a possible solution to "explain a given artwork's success in terms of the underlying influencing artistic styles" [15].

However, this is a challenging task since, in contrast to many other disciplines (math, physics, chemistry), making art "understandable" requires a combination of objective and subjective interpretations to analyze its message. For instance, analyzing the objective features of Guernica, the famous painting by the Spanish artist Pablo Picasso is not enough to understand the background and the interpretations of this painting. The latter information is highly subjective, controversial, and depends on the historical circumstances surrounding the creation of the painting [48]: the Spanish Civil war, Picasso being commissioned by Manuel Azana, the president of the short-lived Spanish Republic, to create a large mural for the Spanish pavilion at the 1937 Paris World's Fair, as well as the bombing of Guernica, a town in the Basque country, on the 26th of April, 1937 by the Condor Legion of the Nazi German air forces, and Picasso's discussion with his friend the poet Juan Larrea who urged him to make the bombing his subject [38], *etc.*

To overcome this challenge, in this paper we propose a human-agent architecture allowing us to foster the needed synergies between the involved parties. Namely, the human end-user, the artist, and the black-box machine learning mechanism. Based on the context, the user preferences, and the artists' recommendations, the agent provides personalized explanations combining machine learning interpretation, agent explainability, as well as the artist's expert analysis. The architecture is discussed, the challenges it raises and are identified and discussed.

The rest of this paper is organized as follows. Section 2 lays out the background for this work. Section 3 introduces the Smart Photobooth project. Section 4 presents the proposed architecture, Sect. 5 identifies the challenges and the open issues, and Sect. 6 concludes this article.

2 Background

2.1 AI & Art

Works combining AI and art in the literature fall into two categories. The first is using AI in the process of creating new art while the second is using AI to analyze existing human-created art.

AI Art Generation. The recent rapid evolution of Deep Neural Networks (DNN) has accelerated the use of AI technologies to create art. In particular, GANs are among the cutting-edge technologies used in this domain. The latter involves a couple of systems of DNNs designed to compete against each other. For instance, in the case of GANs generating visual art, a DNN, called the generator, is trained to generate realistic images whereas the other DNN (known as the discriminator) is trained to classify generated images as fake while identifying real artistic pieces as real art. The training of the GAN is achieved once the generator becomes capable of creating output that cannot be identified as fake by the discriminator (*i.e.,* the generator outperforms the discriminator). Recently, this type of models has been implemented in different configurations (*e.g.,* CycleGAN [60], StyleGAN [31], BigGAN [7]) and has achieved remarkable results in generating human faces (2D [45] and 3D [55]), music [35], as well as furniture [58]. Another application of AI Art generation is transferring some features of the input. Face aging and NST are notable examples.

Understanding AI with Art. The advent of AI has a significant impact on art access ability and understandability. Several works in the literature have explored how AI can help make cultural heritage more accessible for users with special needs. For instance, haptic interfaces have been proposed to be used by museums to help visitors with visual impairments formulate mental pictures of the objects and provide important contextual and navigation information [10, 13,18,56] (*cf.* [47] for a review).

Moreover, in recent decades, thousands of artworks have been digitized and are now available online for access and analysis. OmniArt [53] and Art500 [40] are among the biggest artwork datasets. The former includes about two million artworks allowing for author, style, period, type and iconography retrieval, color classification, and object detection, whereas the latter contains about 550 thousand artworks. The Metropolitan Museum of Art of New York also released in 2017 over 406.000 indexed pictures of public domain artwork [1]. The datasets can retrieve items by authors, genre, styles, events, and historical figures.

Artistic Style Classification. Based on the datasets mentioned above, recent works in the literature propose to classify paintings into their artistic styles (Renaissance, Baroque, Impressionism, *etc.*). Many of these works rely on style patterns and definitions proposals by the Swiss art historian Heinrich Wölfflin [57]. In particular, Wölfflin identifies five key visual principles each defined by two contrasting visual schemes [11,57]:

(i) **Linear *vs.* Painterly.** In the former, elements are clearly outlined and boundaries are clear while in the latter, elements are fused and contours and edges are blurry. *(ii)* **Closed *vs.* Open forms.** In the former, elements are balanced with the frame. Vertical and horizontal compositions are dominant. In the latter, diagonal components are dominants with an impression of the space going beyond the edges of the picture. *(iii)* **Planner** in which elements are organized in successive planes parallel to the picture plane versus **Recessional** which gives an illusion of depth and where elements are arranged on various planes. *(iv)* **Multiplicity.** Elements appear distinct and independent, versus **Unity** where elements are fused into a single whole. *(v)* **Absolute clarity** with explicit and articulated forms versus **relative clarity** with less clearly structured forms avoiding objective clearness in an intended manner. Recent works in the literature suggested that convolutional neural networks trained to classify paintings according to their artistic style, implicitly learn features related to Wölfflin's concepts. For instance, Elgammal et al. [17] used a convolutional neural network to classify paintings into their artistic styles. The results obtained showed that the network managed to smoothly place an artwork into a temporal arrangement based on learning style labels. In a related work, the authors in [11] trained a convolutional neural network to predict the values of the five Wölfflin concepts (or features). The result of this work showed that the proposed network learned to discriminate meaningful features corresponding to the visual characteristics of Wölfflin's concepts.

But, according to other authors, such as Lecoutre, Negrevergne and Yger [39] or Tan et al. [54], identifying the artistic style of a picture in a fully automatic way is a challenging problem since classifying visual styles cannot rely on any definitive feature. This is especially difficult for non-representational artwork.

For these reasons, in this work, we overcome this problem by relying on an autonomous agent capable of combining the explanation obtained from the artist with the interpretation obtained by the machine learning mechanism. The artist provides the broader view, the historical context, and insights on the artist's background influencing his works (*e.g.*, political thought), while the XAI mechanism obtains the values of Wölfflin features and illustrates them on the transformed portrait of the user.

2.2 Styles in Modern and Contemporary Art Painting

The moment a painter takes out his brush and palette and starts painting on a canvas, it is possible that a new style will emerge [2]. Styles in painting had their heyday during the late 19th century and early 20th century. The style displayed or expressed in a painting can become part of an art movement, in which case there is a group of artists who have defined a certain style (in painting or other artistic disciplines). The individual interpretation was the most common impulse that made certain styles evolve to a peak, disseminate and transform into a new style. In the period mentioned above, also called 'modernism', we can count more than 100 different styles among which the Hurufiyya movement (Islamic calligraphy), Peredvizhniki (Russian 'wanderers' protesting against academic restrictions), Letras y figuras (depiction of letters of the alphabet during the Spanish colonial period in the Philippines) and many more [37].

Impressionism became one of the best-known movements in western Europe. The impressionists rejected, like their Russian contemporaries (the Peredvizhniki, Wanderers) classical and imperative aesthetic rules, incorporated by the Salons that detained a monopoly in the field of contemporary art exhibitions. An important innovator within this group was Claude Monet (1840–1926). In the American Magazine of Art (1927), Lilla Cabot Perry reveals the remarkable method Monet used to paint in the typical impressionistic style: He had grooved boxes filled with canvases placed at various points in the garden where there was barely room for him to sit as he recorded the fleeting changes of the light on his water-lilies and arched bridges. He often said that no painter could paint more than one half an hour on any outdoor effect and keep the picture true to nature, and remarked that in this respect he practiced what he preached [46].

Just after the turn of the century in 1907, Pablo Picasso painted *The Young Ladies of Avignon/The Brothel of Avignon*, where Aviñón refers to a street in Barcelona. This proto-cubist work introduces 'primitive style' from Africa in western art. At that time masks and sculptures from African countries became popular and were on sale in shops in Paris but also exhibited in museums. In the 'Autobiography of Alice B. Toklas', Toklas - who was befriended with Picasso - explains his radical turn in style as follows: "In these early days when he created cubism the effect of the African art was purely upon his vision and forms, his imagination remained purely Spanish" [52].

Ten years later. *Fountain*, an artwork of Marcel Duchamp (1887–1968) was submitted to the 1917 exhibition of the American Society of Independent Artists in New York. Sitting on a pedestal, turned upside down and signed with the artist's name R. Mutt. *Fountain* sparked raging controversy after Duchamp's colleagues refused to recognize the item as a legitimate work of art and requested it to be removed. Duchamp coined the term 'readymade' for this work but, while final evidence is missing, it could have been the 'Dada baroness' Elsa von Freytag - Loringhoven who came up with this idea and not Duchamp, who was a friend of her [19].

The utensil, called 'ready made' was an object, not made by an artist but ready to be chosen as an artwork. The case of the *Fountain* had the power to stop the development of all future styles and movements. This did not happen immediately but at the end of the Sixties. In *The conspiracy of Art*, the philosopher Jean Baudrillard describes the consequences of the removal of the artist from the artwork and even from the art world. He points to 'a new reality' that is not about creating styles but about creating interchangeable components that can each serve as 'reality': a procession of models providing autonomy for the virtual, freeing it from reality, and the simultaneous autonomy of reality that we now see functioning for itself - motu proprio - in a hallucinatory perspective, in other words, self - referential ad infinitum. Cast out from its own framework, from its own principle, extraneous, reality has itself become an extreme phenomenon. In other words, we can no longer think of it as reality, but only as otherworldly, as if seen from another world - as an illusion [5].

2.3 Neural Style Transfer

Style Transfer, often called Neural Style Transfer (NST), is the practice of manipulating a piece of data like an image or video to adopt the inherent style of another piece of data [29]. Common uses for this type of application come under the form of Deep Fakes, the image of a given person projected onto someone else's in a video, and the projection of a certain style onto a photograph, similar to a filter.

The main technology used nowadays for this type of functionality is the previously mentioned Generative Adversarial Network [21]. One of the more prominent GAN architectures with the objective of style transfer is the so-called Cycle-Consistent Generative Adversarial Network [61], CycleGAN in short, first proposed by Jun-Yan Zhu, Taesung Park, Phillip Isola, and Alexei A. in 2017. One of the characteristic advantages of the CycleGAN is that it doesn't require paired training data to deliver results. What this means is that when a GAN is trained on two domains, by adding a piece of data to the set of one domain, it is not necessary to add the equivalent data of the opposite domain. This facilitates the creation of CycleGAN datasets since data can be compiled into each domain without further complications.

Just like with a regular GAN, both Generators and Discriminators are competing with each other. The Generators learn to create better fakes and the Discriminators learn to better detect the fakes. Together, the models get incrementally better at their tasks, learning from each other and adapting accordingly [8].

In addition to that, the Generator models are trained to not just create new images in the target domain, but instead reconstruct versions of the input images from the source domain. This is achieved by using generated images as input to the corresponding generator model and comparing the output image to the initial images. Passing an image through both generators is what we designate as a cycle. Together, each pair of generator models get trained to better reproduce the original source image, referred to as cycle consistency.

Another advanced implementation of style transfer can be found in StyleGAN [32]. In contrast to CycleGAN, this one does not need an entire dataset for the desired target-style application, one image proves sufficient. A StyleGAN works similarly to the traditional GAN implementation, however, after a StyleGAN has successfully been trained, it offers much finer control over the generated result than was previously possible. To achieve style transfer, StyleGAN is fed reference data on specific convolution layers. When it reaches those layers in the generation process, it will attempt to map the current sample to that target piece of data, effectively blending the image information into the end result.

2.4 XAI

The recent rapid development of Artificial Intelligence (AI) technology, as well as its widespread use in our daily lives, have raised several concerns about the human understandability of this sophisticated technology. To address this concern, eXplainable AI (XAI) [4] emerged to interpret the sometimes intriguing results of AI and ML learning mechanisms [22] as well as autonomous agents and robots [3]. In future AI systems, it is vital to guarantee a smooth human-agent interaction, as it is not straightforward for humans to understand the agent's state of mind, and explainability is an indispensable ingredient for such interaction [41].

Explanations can be provided by AI for a multitude of purposes including, control and debugging of AI systems, transparency, and accountability, as well as training and education. In the latter, case the main aim of the explanations is to help users understand how the system works and get a glance at its inner workings. Examples of these educational explanations including firefighter training [26], and UAV operation [43,44]. Moreover, recently, the advantages of using XAI in humanities and arts. Yet, this line of research is in its early stages of development with many considerable challenges ahead [47].

3 The Smart Photobooth Project

The Smart Photobooth is a playful and interactive intelligent "machine" where the users can experiment with AI, and learn about the process of how intelligent machines are trained. For this, we have chosen to combine AI and Art. Science & Art have a vibrant and exciting relationship. Thus, they form a perfect combination to engage different audiences - of all ages, genders, and backgrounds - and entice them to learn more and explore computational methods. In particular, we have chosen to use AI to manipulate images to resemble artistic styles for two reasons: *(i)* humans are very visual and drawn to images, which are a powerful media to provoke interaction and enable easy communication – even more in the era of internet and social media - and *(ii)* the Smart Photobooth is similar to an interactive version of the Snapchat filter (which is widely known and popular). The Smart Photobooth is the perfect package to deliver our machine-learning & art message to a wider audience, in particular to teenagers, including those who are not technology drawn.

When the user enters the booth, they can take a portrait of themselves and then select a style to transform their image into. The set of available styles are selected by a professional artist (*e.g.,* impressionism, cubism, *etc.*). The style transfer is conducted using a NST mechanism (*cf.* Sect. 2). The NST mechanism is pre-trained using two training sets representing artistic styles as well as end-user portraits.

Once the user obtains the output (*i.e.,* their portrait pictured in the chosen style), they also obtain a multi-media presentation explaining both *(i)* the basic principles of the chosen style, its historical context, main contributors and most famous paintings, and *(ii)* an illustration of how the style influenced the visual features of the output image. This is obtained by the XAI mechanism which explains what of Wölfflin's features are present in the output image and how to they correspond to the style chosen by the user.

The project will be developed and presented in two different venues: a) Space 1 - workshops in the Scienteens Lab at the University of Luxembourg for STEM (science, technology, engineering, and mathematics) high school students in Luxembourg (April–July 2021); b) Space 2 - exhibition in the Luxembourg Science Center for various types of visitors - children, teenagers, families (August–December 2021). In 2022, we plan to have the Smart Photobooth exhibited for the whole year in the AI & Art Pavilion. The Pavilion, supported by Esch2022 Capital of Culture, will be providing various interactive programs and a series of exhibitions for all types of visitors for the duration of Esch2022.

4 Architecture

Figure 1 depicts our proposed architecture to explain the history of arts to human users by agents. On the bottom side of the figure, a training dataset is used to train the model after preprocessing it to handle any abnormalities. The result of the training model will be stored in the *Art Knowledge Base*, which will be the input to build the deployment machine learning model. The GAN-based NST algorithm takes as input: *(i)* The input image of the human user to be converted; *(ii)* an input dataset of images similar to the human user image; *(iii)* the deployment model features. The output of the algorithm is the output image with a specific art style. This image will be presented to the human and will help along with some features from the deployment model to provide some ML analysis to be used for the explanations. The explanation model in the agent formulates the explanations based on three sources:

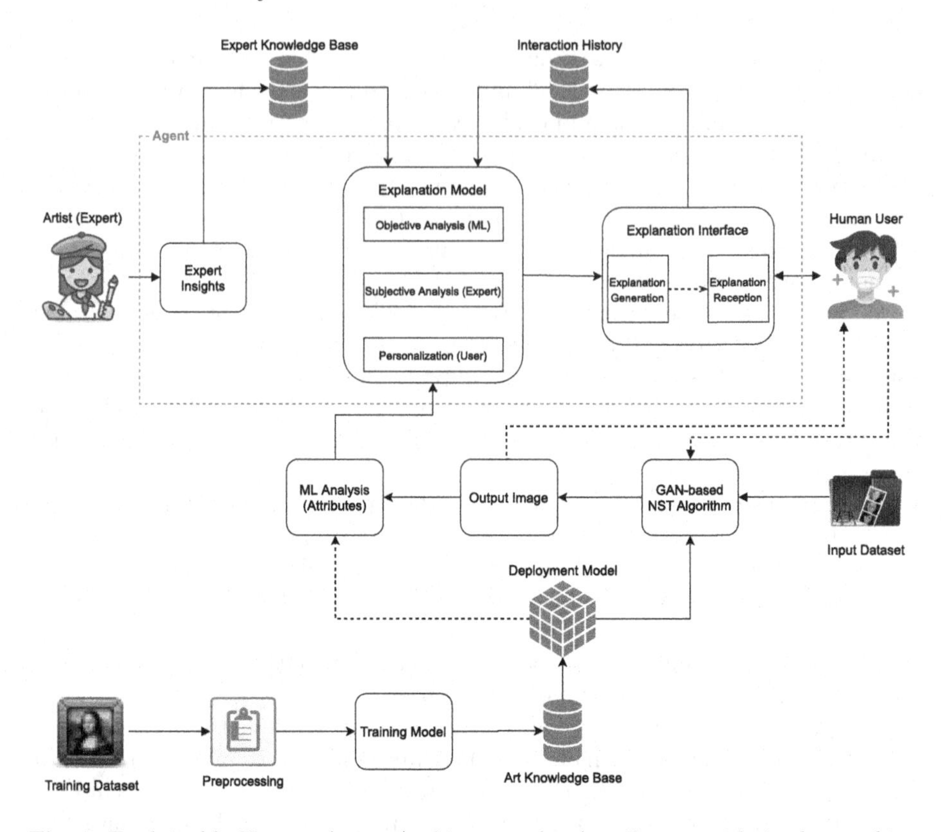

Fig. 1. Explainable Human-Agent Architecture for Art: Bottom side is the machine learning model including the *GAN-based NST Algorithm*. Top side is the agent model responsible for explainability. Two-ways arrow between the *Explanation Interface* and *Human User* to highlight an interaction where the feedback from the user is used to update the *Explanation Model*.

(i) An objective analysis based on the insights provided by the ML algorithm and model; *(ii)* A subjective analysis provided by the expert in the domain (the artist) which will store relevant knowledge in the *Expert Knowledge Base*; *(iii)* Another subjective analysis performed on the human user side to guarantee that the personalized explanations are built based on the preference of the human. For this step, the literature highlights the need to move towards human-centered Models as explanations are subjective [51]. This knowledge about the human user is collected depending on the various interactions with him/her and stored in the *Interaction History Base*, hence the two-ways arrow between the *Explanation Interface* and the *Human User*.

Finally, the formulated explanations are communicated by the agent to the human user through an interface that allows for the following two tasks: *(i)* generating the explanations based on the explanation components formulated in the explanation model. The generation could use templates or be more sophisticated by relying on NLP techniques. *(ii)* Providing the explanation to the human through the explanation reception process that allows the interaction with the human and considers its cognitive load.

5 Open Challenges and Research Directions

Using XAI for AI & Art dissemination is a domain in its early stages of development with most of the pioneering work carried out at the conceptual front [15]. The Smart Photobooth is a work-in-progress aiming to operationalize and test XAI-based art dissemination in a real-life context. As explained in the previous sections, the project proposes an architecture that combines expertise from the artist, input from the end-user, and the output of the most advanced machine learning mechanisms. The intelligent agent proposed by our architecture is in charge of combining these heterogeneous data and making them accessible and understandable by different stakeholders. The project still being in its early implementation phase, this section identifies the following challenges and research directions we will pursue to address them.

(i) **Explaining heterogeneous AI systems:** The Smart Photobooth project involves multiple AI systems. Namely: The NST systems, the style interpretation system, and the agent which is in charge of obtaining the artist's explanation, and the input characterizing the user. Combining these heterogeneous explanations is a challenging task notably because it involves both symbolic knowledge (user data and artist input) and sub-symbolic knowledge within the black-box machine learning mechanism. One potential solution to address this issue is to resort to the latest advances in neuro-symbolic AI which proposes to integrate the symbolic AI systems [20] (*i.e.,* agents knowledge is represented by logic and reasoning) with the sub-symbolic knowledge within the GANs and the convolutional neural networks. Recent works in XAI suggested that this neuro-symbolic integration is highly beneficial to XAI [9]. In particular, compared with the current approach which relies on a simple concatenation of expert explanations with the visual descriptors originating from the neural network, the neuro-symbolic approach allows for a better understanding of the system's output since the symbolic knowledge extracted from the neural network can be reasoned and manipulated by the agent. Moreover, information extracted also be made accessible/understandable by the artist who can also interact with the knowledge to tune the performance of the machine learning mechanism. *(ii)* **Users with Special Needs:** Most of the explanations and insight provided by the Smart Photobooth are communicated audio-visually. For this reason, the content is inaccessible for users with visual and/or audio impairment. One alternative solution is to rely on solutions currently being developed to make cultural heritage accessible for this category of users. For instance, *bind photography* [59] can assist users with low visual acuity take their photos by providing audio feedback that facilitates aiming the camera. Additionally, online accessibility of explanations is needed for those who could not visit the artworks galleries [42] *(iii)* **Gamification:** In recent years, gamification [14,25,28] has been a trending topic in domains as diverse as education, information studies, and

human–computer interaction [50]. It is also considered an interesting means of supporting user engagement and enhancing positive use patterns, such as increasing user activity, social interaction, or quality and productivity of actions [23]. These desired use patterns are considered to emerge as a result of the positive, intrinsically motivating [49], and gameful experiences [28] brought about by game/motivational affordances implemented into a service [24]. Recently, art organizations have also sought to gamify different aspects of their institutions to engage visitors, increase fundraising, or improve marketing objectives [6]. The application of gamification in education and outreach settings like in the context of this paper is still a relatively new trend, but it has gained attention due to its ability to increase student motivation and engagement. According to [16], there are three main concerns when considering gamifying the learning experience: "*(i)* insufficient evidence exists to support the long-term benefits of gamification in educational contexts; *(ii)* the practice of gamifying learning has outpaced researchers' understanding of its mechanisms and methods; *(iii)* the knowledge of how to gamify an activity in accordance with the specifics of the educational context is still limited."

6 Conclusions

This paper presented the Smart Photobooth, an interdisciplinary outreach and knowledge dissemination project organized by the University of Luxembourg. The project relies on XAI to explain the history of art and artistic styles to end-users. Delivering such explaining requires a combination of explanations provided by the human expert (an artist) with interpretation obtained from a machine learning mechanism. To achieve this synergy, the paper proposed an architecture powered by an agent who is in charge of accomplishing this combination. The components of the architecture were identified and explained. Open issues and challenges were identified with their potential solutions. The Smart Photobooth project is a work-in-progress, as the project started in early 2021. Currently, the Photobooth is being installed and the proposed architecture and the XAI mechanism are being implemented. The next step is to evaluate the explanations and assess how well they performed in engaging the end-users and enhancing their knowledge on the history of art. To do so, specific XAI metrics [27] will be defined, several user studies will be conducted, and their results will be statistically validated, studied, and analyzed.

Acknowledgment. The Smart Photobooth project is funded by FNR (Fond National de Recherche), Luxembourg. Reference: PSP-C2020-2/15417971.

References

1. Images of artworks in the public domain. https://www.metmuseum.org/about-the-met/policies-and-documents/image-resources

2. Alpers, S.: Style is what you make it: the visual arts again. In: Lang, B. (ed.) The Concept of Style (Revised and Expanded Edition. Cornell Univ. Pr., Ithaca and London, 1979, 1987), pp. 137–162 (1979)
3. Anjomshoae, S., Najjar, A., Calvaresi, D., Främling, K.: Explainable agents and robots: results from a systematic literature review. In: 18th International Conference on Autonomous Agents and Multiagent Systems (AAMAS 2019), Montreal, Canada, 13–17 May 2019, pp. 1078–1088. International Foundation for Autonomous Agents and Multiagent Systems (2019)
4. Arrieta, A.B., et al.: Explainable artificial intelligence (XAI): concepts, taxonomies, opportunities and challenges toward responsible AI. Inf. Fusion **58**, 82–115 (2020)
5. Baudrillard, J.: The Conspiracy of Art. Manifestos, Interviews, Essays. Semiotext(e) Foreign Agents Series (2003)
6. Bouchard, A.: Gamification in the arts (2013)
7. Brock, A., Donahue, J., Simonyan, K.: Large scale GAN training for high fidelity natural image synthesis. In: International Conference on Learning Representations (2018)
8. Brownlee, J.: How to develop a CycleGAN for image-to-image translation with Keras, August 2019
9. Calegari, R., Ciatto, G., Omicini, A.: On the integration of symbolic and subsymbolic techniques for XAI: a survey. Intell. Artifi. **14**(1), 7–32 (2020)
10. Carrozzino, M., Bergamasco, M.: Beyond virtual museums: experiencing immersive virtual reality in real museums. J. Cult. Herit. **11**(4), 452–458 (2010)
11. Cetinic, E., Lipic, T., Grgic, S.: Learning the principles of art history with convolutional neural networks. Pattern Recogn. Lett. **129**, 56–62 (2020)
12. Cetinic, E., She, J.: Understanding and creating art with AI: review and outlook. arXiv preprint arXiv:2102.09109 (2021)
13. Coelho, A., Cardoso, P., van Zeller, M., Santos, L., Raimundo, J., Vaz, R.: Gamifying the museological experience (2020)
14. Deterding, S., Dixon, D., Khaled, R., Nacke, L.: From game design elements to gamefulness: defining "gamification". In: Proceedings of the 15th International Academic MindTrek Conference: Envisioning Future Media Environments, pp. 9–15 (2011)
15. Díaz-Rodríguez, N., Pisoni, G.: Accessible cultural heritage through explainable artificial intelligence. In: Adjunct Publication of the 28th ACM Conference on User Modeling, Adaptation and Personalization, pp. 317–324 (2020)
16. Dicheva, D., Dichev, C., Agre, G., Angelova, G.: Gamification in education: a systematic mapping study. J. Educ. Technol. Soc. **18**(3), 75–88 (2015)
17. Elgammal, A., Liu, B., Kim, D., Elhoseiny, M., Mazzone, M.: The shape of art history in the eyes of the machine. In: Proceedings of the AAAI Conference on Artificial Intelligence, vol. 32 (2018)
18. Frid, E., Lindetorp, H., Hansen, K.F., Elblaus, L., Bresin, R.: Sound forest: evaluation of an accessible multisensory music installation. In: Proceedings of the 2019 CHI Conference on Human Factors in Computing Systems, pp. 1–12 (2019)
19. Gammel, I.: Baroness Elsa, Gender, Dada and Everyday Modernity. A Cultural Biography. MIT Press, Cambridge (2003)
20. d'Avila Garcez, A.S., Broda, K.B., Gabbay, D.M.: Neural-Symbolic Learning Systems: Foundations and Applications. Springer, London (2002). https://doi.org/10.1007/978-1-4471-0211-3
21. Goodfellow, I.J., et al.: Generative adversarial networks, June 2014

22. Guidotti, R., Monreale, A., Ruggieri, S., Turini, F., Giannotti, F., Pedreschi, D.: A survey of methods for explaining black box models. ACM Comput. Surv. (CSUR) **51**(5), 1–42 (2018)
23. Hamari, J.: Transforming homo economicus into homo ludens: a field experiment on gamification in a utilitarian peer-to-peer trading service. Electron. Commer. Res. Appl. **12**(4), 236–245 (2013)
24. Hamari, J., Koivisto, J., Sarsa, H.: Does gamification work? - a literature review of empirical studies on gamification. In: 2014 47th Hawaii International Conference on System Sciences, pp. 3025–3034. IEEE (2014)
25. Hamari, J., Lehdonvirta, V.: Game design as marketing: how game mechanics create demand for virtual goods. Int. J. Bus. Sci. Appl. Manag. **5**(1), 14–29 (2010)
26. Harbers, M., van den Bosch, K., Meyer, J.-J.: A methodology for developing self-explaining agents for virtual training. In: Dastani, M., El Fallah Segrouchni, A., Leite, J., Torroni, P. (eds.) LADS 2009. LNCS (LNAI), vol. 6039, pp. 168–182. Springer, Heidelberg (2010). https://doi.org/10.1007/978-3-642-13338-1_10
27. Hoffman, R.R., Mueller, S.T., Klein, G., Litman, J.: Metrics for explainable AI: challenges and prospects. arXiv preprint arXiv:1812.04608 (2018)
28. Huotari, K., Hamari, J.: Defining gamification: a service marketing perspective. In: Proceeding of the 16th International Academic MindTrek Conference, pp. 17–22 (2012)
29. Jing, Y., Yang, Y., Feng, Z., Ye, J., Yu, Y., Song, M.: Neural style transfer: a review, October 2018. arXiv: 1705.04058
30. Jing, Y., Yang, Y., Feng, Z., Ye, J., Yizhou, Yu., Song, M.: Neural style transfer: a review. IEEE Trans. Vis. Comput. Graph. **26**(11), 3365–3385 (2019)
31. Karras, T., Laine, S., Aila, T.: A style-based generator architecture for generative adversarial networks. In: Proceedings of the IEEE/CVF Conference on Computer Vision and Pattern Recognition, pp. 4401–4410 (2019)
32. Karras, T., Laine, S., Aila, T.: A style-based generator architecture for generative adversarial networks, March 2019
33. Kostoska, G., Baez, M., Daniel, F., Casati, F.: Virtual, remote participation in museum visits by older adults: a feasibility study. In: 8th International Workshop on Personalized Access to Cultural Heritage (PATCH 2015), ACM IUI 2015, pp. 1–4 (2015)
34. Kostoska, G., Vermeeren, A.P.O.S., Kort, J., Gullström, C.: Video-mediated participation in virtual museum tours for older adults. In: 10th International Conference on Design & Emotion, 27–30 September 2016, Amsterdam. The Design & Emotion Society (2016)
35. Kulkarni, R., Gaikwad, R., Sugandhi, R., Kulkarni, P., Kone, S.: Survey on deep learning in music using GAN. Int. J. Eng. Res. Technol. **8**(9), 646–648 (2019)
36. Lamas, A., et al.: MonuMAI: dataset, deep learning pipeline and citizen science based app for monumental heritage taxonomy and classification. Neurocomputing **420**, 266–280 (2021)
37. Lang, B.: The Concept of Style. Cornell University Press, Ithaca (1987)
38. Latour, G.: Guernica, histoire secrète d'un tableau. Média Diffusion (2013)
39. Negrevergne, B., Lecoutre, A., Yger, F.: Recognizing art style automatically with deep learning. In: Proceedings of Machine Learning Research, no. 77, pp. 327–342 (2017)
40. Mao, H., Cheung, M., She, J.: DeepArt: learning joint representations of visual arts. In: Proceedings of the 25th ACM International Conference on Multimedia, pp. 1183–1191 (2017)

41. Mualla, Y.: Explaining the behavior of remote robots to humans: an agent-based approach. Ph.D. thesis, University of Burgundy - Franche-Comté, Belfort, France (2020). 2020UBFCA023
42. Mualla, Y., Kampik, T., Tchappi, I.H., Najjar, A., Galland, S., Nicolle, C.: Explainable agents as static web pages: UAV simulation example. In: Calvaresi, D., Najjar, A., Winikoff, M., Främling, K. (eds.) EXTRAAMAS 2020. LNCS (LNAI), vol. 12175, pp. 149–154. Springer, Cham (2020). https://doi.org/10.1007/978-3-030-51924-7_9
43. Mualla, Y., Najjar, A., Kampik, T., Tchappi, I., Galland, S., Nicolle, C.: Towards explainability for a civilian UAV fleet management using an agent-based approach. arXiv preprint arXiv:1909.10090 (2019)
44. Mualla, Y., Tchappi, I., Najjar, A., Kampik, T., Galland, S., Nicolle, C.: Human-agent explainability: an experimental case study on the filtering of explanations. In: Proceedings of the 12th International Conference on Agents and Artificial Intelligence - Volume 1: HAMT, pp. 378–385. INSTICC, SciTePress (2020)
45. Xin Ning, Fangzhe Nan, Shaohui Xu, Lina Yu, and Liping Zhang. Multi-view frontal face image generation: A survey. Concurrency and Computation: Practice and Experience, page e6147, 2020
46. Perry, L.C.: Reminiscences of Claude Monet from 1889 to 1909. Am. Mag. Art **XVIII**(3), 123 (1927)
47. Pisoni, G., Díaz-Rodríguez, N., Gijlers, H., Tonolli, L.: Human-centred artificial intelligence for designing accessible cultural heritage. Appl. Sci. **11**(2), 870 (2021)
48. Raento, P., Watson, C.J.: Gernika, Guernica, Guernica?: contested meanings of a Basque place. Polit. Geogr. **19**(6), 707–736 (2000)
49. Ryan, R.M., Deci, E.L.: Self-determination theory and the facilitation of intrinsic motivation, social development, and well-being. Am. Psychol. **55**(1), 68 (2000)
50. Seaborn, K., Fels, D.I.: Gamification in theory and action: a survey. Int. J. Hum.-Comput. Stud. **74**, 14–31 (2015)
51. Singh, R., et al.: Directive explanations for actionable explainability in machine learning applications. arXiv preprint arXiv:2102.02671 (2021)
52. Stein, G.: Autobiography of Alice Toklas 1907–1914. The Library of America (1933)
53. Strezoski, G., Worring, M.: OmniArt: a large-scale artistic benchmark. ACM Trans. Multimed. Comput. Commun. Appl. (TOMM) **14**(4), 1–21 (2018)
54. Tan, W.R., Chan, C.S., Aguirre, H.E., Tanaka, K.: Ceci n'est pas une pipe: a deep convolutional network for fine-art paintings classification. In: 2016 IEEE International Conference on Image Processing (ICIP), pp. 3703–3707. IEEE (2016)
55. Toshpulatov, M., Lee, W., Lee, S.: Generative adversarial networks and their application to 3D face generation: a survey. Image Vis. Comput. **108**, 104119 (2021)
56. Trotta, R., Hajas, D., Camargo-Molina, J.E., Cobden, R., Maggioni, E., Obrist, M.: Communicating cosmology with multisensory metaphorical experiences. J. Sci. Commun. **19**(2) (2020)
57. Wölfflin, H.: Principles of Art History (1915)
58. Zhang, S., Han, Z., Lai, Y.-K., Zwicker, M., Zhang, H.: Stylistic scene enhancement GAN: mixed stylistic enhancement generation for 3D indoor scenes. Vis. Comput. **35**(6), 1157–1169 (2019). https://doi.org/10.1007/s00371-019-01691-w
59. Zhao, Y., Wu, S., Reynolds, L., Azenkot, S.: The effect of computer-generated descriptions on photo-sharing experiences of people with visual impairments. In: Proceedings of the ACM on Human-Computer Interaction, vol. 1, no. CSCW, pp. 1–22 (2017)

60. Zhu, J.-Y., Park, T., Isola, P., Efros, A.A.: Unpaired image-to-image translation using cycle-consistent adversarial networks. In: Proceedings of the IEEE International Conference on Computer Vision, pp. 2223–2232 (2017)
61. Zhu, J.-Y., Park, T., Isola, P., Efros, A.A.: Unpaired image-to-image translation using cycle-consistent adversarial networks, August 2020

Assessing Explainability in Reinforcement Learning

Amber E. Zelvelder(✉), Marcus Westberg, and Kary Främling

Umeå University, Umeå, Sweden
{amber.zelvelder,marcus.westberg,kary.framling}@umu.se

Abstract. Reinforcement Learning performs well in many different application domains and is starting to receive greater authority and trust from its users. But most people are unfamiliar with how AIs make their decisions and many of them feel anxious about AI decision-making. A result of this is that AI methods suffer from trust issues and this hinders the full-scale adoption of them. In this paper we determine what the main application domains of Reinforcement Learning are, and to what extent research in those domains has explored explainability. This paper reviews examples of the most active application domains for Reinforcement Learning and suggest some guidelines to assess the importance of explainability for these applications. We present some key factors that should be included in evaluating these applications and show how these work with the examples found. By using these assessment criteria to evaluate the explainability needs for Reinforcement Learning, the research field can be guided to increasing transparency and trust through explanations.

Keywords: Reinforcement Learning · Explainable AI · XAI · Interpretable Machine Learning

1 Introduction

One key obstacle hindering the full scale adoption of Machine Learning, including Reinforcement Learning (RL), is its inherent opaqueness. This prevents these 'black box' approaches (i.e.; systems that hide their inner logic from users) from becoming more widespread and receiving greater authority and trust in decisions. Being opaque and having no explanations for why the autonomous agent takes an action or makes a decision can cause both practical and ethical issues [18,19]. With the recent increased calls for transparency in computer-based autonomous decision-making, there has been a surge in research into making Machine Learning algorithms more transparent. RL is often thought to not need further transparency if the reward condition is known, but in RL there is a variety of applications, each with their own set of interactions, recommendations etc. These different applications of RL will also have different needs when it comes to explainability.

D. Calvaresi et al. (Eds.): EXTRAAMAS 2021, LNAI 12688, pp. 223–240, 2021.
https://doi.org/10.1007/978-3-030-82017-6_14

This paper will seek to review how different application domains might affect explainability, and assess applications within these domains to determine the amount and type of explainability. We will start by setting out the background of RL, Explainable AI (XAI) and the state of XAI in RL. In this background section, we will go into detail on the main application areas in which RL is being used, or can potentially be used, followed by an overview of the types of explainability and then which types of XAI have been applied to RL. Then we show our methodology to find examples of RL applications. We will then present possible evaluation criteria and key factors for assessing explanations in RL. After that we will make an assessment of the most notable examples of applications we have identified in RL. Finally, we will conclude with a summary of our findings and present what we believe are the remaining challenges that future work could be based on.

2 Background

In RL, an algorithm learns dynamically from its environment and is driven by either a reward or penalty being given when reaching or being in a specific state or states [24,46]. Because of the way RL algorithms learn, through maximising the final reward, it is particularly suited for problems that require a solution that weighs the short term outcome against the long term outcome [46]. Robotics is the application domain that RL is the most prominent in by far [25], but it has gradually started to see more extensive use in control systems [29] and networking [33]. RL has also been showcased publicly as highly proficient in playing and winning a variety of games, such as GO , checkers and video games [4,12].

With the rise of deep learning, it has been made possible to scale RL to attempt tackling decision-making problems on a larger scale [3]. But with this increase in applications being developed using AI such as RL, there has also been an increase in demand for more transparency [50]. In addition to there being legal calls for transparency, there is also the argument that if autonomous agents can be clear about the reasons for their actions, this would help build rapport, confidence and understanding between the AI agents and human operators, thereby increasing the acceptability of the systems and enhancing end-user satisfaction [2,18].

Just like RL, XAI started gaining increased popularity in the 90s with symbolic reasoning systems, such as MYCIN. The interest in XAI remained mostly academic until the rise of Machine Learning (ML) and its involvement in making increasingly important decisions. The interest in explanations really took off after some concerns regarding bias within Machine Learning. Among the more well known and publicised cases of bias are the Amazon recruitment algorithm that advised against hiring women [38], and Flickr's image tagging algorithm that tagged people of some ethnicities as animals or objects [52]. A more common bias within RL is the possibility of model bias, where the learning environment is too different from the intended target environment [10]. Now the interest level in XAI is high due to contemporary trust issues and ethics debate in the field of autonomous AI decision-making and the legal debate and requirements that are being imposed as a result. Despite the increased interest in XAI and the widespread implementation of RL in both research and industry, ways to implement explanations into RL have not been thoroughly researched.

One of the reasons lack of active development in RL explanations is because there is an underlying assumption that knowing the reward for RL is explicable enough. Another reason is that RL is often used in more mechanical situations, rather then conversational, in which case there might not be a user to directly interact with or the user is an expert to whom the actions are explicable when combined with observations of the environment.

In the following sections we will present the most prominent Application Domains where RL is used, the types of XAI that are relevant to RL and the current existing XAI techniques implemented for RL.

2.1 RL Main Application Domains

RL can be considered to be in its early stages of development when it comes to applications. The majority of works in the literature use simple test scenarios that are not always representative of real-life needs, but as some of these could be considered simplified versions of general applications, we will use some simplified examples to illustrate the potential use of RL in each application domain cluster outlined. We will rely on several examples from the recent edition of Sutton and Barto's book [46] on RL in order to identify the main clusters of RL applications. We will also include references to more recent works in literature where appropriate. These will be primarily used to illustrate the potential of RL within specific domain clusters.

Physics. Numerous examples of applied RL involve simplified Physics tasks. This is because we have easy access to mathematical models of the laws of physics, which allow us to build simulated environments in which the algorithm is tested. The simplified and well-known examples of this category (e.g.; cart-pole [46]) can also be moved to a more applied domain. Another example is the mountain-car task, where an under-powered vehicle surrounded by two hills needs to get up a hill but doesn't have the power to ascend it without gathering momentum by backing up the other way first [13,46]. This example could serve in the optimization of fuel use in actual cars in that situation if further developed.

Robotics. Due to it's similarity to the natural learning process, RL is well-suited to train the movement of robots, particularly the optimization of reaching movement goals. Because of this, robotics is an application area where RL has been thoroughly trialled and successfully applied for a variety of purposes [27]. Robotics usually uses a delayed or continuous reward, as it is the set of actions they seek to reward, rather than individual actions. One common application is to use it for path optimization, where the reward is given if the destination is reached (and sometimes a bigger reward if it didn't take long). Many examples of successful applications of RL in Robotics can be found in the survey done by Kober *et al.* [25].

Games. Playing games is one of the ways RL is most well-known to people outside of the Computer Science expertise. RL has been shown to have great potential at learning games, because most games have a clear reward structure of winning the game. There have been cases of RL-trained algorithms being able to beat the (human) masters of the games Go and Chess [4]. There has been further research in making algorithms that can tackle multiple games [45]. Research into RL algorithms that perform video games on a human level (e.g.; With the same tools and at the same or higher skill level as a person) is also being performed, and has had some success [4,12]. Deep RL is also being used in newer research into using machine learning to play games and videogames, including videogames with or against other human players, with promising preliminary results [3].

Autonomous Vehicles and Transport. Although most autonomous vehicles use supervised learning to make sure they learn the correct rules, there has been research into using Reinforcement Learning instead [43]. Also, in the transport sector there are related systems that are using machine learning. For instance, transport systems use route optimization in a way similar to robotics, and there are also urban development applications that use machine learning to manage traffic control [30,35].

Healthcare. The healthcare sector is increasingly utilising ML in their systems, and this includes RL [20]. A lot of the current work on bringing machine learning to the healthcare sector is still in the early stages, but ML algorithms that enhance Computer Vision are already used commercially in diagnostics, medical imaging and surgery, to supplement the medical personnel [17,28]. With ML starting to influence the sector, and the motivation to improve this sector being high, new research is performed continuously to increase the successful use of ML. While ML within healthcare currently mostly supports the experts and professionals, research is also taking more interest in developing algorithms that interact directly with the (potential) patients, to either help them know what doctor they need, whether they are at risk of special ailments, or how to manage their health, either generally or with a specific condition [39].

Finance Predictions. Some areas of finance are doing research into how RL and other machine learning methods could be used to predict developments in aspects of the market [40]. As the finance sector already extensively uses rule- and trend-based models to try to stay ahead of the curve, machine learning is a natural next step for finance. Most applications within finance follow current trends and make estimations based on historical data [6].

Other. There are a few other domains in which RL development is prominent, which will not be as obviously relevant to this paper, but still are worth mentioning as possible domains to look into at a later point. Reinforcement learning is used in industry to try to alert when machines need preventative maintenance [9,15]. It is also used to assist with elevator scheduling [7,53] using a continuous-time Markov chain [16]. Other ways RL is used is in various types of optimization, examples including network communication optimization and general network management [16], optimising/minimising resource consumption [29] and optimising memory control [4]. RL is also being used to automatically optimise the web data shown to users by advertisers [4].

2.2 XAI Types

Sheh [44] proposes a way of categorising the types of explanations that are most commonly required for AI in different contexts. Further research into this has been done by Anjomshoae et al. [2], leading to a distinction of several types of explanations that are currently used in AI.

Teaching explanations aim to teach humans (General users, domain experts, AI experts) about the concepts that the AI has learned. These explanations don't always need to be accompanied by a decision, as the teaching is what is at the core of the explanation. These explanations can take the form of hypotheticals, for example as answers to follow-up questions regarding a previous explanation (i.e. "if parameter X was different, how would this have affected the decision?").

A possible subtype of teaching explanations are **contrastive explanations**, where the hypotheticals involve showing the user why the decision is better by contrasting it with the poorer choice(s). There has already been research into how contrastive explanations can be used in RL [2,49]. In this research, the possible consequences of other actions were generated to show them in contrast to what decision was made [36].

There are **introspective explanations** in the form of **tracing explanations** and **informative explanations**. The former is a trace of internal events and actions taken by the AI, the purpose of which is to provide a complete account (of desired granularity) of the decision process to track down faults or causes behind incorrect decisions. The latter type involves explaining discrepancies between agent decisions and user expectation by looking at the process behind the given decision, the purpose here being to improve human-robot and human-system interaction by either pointing out where an error may have occurred or convince the user that the agent is correct. What both have in common is that they draw directly from the underlying models and decision-making processes of the agent. Due to the nature of ML, these kinds of introspective explanations are often not compatible with such techniques, and in the few cases that they might be, they will not be complete [44].

By contrast, **post-hoc explanations** provide rationalisations of the decision-making process without true introspection. These explanations are helpful in models where tracing underlying processes is not an option, such as with black-box models. These explanations can be derived from a simulation of what the underlying processes might be like, sometimes working from a parallel model that attempts to sufficiently approximate the hidden model.

Execution explanations are the simplest form of explanations, presenting the action or set of actions that the AI agent undertook. Similar to tracing explanations, this type of explanation provides a history of events, but does so by listing the explicit operations undertaken.

Post-hoc and execution explanations, together with teaching explanations, form the three types of explanations most compatible with RL, though their usefulness varies. Post-hoc and execution explanations provide methods of tracing and forming narratives regarding decision-making without having to "look inside the box". In turn, teaching explanations can help users understand the model of the application better.

2.3 Explainable Reinforcement Learning

In RL there is not necessarily a reason for the algorithm to take an action or make a suggestion. The actions or decisions made by the algorithm are often based on the experience that algorithm builds. Because of this there has been a particular interest in contrastive, post-hoc and tracing explanations for RL. Since gaming applications have been a very popular domain for RL, and games are generally a low-risk activity, it is a very popular domain to test new techniques in. Videogames have been used to test the understanding and clarity of saliency maps as explanations [1,21]. A numerical explanation has also been studied using videogames, and these studies also include user feedback to assess the quality of the explanations [11,21,42]. Another method of creating explanations for RL has been to amend the RL algorithm, to maintain an amount of "memory" [8], which provides a form of tracing explanation. This has also been implemented in a way to provide a visual representation of the internal memory of an agent by Jaunet et al.[23], which can be used a combination between a teaching and a tracing explanation. There is also some research looking into transparent or interpretable models, such as PIRL, hierarchical policies and LMUT [41].

3 Methodology

In order to make sure that our proposed evaluation criteria are thorough and applicable in existing application domains, we performed a literature survey of RL survey papers and books to extract as many examples as possible of RL being used in practice, or having a potential use in practice. We have assessed 19 pieces of literature, of which 14 contained viable examples. In total we noted 91 examples of possible applications. The levels of detail of each of these applications varied and some refer to the same application. In Fig. 1 we show a breakdown of the number of examples found in the more prominent sources.

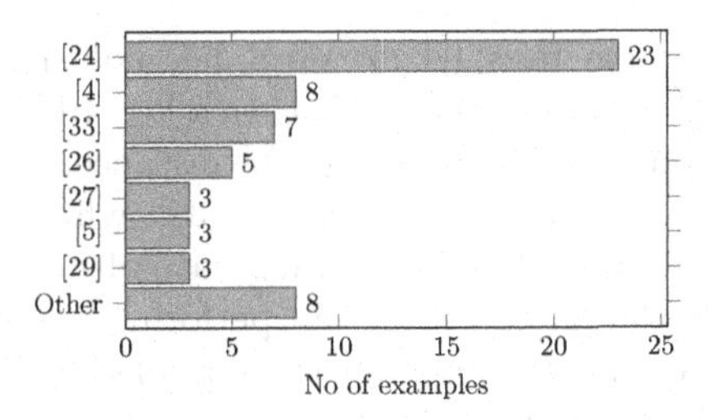

Fig. 1. Number of examples by source

For each survey paper we assessed whether the methods and applications found were applied to any scenario that could be used in practice. For each of these we made note of the survey papers they were mentioned in, their general application domain, a description of the application and, if mentioned, the RL method and the reward setup. After we had listed these, we assessed similarities between different listings and merged them in our data where appropriate, preserving the multiple sources. We were then left with 50 examples that could be evaluated. After these examples were collected, we found that they were in 8 different categories (Fig. 2). In this we found 16 were in the robotics domain, 8 each in the games, control systems and networking domains, 3 in autonomous vehicles and 2 in both finance and healthcare.

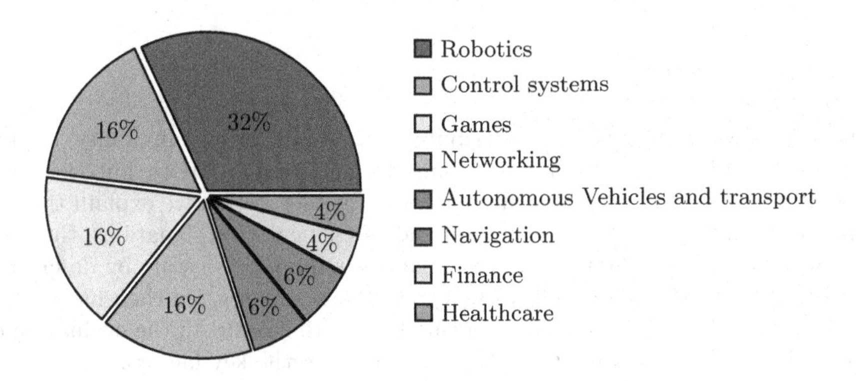

Fig. 2. Application domains of Practical examples

3.1 Analysis of Application Domains

Many of the examples in **robotics** show that RL can be used for a robotic agent to master its understanding of physics by achieving balance of itself or another object, or making some other type of adjustments based on gravity and other laws of physics [4,13,24,26,27]. Another type of robotics that frequently uses RL is robots that move around and perform a task, such as moving objects [4,24,26] or finding an exit [4,31]. There are also robots that use RL to play ball games

[5,26,31] and one example uses RL to do a peg-hole insertion [26]. Another important point to make about robotics is that a lot of RL applications in other domains can also be connected to robotics. For instance, in healthcare one of the applications is gesture reading and replication [32], which would be a combination of computer vision and robotics. **Games** have often been used to showcase how well a machine learning algorithm can perform, as with TD-Gammon and Samuel's Checker player beating human masters of the games of backgammon and checkers [4,24]. Most applications in this domain are one versus one board games. There are also applications that use RL for playing videogames [4,22]. The examples within the **control systems** domain found were primarily in optimization [4,24], resource allocation [4,24,29] and task scheduling [4,24,29]. A lot of these applications are located in factory settings, but there are also examples in smart homes and Internet-of-Things settings. Most of the **networking** applications are related to access control, caching and connection preservation [4,33]. There was one example relating to personalising web services [4] and network security [33]. In **Autonomous vehicles and Transport**, the applications are primarily about avoiding collision and interpreting other traffic [5,26,33,43]. **Navigation** is very similar to the Autonomous vehicles, with the main difference being that traffic is not necessarily a problem [4,14,24]. In **finance** we found two examples, one involved creating economic models [4] and the other was an automated trading application [5]. In **healthcare**, we found the application mentioned before that reads and replicates surgical gestures [32] and an application that detects and maps a person's bloodvessels [54].

4 Key Criteria

There are several factors that contribute to the need for explainability, a large amount of which relate to Human-Computer Interaction (HCI), but there are several other types of factors that contribute. In this section we explain the different criteria that drive the evaluation and assessment of explanations, the need for explanations and the types of explanations required. We start by describing four key factors (Fig. 3) which we will use for assessments, we then have a few sections to highlight notable contributing factors that assist in the evaluation of the key factors and in what way they contribute to the key factors.

Out of the key factors, two are closely related to HCI, User Expectancy and User Expertise, and two of them more related to the consequences of the application, Urgency of the output and Legality. The first key factor, and perhaps the best starting point for assessing the explanations, is the **User Expectancy**. The amount of explanation a user expects is a direct influence on the amount of explanation required, but it is also influenced by the users expectations of the applications actions or outputs [37]. If a user expects little or no explanation, and the application gives the expected output, there is no need for an explanation. But if the user expects an explanation, or if the application is behaving in a way that is outside of the user expectations, a more expansive explanation would be required. Because of the nuances of this factor, it also has a strong

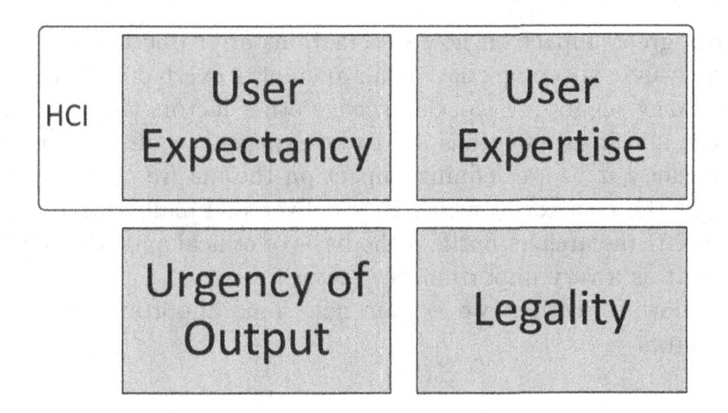

Fig. 3. Four key factors that influence the need for explainability

influence on the type of explanation that would be the most suitable for the application. The other user-connected key factor is, the **User Expertise**. The level of expertise the intended user has varies between every application. This factor mostly dictates the type of explanation to be made available [47]. If the user can be anyone, the explanations will need to be more informative, using general language. If the target users are experts in the application domain, but not on this specific application, an informative explanation that is specific to the domain can be presented. Finally, if the user is an expert on the application and/or any devices the application controls, an explanation that helps the user trace down faults within the system, and explanations of executed actions would be preferred.

A key factor that gives a limitation on the detail and type of explanation given is the **Urgency of the Output**. Longer and detailed explanations also take longer to produce and also take longer for the user to interact with. This is a time-bound scale, which means that it determines if there is time to produce any form of explanation before a decision is taken. If an action needs to be taken immediately, like in an autonomous vehicle, there might not be enough time to explain it to the user, so in this case an explanation can not be expected until later. If there is no urgency, the application can produce a full report and even request the user to approve the decision or action based on the explanation.

Finally, a factor that has gotten more important recently, is the **Legality**. This factor is driven by the laws affecting the application domain, AI in general or the specific case of the application. In this case it matters if it affects or evaluates an individual or group of individuals. If it can do neither there is no legal need for explanations. If it evaluates, then there should be the possibility to produce an explanation of the evaluation, but this can be retrospective. If it affects, it is preferred that an immediate explanation is given, as well as a retrospective explanation being made available.

These key factors were chosen on the grounds that they cover and represent the core areas of XAI concern. Differences in user expertise has shown to affect expectations of output and explanatory content [47], and the context of the audience has great impact on how explanations are to be tailored [37]. In this way, expertise and expectancy have a fair amount of overlap, but in this paper we treat expectancy separately to acknowledge other factors that can also impact expectations, and if explanations are to be expected at all. Urgency of output is chosen because it has a defining impact on the nature of explanation to be provided and the context in which it is delivered. Finally, legality is of great concern for XAI researchers both on the basis of ethical and economic concerns, thus placing it as a very important key factor.

In the following sections we explain how some important criteria feed into these key factors.

4.1 The Intended User

The starting point for any application should be the intended user. This is very important for explanations, as there has to be an individual or group of individuals that the explanation would be intended for. The user affects both HCI-based aspects of the key factors and it also influences both the amount of explanation that should be given, as well as the type of explanations (see Sect. 2.2). The development of an application and explanations for the applications should be started from the user expectations, requirements and expertise.

4.2 Means of Interaction

Any application that includes direct interaction with some kind of user will need some explanation, but this can be limited by the means that the application or the user have to interact. With many RL implementations, there is some limitation to what kind of interaction there can be, and this can limit the possibilities of their communication. For instance, in robotics the application can usually interact by movement or other non-verbal communication, which humans can sometimes intuitively interpret [51]. In applications located in factories or other industrial devices, the means might just be a digital display or a blinking light. In other RL applications the interaction can be done via a monitor, or using audio. This contributing factor has no great influence on the key factors, but is used as a limitation to the type of explanations that can be implemented for an application.

4.3 Industry Sector

The type of industry sector that an application is developed in and for is key in determining the need and nature of the explainability of machine learning and AI systems. The industry often dictates who the intended user is, and what aspects of the system needs automation. For instance, an AI in healthcare can either interact with the medical professionals, or with the patients themselves. But in the manufacturing industry, the user is far more likely to be a person who has expert knowledge on the topic of the algorithm. Although explanations are preferred in most industry sectors, the need for explanations is greater in some than in others. For instance, in the healthcare sector it is very important for AI to explain themselves, because the decision or advice of the RL algorithm could affect the health of an individual. This is the case for all users of AI within healthcare, and it has been shown that people are more likely to trust the health advice of an AI if an explanation and/or motivation is given by the AI that pertains to the patient personally [39]. This is different from the finance sector, where financial predictions are often assumed to be estimates, so a motivation or detailed explanation is not needed as much, but a chart or list of rules as a form of introspective informative explanation to support the prediction would be preferred. This contributing factor strongly influences the HCI aspects of the key factors, as well as the other key factors to a lesser extend, depending on the specific application.

4.4 Urgency/Time-Restraint

The urgency by which an algorithm needs to make a decision has some influence on how much explanation can be expected and/or is needed. If it is, for instance, an AI that drives a car, making the decision to do an emergency brake for a suddenly crossing pedestrian is more important than explaining it. The same goes for a machine stopping production if there is a misalignment in the system. Both of these could be followed up by a longer explanation, after the urgency is reduced. In other examples, such as an algorithm that makes mortgage agreement decisions, time is of less importance than an explanation regarding how a decision was made.

4.5 Legal

This factor is emphasized by recent legislation across the world. As the EU has recently passed a law known as the GDPR [48], which dictates that if an algorithm is fully or partially responsible for any decisions made regarding a person, that person has the right to know the reasons behind the algorithm's conclusion.

4.6 Responsibility

Whether the user, the creator of the application, the manufacturer of a device using the application, or some other individual or organisation has the responsibility for action or decisions chosen by the application is also a driver for wanting or needing explanations of differing types. This contributing factor is tied to the legality key factor.

5 Assessment

Although some of the key factors can be subjective, in Fig. 4 we show an overview of how the unique examples found in Sect. 3 could be classified on a scale of 1–5 on the key factors presented in Sect. 4. This gives an overview of the need for explanations in different sectors and shows how much variability exists in the domains. For a specific application, the details of how the specific key factors are relevant are more important then just their value. Therefore, in this section we will evaluate four of the scenarios we extracted from the examples reviewed in Sect. 3. We are using the key factors from Sect. 4 to perform this evaluation. The examples in this section were chosen because they are from varied application domains and have very different users. They are therefore expected to have very different needs when it comes to explanations.

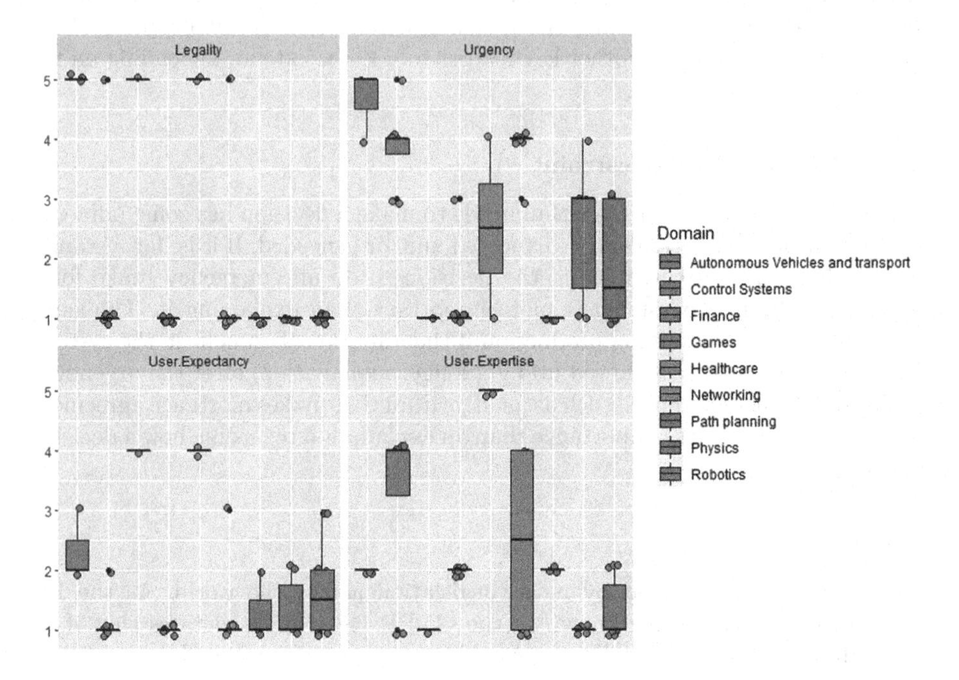

Fig. 4. Figure for classification of known examples on a scale of 1–5

5.1 A Box-Moving Robot

This example was chosen for evaluation because the functionality of the robot is simple and can be kept within the robotics domain, but has the potential of being used in many other domains with minor changes. We will be assuming the robot is an industry-ready adaptation and has therefore the capability to find a (specific) box, move the box, and recover from a position in which the robot itself is stuck in a corner [4,24,26,34]. This application does not have a specific intended user, which presents two options as to who to regard as a user; the person who has given the robot a task to do, or anybody who might be in the area where the robot operates. This implies that the user expertise and expectation of explanations will be low, assuming the robot performs as intended. The urgency in the case of this robot is mostly not time-critical as the robot can wait to make a decision if there is no threat of any kind present. This means that the robot has time to display an indicator or play a sound to facilitate human understanding. Specific design decisions may limit these capacities. For example, the OBELIX robot [34] currently has no means to play sounds or show any kind of display, which means its means of communications are left to physical movements. The robot does not make decisions about people, but has a chance to encounter people within its work space. This means that the only legal requirement for the robot to give an explanation is when it interferes with an individuals actions in any way, such as by bumping into them or obstructing their path with the boxes.

5.2 Cloud Computing Resource Allocation

This example was picked because cloud computing is increasingly popular, and the optimisation of all its procedures is critical to its success. In the cloud computing resource allocation problem, there is a server cluster with a certain amount of physical servers and each of these physical servers can provide a limited number of resources [29]. A job will be processed when enough resources are available, the algorithm is employed to optimise how and when the jobs are allocated and to which machines, to optimise the processing time and minimise the power consumption. In this scenario, a user will be the person who submits a job to the cloud computing. The user will generally have some expertise in cloud computing, but the amount of expertise will vary. A user will typically have little to no expectation of explanations, unless the system has issues with the performance. As one of the goals of the algorithm is to optimise the time, there is no time to explain actions before they are taken. As individuals are not evaluated, there is no legal requirement for explanations. The recommended type of explanation would be a tracing explanation or execution explanation to track down the cause of a fault within the system.

5.3 Frogger Videogame

We are evaluating the frogger videogame as an example from the gaming domain. This example was selected because there already exist an experimental study into explanations [11,42] for this game and because of its iconic reputation. In the game of Frogger you need to guide a frog from one side of a map to the other. In the first part, the frog needs to avoid being hit by a car, and at the second part the frog needs to jump between moving logs to reach the other side of a river. Since this is a game being played by the RL algorithm, there is no user, only observers. We can assume the user knows the rules of the videogame, but has no further expertise. As the objects other than the frog move in real-time, that is the time sensitivity required for the reactions. The game doesn't make decisions on individuals, so there is no legal requirement for explanations. Because of academic interest, the explanations currently being developed for Frogger are to indicate, either numerically or by description, what observations of the AI are being most relevant in its decision at any point during the run. This application is therefore often used to perform benchmarking of explanation techniques and user studies.

5.4 Surgical Robot

The example of a surgical robot has been chosen because it has a strong contrast with the other examples so far. The specific example we use is of a robot performing a suturing task [32]. In the referenced paper this is being done in a simulated environment, but since the goal is to let it perform in a medical environment, we shall evaluate it as such. The robot is performing a medical operation (suture) on a person (the patient), and is being monitored by another person (the doctor). The intended user is the doctor, who will have a high amount of medical expertise to assess that the robot does the correct procedures, and will also be capable of spotting any mistakes made. In the application, a display of certain parameters was included, so this display could be considered a starting point for an explanation. The display in the example indicated how accurate the expected kinaesthetic response was compared to the actual, which can indicate a possible problem if the accuracy is too far off. Since the performance of this robot directly affects the health of an individual, this means that legally there is a strict requirement for explanations to be available prior to use on a patient.

6 Discussion

There are still issues that prevent RL from more widespread application in general, with the core issues being centered around the inability to adapt, lack of correlation, application complexity, increasingly larger and more complex data and the narrow focus of current XAI techniques [27,51]. In RL there is a reward state that can always help clarify the goal that the algorithms are working towards, so one of the important steps towards explainability is to make sure the reward states can be viewed by the user in a way the typical user can understand. As RL algorithms can sometimes use very complicated reasoning, which might be beyond what a human understands, it can be hard for an algorithm to be accompanied by an explanation that a human can easily follow. This is further complicated if the system has to make several actions in succession that each require a longer explanation than a human would be able to keep up with. In deep RL this becomes even more of a problem as the input and the parameters are very expansive, making it harder to tie specific outcomes to specific parameters. Another challenge is that an explanation must be good enough that a human can help keep it accountable. Related to this is the problem of who can be held responsible for any wrong decision that is made as a consequence of an explanation being insufficient. Because XAI is still in early stages of implementation within machine learning and even more so when it comes to RL, due to RL having a more limited application area, there is very little existing research in XAI for RL. In RL most current research into explainability is focusing on 'recommender' systems. However, as presented earlier in this paper, RL is used or will soon be used in many other critical areas where explanations might be required in various degrees.

7 Conclusion

This paper has shown which application domains RL is most used in, and why explainability is important in RL. It has also presented guidelines that can be used to evaluate the explainability needs for specific applications. The guidelines are centered around the HCI aspects of user expectations and expertise as well as the urgency of the output and the legal requirements where applicable. We have assessed various notable applications that use RL algorithms using the guidelines provided. As we continue to work towards Explainable RL, the guidelines set out in this paper will help identify the need for explainability in new RL applications.

Acknowledgement. This work was partially supported by the Wallenberg AI, Autonomous Systems and Software Program (WASP) funded by the Knut and Alice Wallenberg Foundation. We would also like to thank Amro Najjar, for providing guidance on starting this paper.

References

1. Anderson, A., et al.: Explaining reinforcement learning to mere mortals: An empirical study. Proceedings of the Twenty-Eighth International Joint Conference on Artificial Intelligence, August 2019. http://dx.doi.org/10.24963/ijcai.2019/184
2. Anjomshoae, S., Najjar, A., Calvaresi, D., Främling, K.: Explainable agents and robots: Results from a systematic literature review. In: Proceedings of the 18th International Conference on Autonomous Agents and MultiAgent Systems. p. 1078–1088. AAMAS '19, International Foundation for Autonomous Agents and Multiagent Systems, Richland, SC (2019)
3. Arulkumaran, K., Deisenroth, M.P., Brundage, M., Bharath, A.A.: A Brief Survey of Deep Reinforcement Learning. IEEE Signal Processing Magazine, Special Issue on Deep Learning for Image Understanding p. 16 (aug 2017)
4. Barto, A., Thomas, P., Sutton, R.: Some Recent Applications of Reinforcement Learning. Workshop on Adaptive and Learning Systems (2017)
5. Busoniu, L., Cluj-napoca, U.T., Babuska, R., Schutter, B.D.: Innovations in Multi-Agent Systems and Applications - 1, vol. 310. Springer Nature (2010)
6. Choi, J.J., Laibson, D., Madrian, B.C., Metrick, A.: Reinforcement learning and savings behavior. The Journal of Finance **64**(6), 2515–2534 (2009)
7. Crites, R.H., Barto, A.G.: Elevator group control using multiple reinforcement learning agents. Machine Learning **33**(2), 235–262 (1998)
8. Cruz, F., Dazeley, R., Vamplew, P.: Memory-based explainable reinforcement learning. In: Liu, J., Bailey, J. (eds.) AI 2019. LNCS (LNAI), vol. 11919, pp. 66–77. Springer, Cham (2019). https://doi.org/10.1007/978-3-030-35288-2_6
9. Das, T.K., Gosavi, A., Mahadevan, S., Marchalleck, N.: Solving semi-Markov decision problems using average reward reinforcement learning. Manage. Sci. **45**(4), 560–574 (1999)

10. Deisenroth, M., Rasmussen, C.: Reducing model bias in reinforcement learning (12 2010)
11. Ehsan, U., Tambwekar, P., Chan, L., Harrison, B., Riedl, M.: Automated rationale generation: a technique for explainable AI and its effects on human perceptions, pp. 263–274 (03 2019). https://doi.org/10.1145/3301275.3302316
12. Erev, B.I., Roth, A.E.: Predicting how people play games?: reinforcement learning in experimental games with unique, mixed strategy equilibria. Am. Econ. Rev. **88**(4), 848–881 (1998)
13. Främling, K.: Light-weight reinforcement learning with function approximation for real-life control tasks. Proceedings of the 5th International Conference on Informatics in Control, Automation and Robotics, Intelligent Control Systems and Optimization (ICINCO-ICSO) (2008)
14. Garcia, J., Fernandez, F.: A comprehensive survey on safe reinforcement learning. J. Mach. Learn. Res. **16**, 1437–1480 (2015)
15. Gosavi, A.: Reinforcement learning for long-run average cost. Eur. J. Oper. Res. **155**(3), 654–674 (2004). Traffic and Transportation Systems Analysis
16. Gosavi, A.: Reinforcement learning: a tutorial survey and recent advances. INFORMS Journal of Computing **21**, 178–192 (2018)
17. Gupta, M., Konar, D., Bhattacharyya, S., Biswas, S. (eds.): Computer Vision and Machine Intelligence in Medical Image Analysis. AISC, vol. 992. Springer, Singapore (2020). https://doi.org/10.1007/978-981-13-8798-2
18. Hellström, T., Bensch, S.: Understandable robots-what, why, and how. Paladyn J. Behav. Robot. **9**(1), 110–123 (2018)
19. Hendricks, L.A., Akata, Z., Rohrbach, M., Schiele, B., Darrell, T.: Generating Visual Explanations
20. Hinto, G.: Deep learning - a technology with the potential to transform healthcare. JAMA **320**, 1101–1102 (2018)
21. Huber, T., Limmer, B., André, E.: Benchmarking perturbation-based saliency maps for explaining deep reinforcement learning agents. arXiv preprint arXiv:2101.07312 (2021)
22. Jaderberg, M., et al.: Human-level performance in 3D multiplayer games with population-based reinforcement learning. Science **364**(6443), 859–865 (2019)
23. Jaunet, T., Vuillemot, R., Wolf, C.: DRLViz: understanding decisions and memory in deep reinforcement learning. In: Computer Graphics Forum, vol. 39 (2020)
24. Kaelbling, L.P., Littman, M.L., Moore, A.W.: Reinforcement learning: a survey. J. Artif. Intell. Res. **4**, 237–285 (1996)
25. Kober, J., Bagnell, A.J., Peters, J.: Reinforcement learning in robotics: a survey. Reinforcement Learn. **32**, 1238–1274 (2012)
26. Kober, J., Bagnell, J.A., Peters, J.: Reinforcement learning in robotics: a survey. Int. J. Robot. Res. **32**, 1238–1274 (2013)
27. Kormushev, P., Calinon, S., Caldwell, D.: Reinforcement learning in robotics: applications and real-world challenges. Robotics **2**(3), 122–148 (2013)
28. Law, H., Ghani, K., Deng, J.: Surgeon technical skill assessment using computer vision based analysis. In: Doshi-Velez, F., Fackler, J., Kale, D., Ranganath, R., Wallace, B., Wiens, J. (eds.) Proceedings of the 2nd Machine Learning for Healthcare Conference. Proceedings of Machine Learning Research, vol. 68, pp. 88–99. PMLR, Boston, Massachusetts, 18–19 August 2017
29. Li, H., Wei, T., Ren, A., Zhu, Q., Wang, Y.: Deep reinforcement learning: Framework, applications, and embedded implementations: invited paper. In: IEEE/ACM International Conference on Computer-Aided Design, Digest of Technical Papers, ICCAD 2017-November, pp. 847–854 (2017)

30. Liang, X., Du, X., Wang, G., Han, Z.: Deep Reinforcement Learning for Traffic Light Control in Vehicular Networks. arXiv e-prints, March 2018
31. Littman, M.L.: Markov games as a framework for multi-agent reinforcement learning. Mach. Learn. Proc. **1994**, 157–163 (1994)
32. Liu, D., Jiang, T.: Deep reinforcement learning for surgical gesture segmentation and classification. In: Frangi, A.F., Schnabel, J.A., Davatzikos, C., Alberola-López, C., Fichtinger, G. (eds.) MICCAI 2018. LNCS, vol. 11073, pp. 247–255. Springer, Cham (2018). https://doi.org/10.1007/978-3-030-00937-3_29
33. Luong, N.C., Hoang, D.T., Gong, S., Niyato, D., Wang, P., Liang, Y.C., Kim, D.I.: Applications of deep reinforcement learning in communications and networking: a survey. IEEE Commun. Surv. Tutorials **21**(4), 3133–3174 (2019)
34. Mahadevan, S., Connell, J.: Automatic programming of behavior-based robots using reinforcement learning. Artif. Intell. **55**(2), 311–365 (1992)
35. Mannion, P., Duggan, J., Howley, E.: Parallel reinforcement learning for traffic signal control. Procedia Comput. Sci. **52**, 956–961 (2015). The 6th International Conference on Ambient Systems, Networks and Technologies (ANT-2015), the 5th International Conference on Sustainable Energy Information Technology (SEIT-2015)
36. Miller, T.: Contrastive explanation: a structural-model approach. CoRR abs/1811.03163 (2018)
37. Miller, T.: Explanation in artificial intelligence: insights from the social sciences. Artif. Intell. **267**, 1–38 (2019)
38. Mujtaba, D.F., Mahapatra, N.R.: Ethical considerations in AI-based recruitment. In: 2019 IEEE International Symposium on Technology and Society (ISTAS), pp. 1–7. IEEE (2019)
39. Neerincx, M.A., van der Waa, J., Kaptein, F., van Diggelen, J.: Using perceptual and cognitive explanations for enhanced human-agent team performance. In: Harris, D. (ed.) EPCE 2018. LNCS (LNAI), vol. 10906, pp. 204–214. Springer, Cham (2018). https://doi.org/10.1007/978-3-319-91122-9_18
40. Nevmyvaka, Y., Feng, Y., Kearns, M.: Reinforcement learning for optimized trade execution. In: Proceedings of the 23rd International Conference on Machine Learning, ICML 2006, pp. 673–680. ACM, New York (2006)
41. Puiutta, E., Veith, E.M.S.P.: Explainable reinforcement learning: a survey. In: Holzinger, A., Kieseberg, P., Tjoa, A.M., Weippl, E. (eds.) CD-MAKE 2020. LNCS, vol. 12279, pp. 77–95. Springer, Cham (2020). https://doi.org/10.1007/978-3-030-57321-8_5
42. Sequeira, P., Gervasio, M.: Interestingness elements for explainable reinforcement learning: understanding agents' capabilities and limitations. Artif. Intell. **288**, 103367 (2020)
43. Shalev-Shwartz, S., Shammah, S., Shashua, A.: Safe, Multi-Agent, Reinforcement Learning for Autonomous Driving (2016)
44. Sheh, R.: Different XAI for different HRI. In: AAAI Fall Symposium Technical Report, pp. 114–117 (2017)
45. Silver, D., et al.: A general reinforcement learning algorithm that masters chess, shogi, and Go through self-play. Science **362**, 1140–1144 (2018)
46. Sutton, R.S., Barto, A.G.: Reinforcement Learning: An Introduction, 2nd edn. The MIT Press, Cambridge (2018)
47. Szymanski, M., Millecamp, M., Verbert, K.: Visual, Textual or Hybrid: The Effect of User Expertise on Different Explanations, pp. 109–119. Association for Computing Machinery, New York (2021). https://doi.org/10.1145/3397481.3450662

48. Voigt, P., von dem Bussche, A.: The EU General Data Protection Regulation (GDPR). Springer, Cham (2017). https://doi.org/10.1007/978-3-319-57959-7
49. van der Waa, J., van Diggelen, J., van den Bosch, K., Neerincx, M.A.: Contrastive explanations for reinforcement learning in terms of expected consequences. CoRR abs/1807.08706 (2018)
50. Wachter, S., Mittelstadt, B., Russell, C.: Counterfactual explanations without opening the black box: automated decisions and the GDPR. Harv. JL & Tech. **31**, 841 (2017)
51. Westberg, M., Zelvelder, A., Najjar, A.: A historical perspective on cognitive science and its influence on XAI research. In: Calvaresi, D., Najjar, A., Schumacher, M., Främling, K. (eds.) EXTRAAMAS 2019. LNCS (LNAI), vol. 11763, pp. 205–219. Springer, Cham (2019). https://doi.org/10.1007/978-3-030-30391-4_12
52. Yapo, A., Weiss, J.: Ethical implications of bias in machine learning. In: Proceedings of the 51st Hawaii International Conference on System Sciences (2018)
53. Yuan, X., Buşoniu, L., Babuška, R.: Reinforcement learning for elevator control. IFAC Proc. Vol. **41**(2), 2212–2217 (2008). 17th IFAC World Congress
54. Zhang, P., Wang, F., Zheng, Y.: Deep reinforcement learning for vessel centerline tracing in multi-modality 3D volumes. In: Frangi, A.F., Schnabel, J.A., Davatzikos, C., Alberola-López, C., Fichtinger, G. (eds.) MICCAI 2018. LNCS, vol. 11073, pp. 755–763. Springer, Cham (2018). https://doi.org/10.1007/978-3-030-00937-3_86

XAI Logic and Argumentation

Schedule Explainer: An Argumentation-Supported Tool for Interactive Explanations in Makespan Scheduling

Kristijonas Čyras[1,2(✉)], Myles Lee[2], and Dimitrios Letsios[2,3]

[1] Ericsson Research, Stockholm, Sweden
kristijonas.cyras@ericsson.com
[2] Imperial College London, London, UK
myles.lee15@imperial.ac.uk
[3] King's College London, London, UK
dimitrios.letsios@kcl.ac.uk

Abstract. Scheduling is a fundamental optimisation problem that has a wide range of practical applications. Mathematical formulations of scheduling problems allow for development of efficient solvers. Yet, the same mathematical intricacies often make solvers black-boxes: their outcomes are hardly explainable and interactive even to experts, let alone lay users. Still, in real-world applications as well as research environments, lay users and experts likewise require a means to understand why a schedule is reasonable and what would happen with different schedules. Building upon a recently proposed approach to argumentation-supported explainable scheduling, we present a tool, *Schedule Explainer*, that provides interactive explanations in makespan scheduling easily and with clarity.

Keywords: Explainability · Scheduling · Implementation

1 Introduction

Mathematical optimisation affords effective techniques for solving well-defined problems involving resource constraints and objective functions. Makespan scheduling (also known as job shop scheduling) is one such fundamental discrete optimisation problem [6] concerning effective resource allocation. Makespan scheduling underlies numerous real-life applications, ranging from nurse rostering in hospital settings [12] to massive scale industrial sheet production.[1] While highly-effective optimisation solvers, such as CPLEX [7], are readily available to tackle optimisation problems including that of scheduling, they by and large

At the time of writing this paper, all the authors were affiliated with Imperial College London but no longer belong to this institution.

[1] https://www.greycon.com/solutions/planning-scheduling/.

D. Calvaresi et al. (Eds.): EXTRAAMAS 2021, LNAI 12688, pp. 243–259, 2021.
https://doi.org/10.1007/978-3-030-82017-6_15

suffer from the lack of explainability: they are seen as black-boxes unable to interact with and/or justify to the user their solutions or answer queries about those proposed by the user.

Explainability in scheduling includes, but is not necessarily limited to, supporting a user of a system equipped with an optimisation solver in understanding as to why the system behaved as it did and how it would respond to the user's interactions with the system. To be (at least partially) explainable, an optimisation solver should be able to justify why a solution is good and to interact with the user in order to modify the solution and answer whether and why the resulting one is good or not. Explainability is seen to be of critical importance in clinical settings [10] and is arguably an important issue in industrial applications of optimisation. Typical queries in scheduling runs of families of products in manufacturing centres to match incoming orders would be:

- Why not put order X in run A instead of run B?
- Why not swap orders X and Y?
- Why not merge runs A and B?

Similarly, user questions in scheduling applications to nurse rostering could be:

- What if nurse A were to do job Y rather than job X?
- Why is job X assigned to nurse A rather than nurse B?
- Why is the schedule good?

Recently, a novel approach ArgOpt of argumentation for explainable scheduling was proposed [4]. ArgOpt combines (computational) argumentation [9]—a branch of Knowledge Representation and Reasoning within the field of AI—with optimisation to explain the goodness of schedules. Specifically, abstract argumentation (AA) [5] affords an intermediate layer in between an optimisation solver and its user, capturing the makespan scheduling problem and the mathematical conditions underlying the goodness of schedules. The AA representation of the problem and its solutions allows to formulate and extract formal argumentative explanations which are in turn transformed into user-friendly natural language explanations about a given schedule.

The authors in [4] show that ArgOpt meets crucial desiderata in terms of soundness and completeness of explanations as well as cognitive and computational tractability. In particular, they establish that various sound and complete formal AA explanations can be extracted in time at most quadratic in the input size of the problem, and illustrate how those can be turned into template-based natural language explanations. However, the approach is described only in principle, without detailing how its implementation would look in practice.

In this paper we complement the work of [4] by providing a concrete implementation of the AA-based explainable scheduling pipeline. Specifically, we design and develop a system that connects an optimisation solver on the one end with an interactive graphical user interface (GUI) supporting explanations at the other end through an AA layer in the middle. Our tool, *Schedule Explainer*,[2]

[2] https://github.com/kcyras/aes.

allows to instantiate and solve a makespan scheduling problem, interact with the schedules given, indifferently, by an optimisation solver or the user, and provide concise actionable explanations to the goodness of schedules.

In what follows, we first give the necessary background on makespan scheduling and argumentation, and summarise the relevant findings of ArgOpt. In Sect. 3 we present the main algorithms of Schedule Explainer. In Sect. 4 we describe our tool and briefly review some related tools in Sect. 5. We conclude in Sect. 6.

2 Preliminaries

We adopt preliminaries on makespan scheduling [6], background on Abstract Argumentation (AA) [5] and ArgOpt results from [4].

2.1 Makespan Scheduling

An instance I of the *makespan scheduling problem* is a pair $(\mathcal{M}, \mathcal{J})$, where $\mathcal{J} = \{1, \ldots, n\}$ is a set of n *independent* jobs with a vector $\boldsymbol{p} = \{p_1, \ldots, p_n\}$ of processing times which have to be executed by a set $\mathcal{M} = \{1, \ldots, m\}$ of m parallel identical machines. Job $j \in \mathcal{J}$ must be processed by exactly one machine $i \in \mathcal{M}$ for p_j units of time non-preemptively, i.e. in a single continuous interval without interruptions. Each machine may process at most one job per time. The objective is to find a minimum makespan schedule, i.e. to minimise the last machine completion time.

We henceforth assume an instance $I = (\mathcal{M}, \mathcal{J})$ of a makespan scheduling problem with $\mathcal{M} = \{1, \ldots, m\}$ and $\mathcal{J} = \{1, \ldots, n\}$, for $m, n \geqslant 1$, unless stated otherwise.

In a standard mixed integer linear programming formulation, binary decision variable $x_{i,j}$ is 1 if job $j \in \mathcal{J}$ is executed by machine $i \in \mathcal{M}$ and 0 otherwise. Thus, a schedule of $(\mathcal{M}, \mathcal{J})$ can be seen as an $m \times n$ matrix $S \in \{0,1\}^{m \times n}$ with entries $x_{i,j} \in \{0,1\}$ representing job assignments to machines, for $i \in \mathcal{M}$ and $j \in \mathcal{J}$.

Given a schedule S, let C_i be the completion time of machine $i \in \mathcal{M}$ in S and let $C_{\max} = \max_{1 \leqslant i \leqslant m}\{C_i\}$ be the makespan. The problem is formally described by Equations (1a)–(1e) below: (1a) minimises the makespan; (1b)–(1c) define the makespan; constraint (1d) ensures each job is assigned to exactly one machine.

$$\min_{C_{\max}, C_i, x_{i,j}} \quad C_{\max} \tag{1a}$$

$$C_{\max} \geqslant C_i \qquad i \in \mathcal{M} \tag{1b}$$

$$C_i = \sum_{j=1}^{n} x_{i,j} \cdot p_j \qquad i \in \mathcal{M} \tag{1c}$$

$$\sum_{i=1}^{m} x_{i,j} = 1 \qquad j \in \mathcal{J} \tag{1d}$$

$$x_{i,j} \in \{0,1\} \qquad j \in \mathcal{J}, i \in \mathcal{M} \tag{1e}$$

Definition 1. *A schedule S is:*

- feasible *iff S meets constraint (1d);*
- optimal *iff S is feasible and minimises the makespan (1a).*

As finding optimal schedules is generally NP-hard, ArgOpt tractably evaluates schedules that are provably approximately optimal, or *efficient*, as described next.

Definition 2. *Let S be a schedule.*

- *Machine $i \in \mathcal{M}$ is* critical *iff $C_i = C_{\max}$.*
- *Job $j \in \mathcal{J}$ is* critical *iff $x_{i,j} = 1$ and $i \in \mathcal{M}$ is critical.*
- *S satisfies the* single exchange property *(SEP) iff for every critical machine $i \in \mathcal{M}$, for every critical job $j \in \mathcal{J}$ such that $x_{i,j} = 1$, for any $k \neq i$ it holds that $C_i \leqslant C_k + p_j$.*
- *S satisfies the* pairwise exchange property *(PEP) iff for every critical machine $i \in \mathcal{M}$, for every critical job $j \in \mathcal{J}$ such that $x_{i,j} = 1$, for every $k \neq i$ and $l \neq j$ such that $x_{k,l} = 1$ it holds that if $p_j > p_l$, then $C_i + p_l \leqslant C_k + p_j$.*
- *S is* efficient *iff S is feasible and satisfies both SEP and PEP.*

SEP concerns improving a schedule by exchanging critical jobs between machines. PEP concerns exchanges of critical jobs with other jobs on other machines. SEP and PEP are necessary (but not sufficient) optimality conditions.

ArgOpt also accommodates *fixed user decisions* that insist on specific (non-)assignments of jobs to machines.

Definition 3. *Let $D^-, D^+ \subseteq \mathcal{M} \times \mathcal{J}$ be, resp.,* negative *and* positive fixed decisions, *$D = (D^-, D^+)$ be* fixed decisions.[3] *We say D is* satisfiable *iff*

- *$D^- \cap D^+ = \emptyset$,*
- *$\nexists (i,j), (k,j) \in D^+$ with $i \neq k$,*
- *$\forall j \in \mathcal{J}\ \exists i \in \mathcal{M}$ such that $(i,j) \notin D^-$.*

We say that schedule S satisfies

- *D^- iff $(i,j) \in D^-$ implies $x_{i,j} = 0$;*
- *D^+ iff $(i,j) \in D^+$ implies $x_{i,j} = 1$;*
- *$D = (D^-, D^+)$ iff S satisfies both D^- and D^+.*

S violates D^-, D^+, D iff it does not satisfy D^-, D^+, D, resp.

[3] This slightly relaxes the notion of *fixed decisions* from [4] to accommodate for ill-posed user queries, allowing explanations over validations of the user input too, useful in practical applications. For instance, if fixed decisions are not satisfiable, then that should be explained.

2.2 Abstract Argumentation (AA)

An *AA framework* (*AF*) is a directed graph $(Args, \rightsquigarrow)$ with

- a set *Args* of *arguments*, and
- a binary *attack* relation $\rightsquigarrow$ over *Args*.

For $\mathsf{a}, \mathsf{b} \in Args$, $\mathsf{a} \rightsquigarrow \mathsf{b}$ means that a attacks b, and $\mathsf{a} \not\rightsquigarrow \mathsf{b}$ means that a does not attack b. With an abuse of notation, we extend the attack notation to sets of arguments as follows. For $\mathsf{A} \subseteq Args$ and $\mathsf{b} \in Args$:

- $\mathsf{A} \rightsquigarrow \mathsf{b}$ iff $\exists \mathsf{a} \in \mathsf{A}$ with $\mathsf{a} \rightsquigarrow \mathsf{b}$;
- $\mathsf{b} \rightsquigarrow \mathsf{A}$ iff $\exists \mathsf{a} \in \mathsf{A}$ with $\mathsf{b} \rightsquigarrow \mathsf{a}$;

A set $\mathsf{E} \subseteq Args$ of arguments, also called an *extension*, is

- *conflict-free* iff $\mathsf{E} \not\rightsquigarrow \mathsf{E}$;
- *stable* iff E is conflict-free and $\forall \mathsf{b} \in Args \setminus \mathsf{E}$, $\mathsf{E} \rightsquigarrow \mathsf{b}$.

2.3 ArgOpt

ArgOpt maps instances of makespan scheduling problem into AA frameworks and establishes one-to-one correspondences between schedules satisfying certain properties and the stability of extensions naturally corresponding to those schedules. We here recap the main definitions and results concerning the mappings and correspondences.

First, the problem instance is mapped to AA as follows.

The *feasibility AF* $(Args_F, \rightsquigarrow_F)$ is given by

- $Args_F = \{\mathsf{a}_{i,j} \ : \ i \in \mathcal{M}, j \in \mathcal{J}\}$,
- $\mathsf{a}_{i,j} \rightsquigarrow_F \mathsf{a}_{k,l}$ iff $i \neq k$ and $j = l$.

In what follows, unless stated otherwise, $(Args_F, \rightsquigarrow_F)$ is a feasibility AF.

A natural mapping arises between schedules and extensions thus. A schedule S and an extension $\mathsf{E} \subseteq Args_F$ are *corresponding*, denoted $S \approx \mathsf{E}$, when the following invariant holds: $x_{i,j} = 1$ iff $\mathsf{a}_{i,j} \in \mathsf{E}$. Under this natural mapping, the feasibility AF captures the space of feasible schedules: for any $S \approx \mathsf{E}$, S is feasible iff E is stable.

ArgOpt captures schedule efficiency via schedule-specific AA frameworks as follows. We henceforth assume, unless stated otherwise, a fixed but otherwise arbitrary schedule S. The *efficiency*[4] *AF* $(Args_S, \rightsquigarrow_S)$ is given by

- $Args_S = Args_F$,
- $\rightsquigarrow_S = (\rightsquigarrow_F \setminus \{(\mathsf{a}_{i,j}, \mathsf{a}_{i',j}) \ : \ C_i = C_{\max}, x_{i,j} = 1, C_i > C_{i'} + p_j\}) \cup \{(\mathsf{a}_{i',j'}, \mathsf{a}_{i,j}) \ : \ C_i = C_{\max}, x_{i,j} = 1, x_{i',j'} = 1, i' \neq i, j' \neq j, \ p_j > p_{j'}, C_i + p_{j'} > C_{i'} + p_j\}$.

[4] It is called *optimality* in [4], but we rename it efficiency instead, to better match the definitions of optimal and efficient schedules as in Definitions 1 and 2.

We say that $(Args_F, \rightsquigarrow_F)$ is *underlying* $(Args_S, \rightsquigarrow_S)$. Then S being efficient equates to the corresponding extension E being stable in the efficiency AF: if $(Args_S, \rightsquigarrow_S)$ is the efficiency AF and $S \approx \mathsf{E}$, then E is stable in $(Args_S, \rightsquigarrow_S)$ iff S is efficient.

Finally, ArgOpt captures fixed decisions via schedule-specific AA frameworks: given fixed decisions $D = (D^-, D^+)$, the *fixed decision AF* $(Args_D, \rightsquigarrow_D)$ is given by

- $Args_D = Args_F$,
- $\rightsquigarrow_D = (\rightsquigarrow_F \cup \{(\mathsf{a}_{i,j}, \mathsf{a}_{i,j}) : (i,j) \in D^-\}) \setminus \{(\mathsf{a}_{k,l}, \mathsf{a}_{i,j}) : (i,j) \in D^+, (k,l) \in \mathcal{M} \times \mathcal{J}\}$.

Then S is feasible and satisfies D iff $\mathsf{E} \approx S$ is stable in $(Args_D, \rightsquigarrow_D)$.

[4] also show that building all of the above AFs and checking whether the extension corresponding to a given schedule of interest is stable can be done in time at most quadratic, i.e. $\mathcal{O}(n^2m^2)$, in the size of the problem, i.e. nm. This low overhead guarantee ensures that explanations concerning the goodness of schedules can be efficiently extracted from the AFs in question. The argumentative explanations are defined thus.

For $(Args, \rightsquigarrow) \in \{(Args_F, \rightsquigarrow_F), (Args_S, \rightsquigarrow_S), (Args_D, \rightsquigarrow_D)\}$ and $\mathsf{E} \approx S$, *an attack* $\mathsf{a} \rightsquigarrow \mathsf{b}$ *with* $\mathsf{a}, \mathsf{b} \in \mathsf{E}$ *explains why* S:

(i) *is not feasible*, when $(\mathsf{a}, \mathsf{b}) \in \rightsquigarrow_F$;
(ii) *is not efficient*, when $(\mathsf{a}, \mathsf{b}) \in \rightsquigarrow_S \setminus \rightsquigarrow_F$;
(iii) *violates fixed decisions*, when $(\mathsf{a}, \mathsf{b}) \in \rightsquigarrow_D \setminus \rightsquigarrow_F$.

Intuitively, if S is (i) either infeasible due to some job scheduled more than once, (ii) or inefficient as witnessed by some pairwise exchange(s), (iii) or violating some negative fixed decision(s), then this is reflected in the relevant attack relation of the given AF.

Similarly, *a non-attack* $\mathsf{E} \not\rightsquigarrow \mathsf{b}$ *with* $\mathsf{b} \notin \mathsf{E}$ *explains why* S:

(i) *is not feasible*, when $\rightsquigarrow = \rightsquigarrow_F$;
(ii) *is not efficient*, when $\rightsquigarrow = \rightsquigarrow_S$ and $\mathsf{b} \rightsquigarrow_S \mathsf{E}$;
(iii) *violates fixed decisions*, when $\rightsquigarrow = \rightsquigarrow_D$ and b is unattacked.

So if S is (i) either infeasible due to some job being unscheduled, (ii) or inefficient as witnessed by some single exchange(s), (iii) or violating some positive fixed decision(s), then this is reflected by the absence of specific attacks in the relevant attack relation of the given AF.

Given appropriately formalised explanations as attacks and non-attacks, we can proceed to define algorithms for extracting explanations from AFs as well as for translating them into accessible templated natural language explanations as suggested in [4].

3 Algorithms

We here give the (pseudo-)algorithms implemented in Schedule Explainer. To this end, we first introduce some notation. We then advance algorithms for construction of AFs, verification of stable extensions and explanation generation.

3.1 Notation

In our algorithms we use Boolean tensors represented as multi-dimensional arrays to encode data structures for manipulating AFs. For instance, $\mathbf{0}^{d_1,\ldots,d_n}$ is the zero-valued tensor. We omit the dimensions if clear from the context. We also extend matrix subscripts to n dimensions. For example,

$$\mathbf{0}^{2\times 2} = \begin{bmatrix} 0 & 0 \\ 0 & 0 \end{bmatrix}.$$

Adjacency matrices represent directed graphs by indicating edges. If we generalise edges to multiple dimensions, then the graph's corresponding adjacency matrix becomes higher-dimensional, as illustrated in Fig. 1.

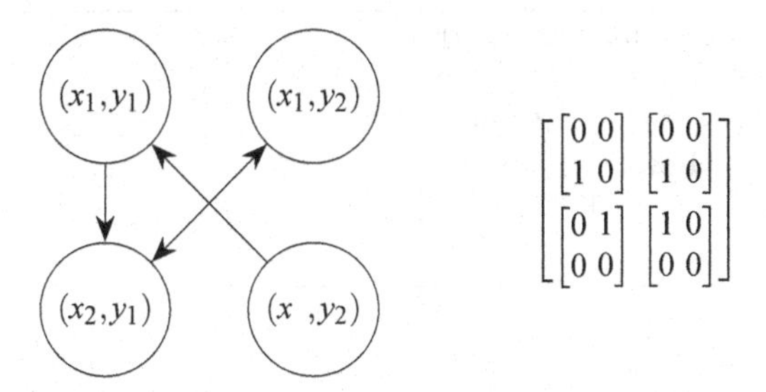

Fig. 1. A directed graph with 2-dimensional nodes and its corresponding 4-dimensional adjacency tensor.

We will interpret attack relations as directed graphs with 2-dimensional nodes themselves, with machines and jobs being the two dimensions.

We use operators over Boolean tensors to manipulate data structures. In particular, $\ominus$, $\oslash$ and $\varoslash$ are, respectively, element-wise negation, conjunction and disjunction operators over tensors. Formally, where $\mathbf{x}$, $\mathbf{y}$ and $\mathbf{z}$ have the same dimensions $d_1, \ldots, d_n$ and $i_1 \in \{1, \ldots, d_1\}, \ldots, i_n \in \{1, \ldots, d_n\}$:

- $\ominus$ is a prefix unary function such that $\ominus \mathbf{x} = \mathbf{y}$ iff $x_{i_1,\ldots,i_n} = 1 - y_{i_1,\ldots,i_n}$;
- $\wedge\!\!\!\bigcirc$ is an infix binary function such that $\mathbf{x} \wedge\!\!\!\bigcirc \mathbf{y} = \mathbf{z}$ iff $x_{i_1,\ldots,i_n} \cdot y_{i_1,\ldots,i_n} = z_{i_1,\ldots,i_n}$;
- $\vee\!\!\!\bigcirc$ is an infix binary function such that $\mathbf{x} \vee\!\!\!\bigcirc \mathbf{y} = \mathbf{z}$ iff $x_{i_1,\ldots,i_n} - x_{i_1,\ldots,i_n} \cdot y_{i_1,\ldots,i_n} + y_{i_1,\ldots,i_n} = z_{i_1,\ldots,i_n}$.

For instance:

$$\ominus \begin{bmatrix} 1 & 0 \\ 0 & 1 \end{bmatrix} = \begin{bmatrix} 0 & 1 \\ 1 & 0 \end{bmatrix}; \quad \begin{bmatrix} 0 & 0 \\ 1 & 1 \end{bmatrix} \wedge\!\!\!\bigcirc \begin{bmatrix} 0 & 1 \\ 0 & 1 \end{bmatrix} = \begin{bmatrix} 0 & 0 \\ 0 & 1 \end{bmatrix}; \quad \begin{bmatrix} 0 & 0 \\ 1 & 1 \end{bmatrix} \vee\!\!\!\bigcirc \begin{bmatrix} 0 & 1 \\ 0 & 1 \end{bmatrix} = \begin{bmatrix} 0 & 1 \\ 1 & 1 \end{bmatrix}.$$

3.2 Constructing AFs

In what follows, we assume as given a fixed set $Args = \mathcal{M} \times \mathcal{J}$ of arguments and will consider as dynamic the various attack relations $\rightsquigarrow \subseteq Args \times Args$. Thus, in delineating the algorithms of Schedule Explainer, we slightly deviate from ArgOpt in simplifying the notation. When appropriate, we also assume a schedule S as given as well as its corresponding extension $\mathsf{E} \approx S$ (slightly abusing the notation).

In our algorithms, we manipulate Boolean tensors as data structures representing the attack relations as follows:

Definition 4. *$\twoheadrightarrow$ is a Boolean tensor data structure to store $\rightsquigarrow$ such that $(i_1, j_1) \rightsquigarrow (i_2, j_2)$ iff $\twoheadrightarrow_{i_1,j_1,i_2,j_2} = 1$.*

We may add various subscripts to $\twoheadrightarrow$ and $\rightsquigarrow$ in describing different algorithms, for instance $\twoheadrightarrow_F$ and $\rightsquigarrow_F$.

Algorithm 1 constructs the feasibility AF.

Algorithm 1

1: **function** CONS-FEASIBILITY(m, n)
2: $\quad \twoheadrightarrow_F \leftarrow \mathbf{0}^{(m \times n)^2}$
3: $\quad$ **for** $j \in \mathcal{J}$, $i_1 \in \mathcal{M}$, $i_2 \in \mathcal{M}$ **do**
4: $\quad\quad$ **if** $i_1 \neq i_2$ **then**
5: $\quad\quad\quad \twoheadrightarrow_{F\, i_1,j,i_2,j} \leftarrow 1$
6: $\quad$ **return** $\twoheadrightarrow_F$

$\twoheadrightarrow_F$ can be constructed trivially in $\mathcal{O}(m^2 n^2)$ computational complexity, due to complexity of zero-initialising $\twoheadrightarrow_F$.

To deal with efficiency we use the following predicates capturing single and pairwise exchange conditions given in Definition 2:

Definition 5. *Define operators SEP and PEP as follows:*

- *SEP*(i_1, i_2, j_1) *iff* $C_{i_1} = C_{\max} \wedge x_{i_1,j_1} = 1 \wedge C_{i_1} > C_{i_2} + p_{j_1}$
- *PEP*(i_1, i_2, j_1, j_2) *iff* $C_{i_1} = C_{\max} \wedge x_{i_1,j_1} = 1 \wedge x_{i_2,j_2} = 1 \wedge i_1 \neq i_2 \wedge j_1 \neq j_2 \wedge p_{j_1} > p_{j_2} \wedge C_{i_1} + p_{j_2} > C_{i_2} + p_{j_1}$.

Algorithm 2 constructs the efficiency AF. There and henceforth, $\mathbf{x}$ encodes a given schedule S as an $m \times n$ matrix.

The construction of $\twoheadrightarrow_S$ is expensive because of the explicit for-loops to iterate over the $\mathcal{M}^2 \times \mathcal{J}^2$ space to compute the edges that satisfy PEP and to copy $\twoheadrightarrow_F$. We added an optimisation by computing SEP outside of the j_2 loop, because PEP is invariant of j_2. Algorithm 2 returns the value of $\mathbf{C}$ to be used later, rather an recompute its value when necessary.

To deal with fixed decisions we use the following predicates capturing the conditions of positive and negative fixed decisions given in Definition 3:

Algorithm 2

1: **function** CONS-EFFICIENCY(m, n, $\boldsymbol{p}$, $\mathbf{x}$, $\twoheadrightarrow_F$)
2: $\quad \mathbf{C} \leftarrow \mathbf{x} \cdot \boldsymbol{p}$
3: $\quad C_{\max} \leftarrow \max(\mathbf{C})$
4: $\quad \twoheadrightarrow_S \leftarrow \twoheadrightarrow_F$
5: $\quad$ **for** $i_1 \in \mathcal{M}$ **do**
6: $\quad\quad$ **if** $C_{i_1} = C_{\max}$ **then**
7: $\quad\quad\quad$ **for** $j_1 \in \mathcal{J}$ **do**
8: $\quad\quad\quad\quad$ **if** $x_{i_1,j_1} = 1$ **then**
9: $\quad\quad\quad\quad\quad$ **for** $i_2 \in \mathcal{M}$ **do**
10: $\quad\quad\quad\quad\quad\quad$ **if** SEP(i_1, j_1, i_2) **then**
11: $\quad\quad\quad\quad\quad\quad\quad \twoheadrightarrow_{S\,i_1,j_1,i_2,j_1} \leftarrow 0$
12: $\quad\quad\quad\quad\quad\quad$ **for** $j_2 \in \mathcal{J}$ **do**
13: $\quad\quad\quad\quad\quad\quad\quad$ **if** PEP(i_1, j_1, i_2, j_2) **then**
14: $\quad\quad\quad\quad\quad\quad\quad\quad \twoheadrightarrow_{S\,i_1,j_2,i_2,j_2} \leftarrow 1$
15: $\quad$ **return** $(\twoheadrightarrow_S, \mathbf{C})$

Definition 6. *Let $D = (D^-, D^+)$ be fixed decisions. Define operators* DP^+ *and* DP^- *as follows:*

- $\mathsf{DP}^+(i_1, i_2, j_1, j_2)$ *iff* $(i_2, j_2) \in D^+$
- $\mathsf{DP}^-(i_1, i_2, j_1, j_2)$ *iff* $(i_1, j_1) \in D^- \wedge i_1 = i_2 \wedge j_1 = j_2$.

Algorithm 3 constructs the fixed decision AF. There and henceforth, D denotes given fixed decisions.

Algorithm 3

1: **function** CONS-FIXEDDECISIONS(m, n, D, $\twoheadrightarrow_F$)
2: $\quad \twoheadrightarrow_D \leftarrow \twoheadrightarrow_F$
3: $\quad$ **for** $(i, j) \in D^-$ **do**
4: $\quad\quad \twoheadrightarrow_{S\,i,j,i,j} \leftarrow 1$
5: $\quad$ **for** $(i_1, j_1) \in D^+$, $i_2 \in \mathcal{M}$, $j_2 \in \mathcal{J}$ **do**
6: $\quad\quad \twoheadrightarrow_{D\,i_2,j_2,i_1,j_1} \leftarrow 0$
7: $\quad$ **return** $\twoheadrightarrow_D$

We are now able to construct the various AFs required for extracting argumentative explanations concerning schedules.

3.3 Verifying Stability

We now need to make sure we can verify stability of extensions, because checking whether the extension corresponding to a given schedule is stable in an AF is a fundamental procedure in ArgOpt. While there exist numerous tools for tackling this task (see e.g. [2] for an excellent overview), they are by and large generic and would require pipeline integration with e.g. ASP or SAT solvers. Instead, we provide a direct efficient algorithm suited for AFs constructed from makespan scheduling instances.

Algorithm 4 returns two tensors: $\mathbf{u}$ encodes unattacked arguments outside the extension and $\mathbf{c}$ encodes attacks among arguments in the extension (thus allowing to check for conflict-freeness). $\bar{\mathbf{u}}$ and $\bar{\mathbf{c}}$ represent nodes and edges to ignore, respectively, useful in tailoring explanations to particular constraints. By default, $\bar{\mathbf{u}} = \mathbf{0}$ and $\bar{\mathbf{c}} = \mathbf{0}$.

Algorithm 4

```
 1: function COMPUTE-STABILITY(x, ↠, ū, c̄)
 2:     u ← COMPUTE-UNATTACKED(x, ↠, ū)
 3:     c ← 0^((m×n)^2)
 4:     for i ∈ M, j ∈ J do
 5:         c_{i,j} ← COMPUTE-CONFLICTS(x, ↠_{i,j}, c̄_{i,j})
 6:     return (u, c)
 7: function COMPUTE-UNATTACKED(x, ↠, ū)
 8:     u ← ⊖ x
 9:     for i ∈ M, j ∈ J do
10:         if x_{i,j} = 1 then
11:             u ← u ⊙∧ ⊖ ↠_{i,j}
12:     u ← u ⊙∧ ⊖ ū
13:     return u
14: function COMPUTE-CONFLICTS(x, ↠_{i,j}, c̄_{i,j})
15:     c_{i,j} ← 0^(m×n)
16:     if x_{i,j} = 1 then
17:         c_{i,j} ← x ⊙∧ ↠_{i,j}
18:     c_{i,j} ← c_{i,j} ⊙∧ ⊖ c̄_{i,j}
19:     return c_{i,j}
```

The following establishes the correctness of Algorithm 4.

Theorem 1. COMPUTE-STABILITY($\mathbf{x}$, $\twoheadrightarrow$, $\mathbf{0}$, $\mathbf{0}$) $=$ $(\mathbf{0}, \mathbf{0})$ *iff* E *is stable in* $(Args, \rightsquigarrow)$.

Proof. First, define a lexicographical ordering $\prec$ on *Args* as follows: $(i_1, j_1) \prec (i_2, j_2)$ iff $i_1 < i_2 \vee i_1 = i_2 \wedge j_1 < j_2$. We proceed with several intermediate lemmas.

Lemma 1. *If* COMPUTE-UNATTACKED($\mathbf{x}$, $\twoheadrightarrow$, $\bar{\mathbf{u}}$) $=$ $\mathbf{u}$, *then* $\forall k \in \mathcal{M}$ $\forall \ell \in \mathcal{J}$ *we have* $u_{k,\ell} = 1$ *iff* $(\neg \exists k' \in \mathcal{M}$ *such that* $\exists \ell' \in \mathcal{J}$ *with* $x_{k,\ell} = 0,\ x_{k',\ell'} = 1,\ \twoheadrightarrow_{k',\ell',k,\ell}= 1,\ \bar{u}_{k,\ell} = 0)$.

Proof. Take arbitrary $\mathbf{x}$, $\twoheadrightarrow$ and $\bar{\mathbf{u}}$. After line 8, we have α : $[\forall k \in \mathcal{M}\ \forall \ell \in \mathcal{J}\ u_{k,\ell} = 1$ iff $x_{k,\ell} = 0]$, by definition of $\ominus$. Define loop invariant β that holds at each iteration of the i and j loops: $\beta : \forall k \in \mathcal{M}\ \forall \ell \in \mathcal{J}$

- $(k,\ell) \prec (i,j) \implies (u_{k,\ell} = 1$ iff $\neg\exists k' \in \mathcal{M}$ s.t. $\exists \ell' \in \mathcal{J}$ with $x_{k,\ell} = 0,\ x_{k',\ell'} = 1,\ \twoheadrightarrow_{k',\ell',k,\ell} = 1)$;
- $(k,\ell) \succeq (i,j) \implies (u_{k,\ell} = 1$ iff $x_{k,\ell} = 0)$.

Then β follows from α because, initially, $i = 1$ and $j = 1$, so there is no $(k,\ell) \prec (i,j)$, whence $u_{k,\ell} = 1$ iff $x_{k,\ell} = 0$ holds trivially given α.

Now, at line 10, if $x_{i,j} = 1$, then we know both β and $x_{i,j} = 1$ hold. At line 11, any argument (k,ℓ) attacked by (i,j) is set to be attacked, so $u_{i,j} = 0$. But at an arbitrary iteration, values i and j are overwritten so we retain $\exists k', \ell'\ (k',\ell') \twoheadrightarrow (k,\ell)$. Then after line 11 we know that $\forall k \in \mathcal{M}\ \forall \ell \in \mathcal{J}\ (k,\ell) \preceq (i,j) \implies (u_{k,\ell} = 1$ iff $\neg\exists k' \in \mathcal{M}$ s.t. $\exists \ell' \in \mathcal{J}$ with $x_{k,\ell} = 0,\ x_{k',\ell'} = 1,\ \twoheadrightarrow_{k',\ell',k,\ell}= 1)$ and that $(k,\ell) \succ (i,j) \implies (u_{k,\ell} = 1$ iff $x_{k,\ell} = 0)$. Thus, the next iteration of (i,j) means that β will follow.

Otherwise, if at line 10 we have $x_{i,j} = 0$, then β trivially holds as only i or j have changed values.

Consequently, at the end of the loop we have $\forall k \in \mathcal{M}\ \forall \ell \in \mathcal{J}\ u_{k,\ell} = 1$ iff $(\neg\exists k' \in \mathcal{M}$ such that $\exists \ell' \in \mathcal{J}$ with $x_{k,\ell} = 0,\ x_{k',\ell'} = 1,\ \twoheadrightarrow_{k',\ell',k,\ell}= 1)$. Since after line 12 $\mathbf{u}$ will be filtered by $\bar{\mathbf{u}}$, we get $\forall k \in \mathcal{M}\ \forall \ell \in \mathcal{J}\ \bar{u}_{k,\ell} = 0 \implies u_{k,\ell} = 1$. As $\bar{\mathbf{u}} = \mathbf{0}$ by default, the statement follows.

Lemma 2. *If* COMPUTE-CONFLICTS$(\mathbf{x}, \twoheadrightarrow_{i,j}, \bar{c}_{i,j}) = c_{i,j}$, *then* $\forall k \in \mathcal{M}\ \forall \ell \in \mathcal{J}$ *we have* $c_{i,j,k,\ell} = 1$ *iff* $x_{i,j} = 1,\ x_{k,\ell} = 1,\ \twoheadrightarrow_{i,j,k,\ell}= 1,\ \bar{c}_{i,j,k,\ell} = 0$.

Proof. At line 15, $c_{i,j,k,l} = 0$ by default. At line 16, if $x_{i,j} = 1$, then (i,j) may be attacked. At line 17, $(i,j) \twoheadrightarrow (k,\ell)$ when $x_{k,\ell} = 1$ and $\twoheadrightarrow_{i,j,k,\ell}= 1$. At line 18, $\bar{c}_{i,j,k,\ell} = 1 \implies c_{i,j,k,\ell} = 0$. Combining these statements the claim follows.

Lemma 3. *If* COMPUTE-STABILITY$(\mathbf{x}, \twoheadrightarrow, \bar{\mathbf{u}}, \bar{\mathbf{c}}) = (\mathbf{u}, \mathbf{c})$, *then*

- $\forall (k_1,\ell_1) \in \mathit{Args} \setminus \mathsf{E}\ \ \bar{u}_{k_1,\ell_1} = 0 \implies (\exists (k_2,\ell_2) \in \mathsf{E} : (k_2,\ell_2) \rightsquigarrow (k_1,\ell_1)$ *iff* $u_{k_1,\ell_1} = 0)$;
- $\forall (k_1,\ell_1), (k_2,\ell_2) \in \mathsf{E}\ \ \bar{c}_{k_1,\ell_1,k_2,\ell_2} = 0 \implies ((k_1,\ell_1) \rightsquigarrow (k_2,\ell_2)$ *iff* $c_{k_1,\ell_1,k_2,\ell_2} = 1)$.

Proof. Assume COMPUTE-STABILITY$(\mathbf{x}, \twoheadrightarrow, \bar{\mathbf{u}}, \bar{\mathbf{c}}) = (\mathbf{u}, \mathbf{c})$. First, let $(k_1,\ell_1) \in \mathit{Args} \setminus \mathsf{E}$ and assume $\bar{u}_{k_1,\ell_1} = 0$. On the one hand, if $u_{k_1,\ell_1} = 0$, then by Lemma 1, $\exists (k_2,l_2)$ with $x_{k_2,l_2} = 1$ and $\twoheadrightarrow_{k_2,\ell_2,k_1,\ell_1}= 1$. I.e. $(k_2,l_2) \in \mathsf{E}$ and $(k_2,\ell_2) \rightsquigarrow (k_1,\ell_1)$. On the other hand, if $(k_2,l_2) \in \mathsf{E}$ with $(k_2,\ell_2) \rightsquigarrow (k_1,\ell_1)$, then $x_{k_2,l_2} = 1$ and $\twoheadrightarrow_{k_2,\ell_2,k_1,\ell_1}= 1$, so by Lemma 1, $u_{k_1,\ell_1} = 0$.

For the second statement, let $(k_1,\ell_1), (k_2,\ell_2) \in \mathsf{E}$. Then $x_{k_1,\ell_1} = 1$ and $x_{k_2,\ell_2} = 1$. Assume $\bar{c}_{k_1,\ell_1,k_2,\ell_2} = 0$. If $(k_1,\ell_1) \rightsquigarrow (k_2,\ell_2)$, then $\twoheadrightarrow_{k_2,\ell_2,k_1,\ell_1}= 1$, whence $c_{k_1,\ell_1,k_2,\ell_2} = 1$ follows from Lemma 2. If $c_{k_1,\ell_1,k_2,\ell_2} = 1$, then $\twoheadrightarrow_{k_1,\ell_1,k_2,\ell_2}= 1$ by Lemma 2 too, whence $(k_1,\ell_1) \rightsquigarrow (k_2,\ell_2)$, as required.

COMPUTE-STABILITY$(\mathbf{x}, \twoheadrightarrow, \mathbf{0}, \mathbf{0}) = (\mathbf{0}, \mathbf{0})$ iff E is stable in $(\mathit{Args}, \rightsquigarrow)$ now follows from Lemma 3.

We next show how to extract user-friendly explanations pertaining to schedules.

3.4 Generating Explanations

As suggested in [4], we generate templated natural language explanations from the argumentative explanations identifiable in the various AFs (see Sect. 2.3). In the following algorithms, the generated *explanations are given in italics*, to be accordingly instantiated with the job and machine names in the user interface.

Algorithm 5 generates explanations regarding feasibility. As [4] do not state explanations for trivial cases when $m = 0$ or $n = 0$, to complement their work, we handle these cases with additional explanations.

Note that a problem with the naive implementation of generating an explanation for each conflict in $\mathbf{c}$ results in k^2 explanations for k conflicting machines for a job. This results in superfluous text for the user. To summarise these explanations, Algorithm 5 constructs a pseudo-schedule $\mathbf{z}$, interpreted as $\mathbf{x}$ transposed and rows filtered if $\sum_{i \in \mathcal{M}} x_{i,j} > 1$ for all jobs j. Then non-zero indices of $\mathbf{z}$ referring to the machines that cause over-allocation are used.

Algorithm 5

```
1: function EXPLAIN-FEASIBILITY(u, c)
2:     if m = 0 then
3:         if n = 0 then
4:             There are no jobs, so the schedule is feasible.
5:         else
6:             There are no machines to allocate jobs.
7:     else
8:         y ← 0^n,  z ← 0^{n×m}
9:         for j ∈ J, i_1 ∈ M, i_2 ∈ M do
10:            if c_{i_1,j,i_2,j} = 1 then
11:                y_j ← 1,  z_{j,i_1} ← 1,  z_{j,i_2} ← 1
12:        if u_0^T = 0 ∧ y = 0 then
13:            Each job is allocated to exactly one machine.
14:        else
15:            for j ∈ J do
16:                if u_{0,j} = 1 then
17:                    Job j is not allocated to any machine.
18:                if y_j ≠ 0 then
19:                    Job j is over-allocated to machines {i | i ∈ M, z_{j,i} = 1}.
```

Algorithm 6 generates explanations regarding efficiency.

In addition to finding explanations as proposed in ArgOpt, Algorithm 6 sorts in decreasing order the possible exchanges by improvements in the makespan of the schedule. While such exchanges do not guarantee the quickest convergence to an efficient schedule, it may be reasonable for the user to consider first the biggest problems.

A key limitation with sorting by reduction pertains to multiple critical machines. In such cases, all reductions are zero, as single or pairwise exchanges

Algorithm 6

1: **function** EXPLAIN-EFFICIENCY($\boldsymbol{p}$, $\mathbf{C}$, $\mathbf{u}$, $\mathbf{c}$)
2: $i_1 \leftarrow$ first argmax of $\mathbf{C}$
3: reasons $\leftarrow$ empty list
4: **for** $j_1 \in \mathcal{J}$, $i_2 \in \mathcal{M}$ **do**
5: **if** $u_{i_2,j_1} = 1$ **then**
6: reason $\leftarrow$ *Job* j_1 *can be allocated to machine* i_2.
7: append reason to reasons
8: **for** $j_2 \in \mathcal{J}$ **do**
9: **if** $c_{i_1,j_1,i_2,j_2} = 1$ **then**
10: reason $\leftarrow$ *Job* j_1 *and* j_2 *can be swapped on machines* i_1 *and* i_2.
11: append reason to reasons
12: sort reasons by $\langle$reduction, processing time$\rangle$
13: **if** reasons is empty **then**
14: *All jobs satisfy SEP and PEP.*
15: **else**
16: output reasons

result in local optimisations of the same objective value. To find a strictly more efficient schedule, we look k steps ahead, with k the number of critical machines. To this end, we would need to generate instructions of k actions, which may cause an exponential explosion in k on the explanation length. Instead, we restrict the explanation space by considering only one critical machine. This reduces the computational complexity by a factor of m, which is significant since efficiency is the most expensive schedule property to explain.

Algorithm 7 generates explanations regarding fixed decisions. (There $\mathbf{y}$ refers to allocations violating positive fixed decisions D^+.) In contrast to [4], we assume that fixed decisions $D = (D^-, D^+)$ need not be satisfiable (see Definition 3), whence we must check the conditions for this and generate explanations accordingly (lines 2–8).

Overall, Schedule Explainer executes Algorithm 8 to explain the goodness of schedules.

4 Tool

We here detail the overall design of Schedule Explainer and illustrate its interface.

Schedule Explainer integrates the algorithms described in Sect. 3 with, on the one end, an optimisation solver of choice, and a GUI on the other end. Specifically, the tool allows to formulate an instance of a makespan scheduling problem, which can then be either passed to the optimisation solver to obtain a schedule, or a candidate schedule can be manually generated (e.g. randomised). The working schedule as well as fixed decisions can be revised at any time.

The tool then allows for explaining whether and why the schedule is good in terms of feasibility, efficiency and fixed decisions, presenting natural language explanations to the user. Importantly, the explanations are *actionable* in that

Algorithm 7

```
1: function EXPLAIN-FIXEDDECISIONS(D, u, c)
2:     for j ∈ J do
3:         if ∃i ∈ M. (i, j) ∉ D⁻ then
4:             Job j cannot be allocated to any machine.
5:         if D⁻ and D⁺ are not disjoint then
6:             Job j subject to conflicting negative and positive fixed decisions.
7:         if |{i ∈ M | (i, j) ∈ D⁺}| > 1 then
8:             Job j cannot be allocated to multiple machines.
9:     y ← 0^{m×n}
10:    for i ∈ M, j ∈ J do
11:        y ← y ⓥ c_{i,j}
12:    if u = 0 ∧ y = 0 then
13:        All jobs satisfy user fixed decisions.
14:    else
15:        for i ∈ M, j ∈ J do
16:            if u_{i,j} then
17:                Job j must be allocated to machine i.
18:            if y_{i,j} then
19:                Job j can't be allocated to machine i.
```

the user can apply the actions suggested by the tool, such as reassigning jobs, to improve the schedule. Upon an action being applied, the tool immediately recomputes whether the schedule is good and gives explanations accordingly.

Schedule Explainer is therefore an *interactive* tool: it allows the user to easily engage with the scheduling problem, its solutions and explanations thereof, via an accessible GUI. The GUI keeps in view the definition of the problem and its parameters and at the same time presents the actionable explanations in natural language, alongside a standard cascade chart schedule visualisation. When the user applies any of the actions suggested, the chart indicates pre- and post-action situations and the tool refreshes the explanations and actions accordingly. Figure 2 shows a screenshot of the Schedule Explainer's GUI. We note that the GUI is proof-of-concept and specific applications of the tool may require appropriate adaptations, as in e.g. [3].

The tool is implemented in Python, which allows for easy integration with several optimisation solvers, particularly GLPK [1] via Pyomo, and offers a wide range of libraries for various functionalities, including graphical interfaces. The tool easily runs on a personal laptop, with computation of explanations taking well under 1 s for problems with $mn \leqslant 1000$. (Optimisation solver running times can be much greater, but this is irrelevant from the explanation point of view.) Schedule Explainer is publicly available at https://github.com/kcyras/aes.

Algorithm 8

1: **function** EXPLAIN(m, n, $\boldsymbol{p}$, D, $\mathbf{x}$)
2: $\twoheadrightarrow_F \leftarrow$ CONS-FEASIBILITY(m, n)
3: $(\mathbf{u_F}, \mathbf{c_F}) \leftarrow$ COMPUTE-STABILITY($\mathbf{x}$, $\twoheadrightarrow_F$, $\mathbf{0}$, $\mathbf{0}$)
4: EXPLAIN-FEASIBILITY($\mathbf{u_F}$, $\mathbf{c_F}$)
5: $(\twoheadrightarrow_S, \mathbf{C}) \leftarrow$ CONS-EFFICIENCY(m, n, $\boldsymbol{p}$, $\mathbf{x}$, D, $\twoheadrightarrow_F$)
6: $(\mathbf{u_S}, \mathbf{c_S}) \leftarrow$ COMPUTE-STABILITY($\mathbf{x}$, $\twoheadrightarrow_S$, $\mathbf{u_F}$, $\mathbf{c_F}$)
7: EXPLAIN-EFFICIENCY($\boldsymbol{p}$, $\mathbf{C}$, $\mathbf{u_S}$, $\mathbf{c_S}$)
8: $\twoheadrightarrow_D \leftarrow$ CONS-FIXEDDECISIONS(m, n, $\mathbf{x}$, $\twoheadrightarrow_F$)
9: $(\mathbf{u_D}, \mathbf{c_D}) \leftarrow$ COMPUTE-STABILITY($\mathbf{x}$, $\twoheadrightarrow_D$, $\mathbf{u_F}$, $\mathbf{c_F}$)
10: EXPLAIN-FIXEDDECISIONS($\mathbf{u_D}$, $\mathbf{c_D}$)

5 Related Work

Explainable scheduling is a relatively new paradigm within either explainable AI or optimisation: there has been some recent activity in, for instance, Workshops on Explainable Planning (XAIP).[5] However, to the best of our knowledge no implemented systems for explaining makespan scheduling exist, other than the application front-end [3] which itself uses the back-end presented in this paper.

We are likewise not aware of commercial optimisation solvers' capabilities in terms of explaining the goodness of schedules. We will nevertheless briefly discuss two relevant and readily available scheduling tools, namely Setmore [11] and LEKIN [8], to indicate the possible limitations in terms of explainability.

Setmore is a commercial online application that records appointments, schedules and employees, designed for small businesses where managers can organise appointments on a calender. Makespan schedules are formulae where employees are machines and appointments are jobs. The graphical interface enables users to quickly glance at appointments and their times, clearly indicating overlapping appointments. Modification of an existing schedule is well-facilitated within the interface, where appointments can be moved or swapped between employees. However, the application does not offer explainability as such: it is limited to data input verification and spawning validation error messages for infeasible schedules. Furthermore, no notion of optimality is at play in Setmore, perhaps naturally because appointment optimality is not well-defined for arbitrary businesses.

LEKIN is an academic-oriented scheduling application to teach students scheduling theory and its applications. The application features numerous optimisation algorithms for scheduling and supports flexible makespan scheduling settings. The application validates a schedule's feasibility at input, whence infeasibility results in error messages. The application computes optimal schedules and produces common scheduling performance metrics such as makespan completion time and tardiness. However, these metrics are global across all machines

[5] http://xaip.mybluemix.net/.

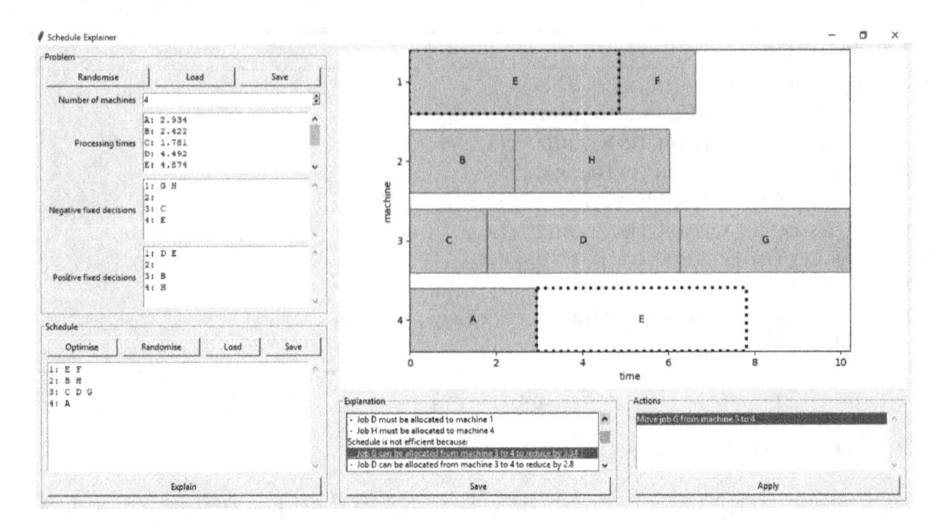

Fig. 2. A screenshot of Schedule Explainer's GUI. Machines are named by integers 1 to 4 and jobs by capital letters A to H. Denotation of job processing times, fixed decisions and working schedule is self-explanatory. The schedule together with the last action applied to it is visualised as a cascade chart: job E was moved from machine 4 to machine 1 so as to satisfy both the positive and negative fixed decisions.

and give no indication as to improving schedules. Moreover, no explanation is presented regarding the pre- and post-optimised schedules.

It is clear that both applications offer limited explanations, explicitly via error messages and implicitly via cascade charts. Similar considerations apply to very advanced, commercial scheduling tools, such as Greycon's OptStudio or OptaPlanner.[6] Instead, for effective knowledge transfer, our Schedule Explainer provides actionable textual explanations accompanied by cascade charts to help the user to understand why a particular schedule is good or not and how to possibly improve it.

6 Conclusions and Future Work

We presented Schedule Explainer, an interactive tool for argumentation-supported explanations in the makespan scheduling setting. The tool realises the recently proposed ArgOpt approach [4] by integrating an optimisation solver with an argumentation-driven engine to provide actionable textual explanations pertaining to the goodness of schedules in a visually-supporting user interface.

In the future it would be interesting to evaluate our tool in various settings: for instance with computer science students to gauge whether it can help to better understand makespan scheduling and optimisation techniques, as well as with generic audiences to see if Schedule Explainer with an appropriately

[6] https://www.optaplanner.org/.

enhanced graphical interface such as the one in [3] can be of educational value in real-life applications such as doctor appointment scheduling. Providing personalised explanations would also be desirable, as well as accounting for a potentially exponential blow-up in the number of explanations. We stipulate that such explorations would also entail some user studies and for easier accessibility in such settings it may as well be useful to develop a web interface for Schedule Explainer. It would also be beneficial to extend the tool to more complicated makespan scheduling problems such as flow- and open-shop scheduling, interval scheduling, scheduling with user preferences.

References

1. Makhorin, A.: GLPK (GNU Linear Programming Kit) (2000). https://www.gnu.org/software/glpk/. Accessed 28 Feb 2021
2. Charwat, G., Dvořák, W., Gaggl, S., Wallner, J.P., Woltran, S.: Methods for solving reasoning problems in abstract argumentation - a survey. Artif. Intell. **220**, 28–63 (2015). https://doi.org/10.1016/j.artint.2014.11.008
3. Čyras, K., Karamlou, A., Lee, M., Letsios, D., Misener, R., Toni, F.: AI-assisted Schedule Explainer for Nurse Rostering. In: Seghrouchni, A.E.F., Sukthankar, G., An, B., Yorke-Smith, N. (eds.) 19th International Conference on Autonomous Agents and MultiAgent Systems - Demo Track, pp. 2101–2103. IFAAMAS, Auckland (2020)
4. Čyras, K., Letsios, D., Misener, R., Toni, F.: Argumentation for explainable scheduling. In: 33rd AAAI Conference on Artificial Intelligence. AAAI Press, Honolulu (2019)
5. Dung, P.M.: On the acceptability of arguments and its fundamental role in non-monotonic reasoning, logic programming and n-person games. Artif. Intell. **77**, 321–357 (1995). https://doi.org/10.1016/0004-3702(94)00041-X
6. Graham, R.L.: Bounds on multiprocessing timing anomalies. SIAM J. Appl. Math. **17**(2), 416–429 (1969)
7. Laborie, P., Rogerie, J., Shaw, P., Vilím, P.: IBM ILOG CP optimizer for scheduling. Constraints **23**(2), 210–250 (2018). https://doi.org/10.1007/s10601-018-9281-x
8. Pinedo, M.L., Chao, X., Leung, J., Feldman, A.: Lekin scheduling system. http://web-static.stern.nyu.edu/om/software/lekin/index.htm (2010). Accessed 28 Feb 2021
9. Rahwan, I., Simari, G.R.: Argumentation in Artificial Intelligence. Springer Springer, New York (2009). https://doi.org/10.1007/978-0-387-98197-0
10. Sacchi, L., et al.: From decision to shared-decision: introducing patients' preferences into clinical decision analysis. Artif. Intell. Med. **65**(1), 19–28 (2015). https://doi.org/10.1016/j.artmed.2014.10.004
11. Setmore: Free online appointment scheduling calender software (2019). https://www.setmore.com/. Accessed 28 Feb 2021
12. Warner, D.M., Prawda, J.: A Mathematical programming model for scheduling nursing personnel in a hospital. Manage. Sci. **19**(4), 411–422 (1972). http://www.jstor.org/stable/2629025

Towards Explainable Practical Agency

A Logical Perspective

Nourhan Ehab[2](✉) and Haythem O. Ismail[1,2]

[1] Department of Engineering Mathematics, Cairo University, Giza, Egypt
[2] Department of Computer Science and Engineering, German University in Cairo, Cairo, Egypt
{nourhan.ehab,haythem.ismail}@guc.edu.eg

Abstract. Practical reasoning is such an essential cornerstone of artificial intelligence that it is impossible to see how autonomous agents can be realized without it. As a first step of practical reasoning, an autonomous agent is required to form its intentions by choosing amongst its motivations in light of its beliefs. An autonomous agent is also expected to seamlessly revise its intentions whenever its beliefs or motivations change. In the modern world, it becomes an impelling priority to endow agents with explainable practical reasoning capabilities in order to foster the trustworthiness of artificial agents. An adequate framework of practical reasoning must be able to (i) capture the process of intention formation, (ii) model the joint revision of beliefs and intentions, and (iii) provide explanations for the chosen beliefs and intentions. Despite the abundance of approaches in the literature for modelling practical reasoning, such approaches fail to possess at least one of the previously mentioned capabilities. In this paper, we present formal algebraic semantics for a logical language that can be used for practical reasoning. We demonstrate how our language possesses all of the aforementioned capabilities providing an adequate framework for explainable practical reasoning.

Keywords: Explainable agency · Practical reasoning · Algebraic semantics

1 Introduction

"What should I do?" is a question that an artificial agent is compelled to answer if it is fully autonomous. The process of deliberation the agent undergoes in order to answer this question is referred to as practical reasoning [6]. In this way, practical reasoning is reasoning about *what to do*. According to Bratman, practical reasoning is a matter of weighing reasons for and against competing actions, where the relevant reasons are provided by the agent's motivations and beliefs [29]. Modelling practical reasoning has posed a difficult challenge over the years for philosophers, logicians, and computer scientists alike and has received less attention than theoretical reasoning or reasoning about what to believe. The process of practical reasoning can be thought of as comprising two separate subprocesses: (i) a process of deliberation where the agent forms its intentions by choosing among its possibly conflicting motivations in light of its possibly uncertain beliefs, and (ii) a process of means-end reasoning to decide on a sequence of actions to achieve the selected intentions [40]. As an example of a typical practical reasoning scenario, consider the following situation.

D. Calvaresi et al. (Eds.): EXTRAAMAS 2021, LNAI 12688, pp. 260–279, 2021.
https://doi.org/10.1007/978-3-030-82017-6_16

Example 1. Doc is a medical rescue agent operating in a disaster zone. Doc's task is to assess the victims' injuries and accordingly decide either to treat them on the spot or take them to the hospital. For most cases, a victim with any open wounds must be taken to the hospital. Due to the hospital's limited capacity, Doc should not take a victim it can treat to the hospital. For this reason, it is awarded points for treating victims that it can treat on the spot. Doc is assigned a case of a little girl with what appears to Doc to be an open wound. Doc thinks it can treat her on the spot, and actually desires to treat her to earn more reward points. At the same time, Doc feels obliged to take the girl to the hospital as it is not very certain that it can treat her. What should Doc do? □

In the above example, Doc needs to do practical reasoning in order to either form an intention to treat the girl or take her to the hospital. However, it can not intend both as it believes that it can not do both. Practical reasoning is thus a process of filtering out an agent's motivations to form its intentions guided by its beliefs. In practical settings, the agent's beliefs and motivations are being continuously revised. Consequently, the revision of beliefs and motivations must be reflected on the agent's intentions. In our example, for instance, if Doc drops its belief that it can not treat a victim on the spot and take them to the hospital, Doc can now have both intentions. Similarly, if it drops its obligation to take the girl to the hospital, it will only have the intention to treat the girl on the spot to earn more reward points. This process of revising the agent's beliefs and motivations to preserve the consistency of its beliefs and intentions will be referred to as *the joint revision of beliefs and intentions*. Unlike the tremendous body of literature on belief revision, this problem of the joint revision of beliefs and intentions has been scarcely investigated [22].

As autonomous artificial agents become more prevalent in a plethora of domains, it becomes an impelling priority to endow agents with explainable practical reasoning capabilities if we are to trust them to perform critical tasks [27]. A rescue agent like the one presented in the previous example, for instance, must be able to explain why it chose to treat a victim rather than take them to the hospital. For this reason, the topic of eXplainable Artificial Intelligence (XAI) has attracted a lot of research interest in both academia and industry with the main objective of developing transparent and understandable AI systems. However, most of the recent work on the topic focuses, almost exclusively, on explaining the behaviour of black-box machine learning algorithms such as deep neural networks (DNNs) [41, for instance]. Explaining the behaviour of goal-driven logical agents has received much less attention in comparison [1]. The existing approaches for explaining the behaviour of logic-based agents predominantly focus on keeping track of the chain of reasoning that led the agent to hold a belief or adopt an intention. Such explanations are not sufficient in a typical dynamic world where the agent's beliefs and intentions are always being revised. Explaining the reasons why beliefs and intentions were accepted or rejected after the revision process is definitely needed if the agent's reasoning is truly transparent. In our rescue agent example, Doc must be able to explain why it believes the girl has an open wound. Further, if Doc drops this belief and accordingly does not intend to take her the hospital anymore, it must be able to also explain why it dropped its belief and intention.

Several attempts have been made over the years to come up with theories of practical reasoning, however, an adequate theory remains missing [39]. An adequate framework of practical reasoning must be able to (i) capture the process of intention formation, (ii) model the joint revision of beliefs and intentions, and (iii) provide explanations for the chosen beliefs and intentions after the revision. Unfortunately, the numerous approaches in the literature for modelling practical reasoning fail to possess at least one of the aforementioned capabilities. With the objective of addressing this gap, we present in this paper formal algebraic semantics for a logical language we refer to as $Log_A\mathbf{PR}$ ("*Log*" stands for logic, "*A*" for algebraic, and "**PR**" for practical reasoning) that can be utilized for *first-person* practical reasoning with its beliefs and different motivations while possessing the above mentioned three capabilities all in one framework. We do not restrict the agent's motivations to only desires or goals. Rather, in $Log_A\mathbf{PR}$, reasoning with any number of arbitrary motivations can be modelled such as desires, obligations, norms, promises, ... etc. The utility of adopting the first person perspective has been investigated in the literature [2, 13, 32]. Following these approaches, we use $Log_A\mathbf{PR}$ to model the internal reasoning of an agent *with* its beliefs and motivations rather than model an external observer's reasoning *about* the beliefs and motivations of an agent. We also follow [9, 11] and treat intentions as a derived attitude from beliefs and motivations rather than treating them as a basic mental attitude.

In giving the semantics of $Log_A\mathbf{PR}$, we depart from the mainstream modal approaches and take the *algebraic* route, which has several merits that we presented in previous work [19,24,25]. $Log_A\mathbf{PR}$ is a recent addition to our growing family of algebraic logics. Hence, independent motivations for the algebraic approach are also motivations for $Log_A\mathbf{PR}$. The algebraic approach is based on an ontological commitment to propositions as first-class individuals in the universe of discourse; this leads to a language with no sentences, but with functional terms taken to denote propositions. Though non-standard, the inclusion of propositions in the ontology has been suggested by several authors [3, 10, 31, 36]. What does this buy us? Take $Log_A\mathbf{B}$ [24] for example. As an algebraic language for reasoning about beliefs, $Log_A\mathbf{B}$ strikes a middle ground between two major approaches to doxastic logic: the dominant, modal approach [15, for example] and the (now relatively out of fashion) first-order syntactical approach [26, 33, for instance]. This allows $Log_A\mathbf{B}$ to avoid problems of logical omniscience, which mar the classical modal approach, while staying immune to paradoxes of self-reference plaguing the syntactical approach. Another example is $Log_A\mathbf{G}$ [19], an algebraic logic for non-monotonic reasoning about graded beliefs. As proven in [19], $Log_A\mathbf{G}$ can capture a wide array of non-monotonic reasoning formalisms providing a unified framework for non-monotonicity. Last but not least, $Log_A\mathbf{C}_n$ [25], which is an algebraic logic for reasoning about preference, desire, and obligation, avoids the so-called paradoxes of deontic logic [28] by, again, abandoning classical possible-worlds semantics. These different motivations for the algebraic approach suggest that it is only natural to consider a language like $Log_A\mathbf{PR}$ if one is to model practical reasoning the algebraic way.

The rest of the paper is structured as follows. In Sect. 2, we present the algebraic foundations on which our proposed logic is based. Next, in Sect. 3, we present the syntax and semantics of $Log_A\mathbf{PR}$. In Sect. 4, we present $Log_A\mathbf{PR}$ theories and our monotonic logic consequence relations for beliefs and motivations. Next, in Sect. 5, we present corresponding non-monotonic consequence relations for reasoning with and jointly revising *graded* beliefs and motivations to form a consistent set of *intentions*. In Sect. 6, we present our account of explanations for both the chosen and rejected beliefs and intentions after the revision. In Sect. 7, we show how our proposed framework addresses several gaps in the related work. Finally, in Sect. 8, we make some concluding remarks. All the proofs of the theorems presented in this paper has been omitted for space limitations but can be found at [18].

2 Algebraic Foundations

$Log_A\mathbf{PR}$ is based on Boolean algebras. We start by reviewing Boolean algebra and then extend the classical Boolean algebraic notion of *filter* to what we refer to as a *multifilter*. In the next section, multifilters are used to define a logical consequence relation for $Log_A\mathbf{PR}$ accommodating practical reasoning with multiple mental attitudes.

Definition 1. *A Boolean algebra is a sextuple $\mathfrak{A} = \langle \mathcal{P}, +, \cdot, -, \bot, \top \rangle$ where $\mathcal{P}$ is a non-empty set with $\{\bot, \top\} \subseteq \mathcal{P}$. $\mathcal{P}$ is closed under the two binary operators $+$ and $\cdot$ and the unary operator $-$ with commutativity, associativity, absorption, and complementation properties as detailed in [35].*

The partial order $\leq$ on $\mathcal{P}$ is such that, for any $p, q \in \mathcal{P}$, $p \leq q$ if $p \cdot q = p$. Since we are using Boolean algebras for reasoning purposes, we intuitively take elements of $\mathcal{P}$ to be propositions and the operators $+$, $\cdot$, and $-$ to be disjunction, conjunction, and negation respectively. The order $\leq$ is provably a Tarskian logical consequence relation [24]. On this interpretation, classical Boolean-algebraic *filters* [35] are logically-closed sets of propositions. Henceforth, an arbitrary Boolean algebra $\mathfrak{A} = \langle \mathcal{P}, +, \cdot, -, \bot, \top \rangle$ is assumed.

Definition 2. *A filter of $\mathfrak{A}$ is a subset F of $\mathcal{P}$ where:*

1. *$\top \in F$;*
2. *if $p, q \in F$, then $p \cdot q \in F$; and*
3. *if $p \in F$ and $p \leq q$, then $q \in F$.*

The filter generated by $\mathcal{Q} \subseteq \mathcal{P}$ is the smallest filter $F(\mathcal{Q})$ of which $\mathcal{Q}$ is a subset.

As practical reasoning involves joint reasoning with *multiple* mental attitudes, we are interested, not in sets of propositions, but in *tuples* of such sets. Hence, we generalize filters—logically-closed sets of propositions—to what we refer to as *multifilters*. The generalization is twofold. First, multifilers are *closed tuples* of sets of propositions. Second, since the propositions in different mental attitudes are interdependent and need not be *logically*-closed (motivations, for example [11]), we generalize the classical order $\leq$ on propositions, on which filters are based, to a more liberal order on *tuples of sets of propositions*. For a positive integer k, a k-preorder on $\mathfrak{A}$ is a preorder $\preceq_k$ on $(2^{\mathcal{P}})^k$.

Definition 3. *Let $\mathcal{Q} = \langle \mathcal{Q}_1, \ldots, \mathcal{Q}_k \rangle$ be a tuple of subsets of $\mathcal{P}$ and $\preceq$ be a k-preorder on $\mathfrak{A}$. The $\preceq$-multifilter generated by $\mathcal{Q}$, denoted $\mathcal{F}_{\preceq}(\mathcal{Q})$, is a tuple $\langle FQ_1, \ldots, FQ_k \rangle$ where FQ_i, for all $1 \leq i \leq k$, is the smallest set containing $\mathcal{Q}_i$ such that $\top \in FQ_i$ and if $\langle P_1, \ldots, P_k \rangle \preceq \langle P_1', \ldots, P_k' \rangle$ and $P_i \subseteq FQ_i$, then $P_i' \subseteq FQ_i$.*

The above properties of multifilters are intended generalizations of the three conditions on filters presented in Definition 2 to accommodate reasoning with multiple sets of propositions rather than a single set by utilising a k-preorder. To accommodate reasoning with mental states with attitudes that have different logical properties, the only required closure condition on multifilters is based entirely on the k-preorder on tuples of sets.

3 Log_APR Languages

The syntax of Log_A**PR** consists of terms constructed algebraically from function symbols. There are no sentences; instead, we use terms of a distinguished syntactic type to denote propositions. Propositions are included as first-class individuals in the Log_A**PR** ontology and are structured in a Boolean algebra. *Grades* are also taken to be first-class individuals. As a result, propositions *about* graded beliefs and motivations can be constructed, which are themselves recursively gradable. A Log_A**PR** language $\mathcal{L}$ is a many-sorted language composed of a set of terms partitioned into three base sorts: σ_P is a set of terms denoting propositions, σ_G is a set of terms denoting grades, and σ_I is a set of terms denoting anything else. A Log_A**PR** *signature* Ω includes a non-empty, countable set of constant and function symbols each having a syntactic sort from the set $\sigma = \{\sigma_P, \sigma_G, \sigma_I\} \cup \{\tau_1 \longrightarrow \tau_2 \mid \tau_1 \in \{\sigma_P, \sigma_G, \sigma_I\} \text{ and } \tau_2 \in \sigma\}$ of syntactic sorts. Intuitively, $\tau_1 \longrightarrow \tau_2$ is the syntactic sort of function symbols that take a single argument of sort σ_P, σ_G, or σ_I and produce a symbol of sort τ_2. In addition, an alphabet Ω includes a countably infinite set of variables of the three base sorts; a set of syncategorematic symbols including the comma, various matching pairs of brackets and parentheses, and the symbol $\forall$; and a set of logical symbols defined as the union of the following sets: (i) $\{\neg\} \subseteq \sigma_P \longrightarrow \sigma_P$, (ii) $\{\wedge, \vee\} \subseteq \sigma_P \times \sigma_P \longrightarrow \sigma_P$, (iii) $\{<, \doteq\} \subseteq \sigma_G \times \sigma_G \longrightarrow \sigma_P$, and (iv) $\{\mathbf{A}_i\}_{i=1}^{k} \subseteq \sigma_P \times \sigma_G \longrightarrow \sigma_P$. $\mathbf{A}_i(p, g)$ denotes the proposition that the agent has *attitude* i towards proposition p with a grade of g. This attitude could be a belief or some kind of a motivation. In the remainder of this paper, we assume a Log_A**PR** language $\mathcal{L}$ with a signature Ω.[1]

The basic ingredient of the Log_A**PR** semantic apparatus is a Log_A**PR** *structure*.

Definition 4. *A* Log_A**PR** *structure is a quintuple* $\mathfrak{S}_k = \langle \mathcal{D}, \mathfrak{A}, \mathfrak{a}, \ll, \mathfrak{e} \rangle$ *where:*

- $\mathcal{D}$*, the domain of discourse, is a countable set comprising three disjoint sets: (i) a set of propositions* $\mathcal{P}$*, (ii) a set of grades* $\mathcal{G}$*, and (iii) the set* $\overline{\mathcal{P} \cup \mathcal{G}}$ *of other entities;*
- $\mathfrak{A} = \langle \mathcal{P}, +, \cdot, -, \bot, \top \rangle$ *is a complete, non-degenerate Boolean algebra [35].*
- $\mathfrak{a} = \{\mathfrak{a}_i \mid 1 \leq i \leq k \text{ and } \mathfrak{a}_i : \mathcal{P} \times \mathcal{G} \longrightarrow \mathcal{P}\}$ *is a set of k grading functions;*
- $\ll: \mathcal{G} \times \mathcal{G} \longrightarrow \mathcal{P}$ *is a ordering function imposing a total order on* $\mathcal{G}$*;*
- $\mathfrak{e} : \mathcal{G} \times \mathcal{G} \longrightarrow \{\bot, \top\}$ *is an equality function, where for every* $g_1, g_2 \in \mathcal{G}$*:* $\mathfrak{e}(g_1, g_2) = \top$ *if* $g_1 = g_2$*, and* $\mathfrak{e}(g_1, g_2) = \bot$ *otherwise.*

A *valuation* $\mathcal{V}$ of a Log_A**PR** language is a triple $\langle \mathfrak{S}_k, \mathcal{V}_F, \mathcal{V}_X \rangle$, where $\mathfrak{S}_k$ is a Log_A**PR** structure, $\mathcal{V}_F$ is a function that assigns to each function symbol an appropriate

[1] Terms involving '$\Rightarrow$' (material implication), '$\Leftrightarrow$' (equivalence), and '$\exists$' are abbreviations defined in the standard way.

function on $\mathcal{D}$, and $\mathcal{V}_X$ is a function mapping each variable to a corresponding element of the appropriate block of $\mathcal{D}$. For a valuation $\mathcal{V} = \langle \mathfrak{S}, \mathcal{V}_F, \mathcal{V}_X \rangle$ with variable x and $a \in \mathcal{D}$, $\mathcal{V}[a/x] = \langle \mathfrak{S}, \mathcal{V}_F, \mathcal{V}_X[a/x] \rangle$, where $\mathcal{V}_X[a/x](x) = a$, and $\mathcal{V}_X[a/x](y) = \mathcal{V}_X(y)$ for every $y \neq x$. An *interpretation* of $Log_A\mathbf{PR}$ terms is given by a function $[\![\cdot]\!]^{\mathcal{V}}$. The behaviour of the interpretation function is depicted in Fig. 1.

Definition 5. *Let $\mathcal{L}$ be a $Log_A\mathbf{PR}$ language and let $\mathcal{V}$ be a valuation of $\mathcal{L}$. An* interpretation *of the terms of $\mathcal{L}$ is given by a function $[\![\cdot]\!]^{\mathcal{V}}$:*

- $[\![x]\!]^{\mathcal{V}} = \mathcal{V}_X(x)$, *for a variable x.*
- $[\![c]\!]^{\mathcal{V}} = \mathcal{V}_F(c)$, *for a constant c.*
- $[\![f(t_1, \ldots, t_m)]\!]^{\mathcal{V}} = \mathcal{V}_F(f)([\![t_1]\!]^{\mathcal{V}}, \ldots, [\![t_m]\!]^{\mathcal{V}})$, *for an m-adic ($m \geq 1$) function f.*
- $[\![(t_1 \wedge t_2)]\!]^{\mathcal{V}} = [\![t_1]\!]^{\mathcal{V}} \cdot [\![t_2]\!]^{\mathcal{V}}$.
- $[\![(t_1 \vee t_2)]\!]^{\mathcal{V}} = [\![t_1]\!]^{\mathcal{V}} + [\![t_2]\!]^{\mathcal{V}}$.
- $[\![\neg t]\!]^{\mathcal{V}} = -[\![t]\!]^{\mathcal{V}}$.
- $[\![\forall x(t)]\!]^{\mathcal{V}} = \prod_{a \in \mathcal{D}} [\![t]\!]^{\mathcal{V}[a/x]}$.
- $[\![\mathbf{A}_i(t_1, t_2)]\!]^{\mathcal{V}} = \mathfrak{a}_i([\![t_1]\!]^{\mathcal{V}}, [\![t_2]\!]^{\mathcal{V}})$.
- $[\![t_1 \lessdot t_2]\!]^{\mathcal{V}} = [\![t_1]\!]^{\mathcal{V}} \ll [\![t_2]\!]^{\mathcal{V}}$.
- $[\![t_1 \doteq t_2]\!]^{\mathcal{V}} = \mathfrak{e}([\![t_1]\!]^{\mathcal{V}}, [\![t_2]\!]^{\mathcal{V}})$.

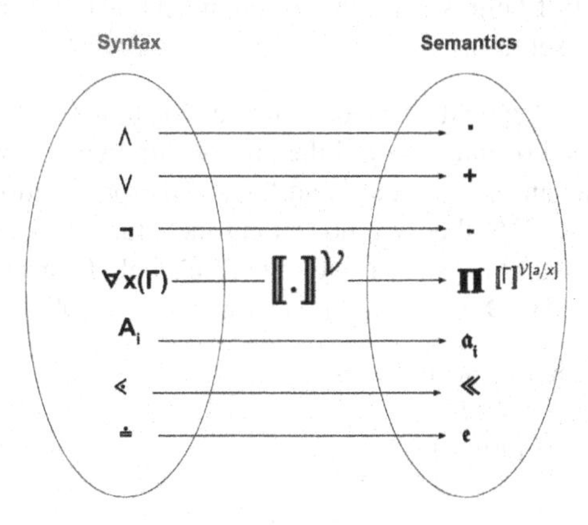

Fig. 1. The interpretation of $Log_A\mathbf{PR}$ terms.

4 Monotonic Logical Consequence

Having defined the syntax and semantics of $Log_A\mathbf{PR}$, we proceed to defining $Log_A\mathbf{PR}$ theories and presenting a monotonic logical consequence relation for $Log_A\mathbf{PR}$. In defining this relation, we employ our notion of multifilters from Sect. 2.

Definition 6. *A* $Log_A\mathbf{PR}$ *theory* $\mathbb{T}$ *is a quadruple* $\langle \mathbb{A}, \mathbb{R}, \mathbb{C}, b \rangle$*, where:*

- $\mathbb{A} = \langle \mathbb{A}_1, \ldots, \mathbb{A}_k \rangle$ *is a k-tuple of subsets of* σ_P *representing the agent's beliefs and* $k-1$ *motivation types;*
- $\mathbb{R}$ *is a set of bridge rules each of the form* $A_1, \ldots, A_k \longmapsto A'_1, \ldots, A'_k$ *where* $A_1, \ldots, A_k \subseteq \sigma_P$ *and* $A'_1, \ldots, A'_k \subseteq \sigma_P$*;*
- $\mathbb{C} \subseteq \{1, \ldots, k\}$*; and*
- $b \in \mathbb{C}$ *is the position of the set of beliefs in* $\mathbb{A}$*.*

Each set $\mathbb{A}_1, ..., \mathbb{A}_k$ in $\mathbb{A}$ represents a separate mental attitude of the agent including its beliefs and different types of motivations. If a propositional term $\phi \in \mathbb{A}_i$, then the agent has the attitude i towards ϕ. The existence of $\mathbf{A}_i(\phi, g)$ in $\mathbb{A}_j$ for some $1 \leq i, j \leq k$ reflects that the agent has the attitude j towards $\mathbf{A}_i(\phi, g)$. This facilitates the representation of introspection and higher-order attitudes. The order of the attitudes in $\mathbb{A}$ represents the agent's character which will be utilized later for conflict resolution. The bridge rules serve to "bridge" propositions across the different mental attitudes. Intuitively, a bridge rule $A_1, \ldots, A_k \longmapsto A'_1, \ldots, A'_k$ means that if every A_i follows from $\mathbb{A}_i$, then every $\mathbb{A}'_j$ may be expanded with A'_j. The set $\mathbb{C}$ specifies the positions of the sets in $\mathbb{A}$ that are closed under conjunction. In typical cases, only the set of beliefs will be closed under conjunction. However, we do not enforce this for generality since we do not restrict $Log_A\mathbf{PR}$ theories to include particular types of motivations. Some sets of motivations can be closed under conjunction as well if their indices are in $\mathbb{C}$. To distinguish the set of beliefs from the different sets of motivations, b indicates the position of the beliefs set in $\mathbb{A}$.

Example 2. Let "$t(x)$" denote the proposition that Doc treats x, "$h(x)$" the proposition that Doc takes x to the hospital, "$w(x)$" the proposition that x is wounded, "$o(x)$" the proposition that x has an open wound, and "$p(x)$" the proposition that x earns more points. A possible $Log_A\mathbf{PR}$ theory representing the situation described in Example 1 is $\mathbb{T} = \langle \langle \mathbb{A}_1, \mathbb{A}_2, \mathbb{A}_3 \rangle, \mathbb{R}, \{1\}, 1 \rangle$. $\mathbb{A}_1$ represents Doc's beliefs, $\mathbb{A}_2$ represents its desires, and $\mathbb{A}_3$ its obligations. **B**, **D**, and **O** are mnemonics for $\mathbf{A}_1$, $\mathbf{A}_2$, and $\mathbf{A}_3$, and:

- $\mathbb{A}_1$ is made up of the following terms:
  b1. $\mathbf{B}(\forall x[o(x) \Rightarrow h(x)], 2)$
  b2. $\mathbf{B}(\forall x[t(x) \Leftrightarrow \neg h(x)], 5)$
  b3. $w(G)$
  b4. $\mathbf{B}(o(G), 6)$
  b5. $\mathbf{B}(t(G), 4)$
  in addition to beliefs about the natural order of the grades (interpreted as rational numbers).
- $\mathbb{A}_2 = \{\mathbf{D}(p(Doc), 10)\}$ and $\mathbb{A}_3 = \{\}$.
- $\mathbb{R}$ is the set of instances of the following rule schema where x and g are variables.
  r1. $\{w(x)\}, \{\mathbf{D}(p(Doc), g)\}, \{\} \longmapsto \{\}, \{\mathbf{D}(t(x), g)\}, \{\}$.
  r2. $\{\mathbf{B}(t(x), g) \wedge g < 5\}, \{\}, \{\} \longmapsto \{\}, \{\}, \{\mathbf{O}(h(x), 10)\}$.

b1 represents Doc's uncertain belief that if x is with an open wound, then Doc takes x to the hospital; **b2** represents Doc's uncertain belief that Doc treats x if and only if Doc does not take x to the hospital; **b3** represents Doc's certain belief that G is wounded;

and **b4** and **b5** represent Doc's uncertain beliefs that G has an open wound and that Doc can treat G respectively. $\mathbb{A}_2$ contains only Doc's desire to earn more points with a degree of 10. **r1** represents the bridge rule that if Doc desires to earn points with a grade of g while believing that x is wounded, then Doc desires to treat x with the same grade g. **r2** represents that if Doc believes that he treats x with a grade smaller than 5, then Doc is obliged to take x to the hospital with a higher grade of 10. □

In Sect. 2, we defined multifilters based on an arbitrary partial order $\preceq$. We now define how to construct such an order for the tuples of sets propositions in $\mathcal{P}$ in light of a $Log_A\mathbf{PR}$ theory. $\circ$ is used to denote sequence concatenation in what follows.

Definition 7. *Let $\mathbb{T} = \langle \mathbb{A}, \mathbb{R}, \mathbb{C}, b \rangle$ be a $Log_A\mathbf{PR}$ theory and $\mathcal{V}$ a valuation. A $\mathbb{T}^{\mathcal{V}}$-induced order, denoted $\preceq_{\mathbb{T}^{\mathcal{V}}}$, is a preorder over $(2^{\mathcal{P}})^k$ such that:*

1. *if $A_1, \ldots, A_k \longmapsto A'_1, \ldots, A'_k \in \mathbb{R}$, then $\langle [\![A_1]\!]^{\mathcal{V}}, \ldots, [\![A_k]\!]^{\mathcal{V}} \rangle \preceq_{\mathbb{T}^{\mathcal{V}}} \langle [\![A'_1]\!]^{\mathcal{V}}, \ldots, [\![A'_k]\!]^{\mathcal{V}} \rangle$.*
2. *If $i \in \mathbb{C}$ and $P_i \subseteq \mathcal{P}$, then $\langle \{\} \rangle^{i-1} \circ P_i \circ \langle \{\} \rangle^{k-i} \preceq_{\mathbb{T}^{\mathcal{V}}} \langle \{\} \rangle^{i-1} \circ \langle \{ \prod_{p \in P_i} p \} \rangle \circ \langle \{\} \rangle^{k-i} \rangle$.*
3. *if $p, q \in \mathcal{P}$ and $p \leq q$, then $\langle \{\} \rangle^{b-1} \circ \langle \{p\} \rangle \circ \langle \{\} \rangle^{k-b} \preceq_{\mathbb{T}^{\mathcal{V}}} \langle \{\} \rangle^{b-1} \circ \langle \{q\} \rangle \circ \langle \{\} \rangle^{k-b}$.*

The intuition is that $\preceq_{\mathbb{T}^{\mathcal{V}}}$ is induced by the bridge rules in a $Log_A\mathbf{PR}$ theory (Condition 1), the set $\mathbb{C}$ (Condition 2), and the natural order $\leq$ among the beliefs (Condition 3). The natural order is only considered for the set of beliefs as motivations are typically not closed under logical implication.

We next utilise a multifilter based on a $\mathbb{T}^{\mathcal{V}}$-induced order to define a set of logical consequence relations one for beliefs and one for each type of motivation. The intuition is that ϕ is an i-consequence if its denotation is in the ith set of the multifilter based on the $\mathbb{T}^{\mathcal{V}}$-induced order.

Definition 8. *Let $\mathbb{T} = \langle \mathbb{A}, \mathbb{R}, \mathbb{C}, b \rangle$ be a $Log_A\mathbf{PR}$ theory. For any $\phi \in \sigma_P$, ϕ is an i-consequence of $\mathbb{T}$, denoted $\mathbb{T} \models_i \phi$, if for every valuation $\mathcal{V}$, $\mathcal{F}_{\preceq_{\mathbb{T}^{\mathcal{V}}}}([\![\mathbb{A}]\!]^{\mathcal{V}}) = \langle FQ_1, \ldots, FQ_k \rangle$ and $[\![\phi]\!]^{\mathcal{V}} \in FQ_i$.*

Example 3. Recall the $Log_A\mathbf{PR}$ theory presented in Example 2, the following are examples of the possible belief and motivation consequences. We use $\models_B, \models_D$, and $\models_O$ instead of $\models_1, \models_2$, and $\models_3$ respectively for readability.

- $\mathbb{T} \models_B \mathbf{B}(o(G), 6) \wedge \mathbf{B}(t(G), 4)$
- $\mathbb{T} \models_D \mathbf{D}(p(D), 10), \mathbb{T} \models_D \mathbf{D}(t(G), 10), \mathbb{T} \models_O \mathbf{O}(h(G), 10)$

The first belief consequence is a classical logical consequence of the set of beliefs since only the set of beliefs is closed under conjunction. $\mathbf{D}(p(D), 10)$ is a desire consequence as it is in $\mathbb{A}_2$. $\mathbf{D}(t(G), 10)$ and $\mathbf{O}(h(G), 10)$ are desire and obligation consequences respectively as a result of applying the bridge rules **r1** and **r2**. Note that the graded belief $o(G)$, for example, is not a belief consequence of $\mathbb{T}$. This, in addition to defining intention consequences from the different motivations consequences, will be addressed in the next section. □

We next examine the properties of our just defined consequence relations. Each $\models_i$ observes modified versions of the distinctive properties of Tarskian logical consequence (reflexivity, monotony, and cut). Further, each $\models_i$ with $i \neq b$ is non-explosive. This last property is very important for motivations as, for example, an agent can have inconsistent desires without desiring everything.

Theorem 1. *Let* $\mathbb{T} = \langle \mathbb{A}, \mathbb{R}, \mathbb{C}, b \rangle$ *and* $\mathbb{T}' = \langle \mathbb{A}', \mathbb{R}', \mathbb{C}, b \rangle$. *The following properties are true.*

1. **(Reflexivity)** *If* $\phi \in \mathbb{A}_i$, *then* $\mathbb{T} \models_i \phi$.
2. **(Monotony)** *Let* $\mathbb{A}_j \subseteq \mathbb{A}'_j$ *for all* $1 \leq j \leq k$, *and* $\mathbb{R} \subseteq \mathbb{R}'$. *If* $\mathbb{T} \models_i \phi$, *then* $\mathbb{T}' \models_i \phi$.
3. **(Cut)** *Let* $\mathbb{A}'_i = \mathbb{A}_i \cup \{\psi\}$ *for some* i, $1 \leq i \leq k$, *and* $\mathbb{A}'_j = \mathbb{A}_j$ *for* $j \neq i$, *and* $\mathbb{R}' = \mathbb{R}$. *If* $\mathbb{T} \models_i \psi$ *and* $\mathbb{T}' \models_i \phi$, *then* $\mathbb{T} \models_i \phi$.
4. **(Non-Explosion of Motivations)** *If* $\mathbb{T} \models_i \phi$ *and* $\mathbb{T} \models_i \neg\phi$ *for* $i \neq b$, *it is not necessarily the case that* $\mathbb{T} \models_i \psi$ *for all* $\psi \in \sigma_P$.

5 Graded Consequence

As shown in the previous section, multifilters do not account for reasoning with graded propositions or with the process of intention formation. In this section, we extend multifilters to handle both issues. Consider the *Log*$_A$**PR** theory presented in Example 2. Given that Doc believes $\mathbf{B}(o(G), 4)$, and does not believe $\neg o(G)$, it makes sense for it to accept $o(G)$ despite its uncertainty about it. (Who is ever absolutely certain of their beliefs?) Similarly, it makes sense for Doc to add its desire $p(Doc)$ to its intentions if they do not conflict with other intentions or beliefs. However, if we only use multifilters, we will never be able to reason with those graded beliefs and motivations as they are not themselves in the agent's theory but only grading propositions thereof (as shown in Example 3). For this reason, we extend our notion of multifilters into a more liberal notion of *graded multifilters* to enable the agent to conclude, in addition to the consequences of the initial theory, beliefs and motivations graded by the initial beliefs and motivations (like $o(G)$).

Henceforth, we consider a special class of *Log*$_A$**PR** theories to simplify the presentation of the upcoming formal semantics.[2] We consider only *Log*$_A$**PR** theories $\langle \mathbb{A}, \mathbb{R}, \mathbb{C}, b \rangle$ where, for every $1 \leq i, j \leq k$, if $\mathbf{A_i}(\phi, g) \in \bigcup_{l=1}^{k} \mathbb{A}_l$, then (i) ϕ does not contain any occurrence of $\mathbf{A_j}$ and (ii) there is no $g' \neq g$ with $\mathbf{A_i}(\phi, g') \in \bigcup_{l=1}^{k} \mathbb{A}_l$. Since we are modelling joint reasoning with beliefs and different types of motivations, we assume a tuple $\mathcal{Q} = \langle \mathcal{Q}_1, \ldots, \mathcal{Q}_k \rangle$ of subsets of $\mathcal{P}$. If $\mathfrak{a}_i(p, g) \in \mathcal{Q}_i$ we say that p is *graded* in $\mathcal{Q}_i$. The rest of this section is dedicated to formally defining graded multifilters and the non-monotonic consequence relation based on them.

Definition 9. *The tuple of embedded graded propositions in* $\mathcal{Q}$, *denoted* $E(\mathcal{Q})$, *is a tuple* $\langle E(\mathcal{Q}_1), \ldots, E(\mathcal{Q}_k) \rangle$ *where* $E(\mathcal{Q}_i) = \mathcal{Q}_i \cup \{p \mid \mathfrak{a}_i(p, g) \in \mathcal{Q}_i\}$ *for all* $1 \leq i \leq k$.

Example 4. Recall the *Log*$_A$**PR** in Example 2. Given a valuation $\mathcal{V}$, let $\mathcal{Q} = [\![\mathbb{A}]\!]^{\mathcal{V}}$. We use $\mathbf{d}$ and $\mathbf{o}$ as mnemonics for $\mathfrak{a}_2$ and $\mathfrak{a}_3$ respectively. We use the syntactic $\wedge$ and $\Leftrightarrow$ operators here rather than their semantic counterparts for readability.

[2] The complete semantics of *Log*$_A$**PR** without the imposed restrictions can be found in [17].

- $\mathcal{F}_{\preceq_{\mathbb{T}^\mathcal{V}}}(\mathcal{Q}) = \langle F\mathcal{Q}_1, F\mathcal{Q}_2, F\mathcal{Q}_3\rangle$ where $F\mathcal{Q}_1 = F([\![\mathbb{A}_1]\!]^\mathcal{V})$, $F\mathcal{Q}_2 = \{\mathbf{d}(p(Doc), 10), \mathbf{d}(t(G), 10)\}$, and $F\mathcal{Q}_3 = \{\mathbf{o}(h(G), 10)\}$.
- $E(\mathcal{F}_{\preceq_{\mathbb{T}^\mathcal{V}}}(\mathcal{Q})) = \langle E(\mathcal{Q}_1), E(\mathcal{Q}_2), E(\mathcal{Q}_3)\rangle$ where:
  - $E(\mathcal{Q}_1) = F\mathcal{Q}_1 \cup \{\forall x[o(x) \Rightarrow h(x)], \forall x[t(x) \Leftrightarrow \neg h(x)], w(G), o(G), t(G)\}$.
  - $E(\mathcal{Q}_2) = F\mathcal{Q}_2 \cup \{p(Doc), t(G)\}$, $E(\mathcal{Q}_3) = F\mathcal{Q}_3 \cup \{h(G)\}$. □

A problem that we have to circumvent is that extracting the embedded graded propositions can introduce inconsistencies among the agent's beliefs and motivations. As shown in Example 4, Doc believes that he can not treat the girl on the spot and take her to the hospital, desires to treat the girl, and is obliged to take her to the hospital. To resolve the introduced inconsistencies, we allude to the agent's *character* and the grades of the inconsistent propositions to resolve them resulting in a consistent set of beliefs and motivations. Given a Log_A**PR** theory $\mathbb{T} = \langle \mathbb{A}, \mathbb{R}, \mathbb{C}, b\rangle$, the agent's character is represented by the order of the attitudes in $\mathbb{A}$. In Example 2, for instance, since beliefs appears before desires and desires before obligations, then Doc prefers to give up beliefs then desires then obligations to resolve any inconsistencies. After resolving all the inconsistencies, the union of all the motivations sets makeup the agent's *intentions*. From now on, we will refer to extracting the embedded graded propositions in $\mathcal{Q}$ and handling the inconsistencies in the resulting tuple as the *telescoping* of $\mathcal{Q}$. A central tool that will be used in telescoping $\mathcal{Q}$ is *telescoping structures*.

Definition 10. *Let $\mathfrak{S}_k$ be a Log_A**PR** structure. A telescoping structure for $\mathfrak{S}_k$ is a triple $\mathfrak{T} = \langle \mathcal{T}, \mathfrak{O}, \mathfrak{s}\rangle$ where:*

- *$\mathcal{T} = \langle \mathcal{T}_1, \ldots, \mathcal{T}_k\rangle$, with $\mathcal{T}_1, \ldots, \mathcal{T}_k \subseteq \mathcal{P}$, is the tuple of top propositions;*
- *$\mathfrak{O}$ is an ultrafilter of the subalgebra induced by $Range(\ll)$ (an ultrafilter is a maximal filter with respect to not including $\perp$ [35]); and*
- *$\mathfrak{s} = \langle \mathfrak{s}_1, \mathfrak{s}_2\rangle$ is a pair of selection functions with $\mathfrak{s}_1 : 2^\mathcal{P} \longrightarrow \mathcal{P}$ and $\mathfrak{s}_2 : 2^{(2^\mathcal{P})^k} \longrightarrow (2^\mathcal{P})^k$.*

The telescoping structure provides the tuple of top propositions $\mathcal{T}$ that will never be given up together with their consequences. $\mathcal{T}$ is the semantic counterpart to $\mathbb{A}$ in a Log_A**PR** theory. Thus $\mathcal{T}_b$ is the set of top beliefs and $\mathcal{T}_i$, where $i \neq b$, is a set of top motivations of type i. Just like we did with $\mathbb{A}$, we assume that the order of the sets in $\mathcal{T}$ represents the agent character from the least preferred to the most preferred. The ultrafilter $\mathfrak{O}$ provides a total order over grades to enable comparing them. The selection function $\mathfrak{s}_1$ picks a proposition from a set of propositions, and $\mathfrak{s}_2$ picks a tuple of sets of propositions from a set of k-tuples of sets of propositions. The order of the sets in $\mathcal{T}$, the ultrafilter, and the selection functions will all be used when picking a consistent set of beliefs and a consistent set of motivations making up the agent's intentions.

We now proceed to defining precisely how the inconsistencies are resolved. To fulfill this aim, we generalize the notion of a kernel of a belief base [20] to suit reasoning with multiple sets of propositions. The intuition is that a $\perp$-kernel is a tuple of sets, one for each mental attitude, where the union of the sets is a subset-minimal inconsistent set. This means that if we remove a single proposition from one of the sets in the $\perp$-kernel, the union becomes consistent. In what follows, we say that a set $\mathcal{R} \subseteq \mathcal{P}$ is inconsistent whenever the classical filter of $\mathcal{R}$ is proper ($F(\mathcal{R}) = \mathcal{P}$). Otherwise, $\mathcal{R}$ is inconsistent.

Definition 11. *Let* $\mathcal{X} = \langle \mathcal{X}_1, \ldots, \mathcal{X}_k \rangle$*, and the multifilter* $\mathcal{F}_{\preceq}(\mathcal{X}) = \langle F\mathcal{X}_1, \ldots, F\mathcal{X}_k \rangle$*.* $\mathcal{X}$ *is a* $\perp$*-kernel of* $\mathcal{Q}$ *if and only if each* $\mathcal{X}_i \subseteq \mathcal{Q}_i$*, and* $\bigcup_{i=1}^{k} F\mathcal{X}_i$ *is a subset-minimal inconsistent set. The set of* $\perp$*-kernels in* $\mathcal{Q}$ *will be referred to as* $\mathcal{Q}_{\perp}$*.*

Example 5. We go back to Example 4. The following are the $\perp$-kernels in $E(\mathcal{F}_{\preceq_{\mathrm{T}\mathcal{V}}}(\mathcal{Q}))$. The first set in each $\perp$-kernel represents beliefs of Doc's, the second represents desires thereof, and the third obligations. Recall that $\forall x[t(x) \Leftrightarrow \neg h(x)]$ has a grade of 5, $t(G)$ has a grade of 4, $o(G)$ has a grade of 6, and $\forall x[o(x) \Rightarrow h(x)]$ has a grade of 2.

1. $\langle \{\forall x[t(x) \Leftrightarrow \neg h(x)], t(G), o(G), \forall x[o(x) \Rightarrow h(x)]\}, \{\}, \{\} \rangle$.
2. $\langle \{\forall x[t(x) \Leftrightarrow \neg h(x)]\}, \{t(G)\}, \{h(G)\} \rangle$.
3. $\langle \{\forall x[t(x) \Leftrightarrow \neg h(x)], o(G), \forall x[o(x) \Rightarrow h(x)]\}, \{t(G)\}, \{\} \rangle$.
4. $\langle \{\forall x[t(x) \Leftrightarrow \neg h(x)], t(G)\}, \{\}, \{h(G)\} \rangle$

The first kernel shows an inconsistency between Doc's beliefs, the second between Doc's beliefs, desires, and obligations, the third between Doc's beliefs and desires, and the fourth between Doc's beliefs and obligations. □

But how do we choose propositions to give up from a $\perp$-kernel and resolve inconsistency? The intuition is this: the proposition to be given up must be the proposition with the *least grade from the least preferred set* in the mental state. Recall that the mental state is ordered from the least preferred to the most preferred set. The role of the selection function $\mathfrak{s}_1$ becomes evident here. It selects a proposition with the least grade from several propositions with the minimum grades in the least preferred set according to the agent's character. Utilizing $\mathfrak{s}_1$ rather than just giving up all the propositions with the least grade is motivated by giving up as few beliefs and motivations as possible. This is called for as in practical reasoning we typically want to retain as much beliefs and motivations as possible. In what follows, given a $\perp$-kernel $\mathcal{X} = \langle \mathcal{X}_1, \ldots, \mathcal{X}_k \rangle$, we say that $\mathcal{X}_i$ is the *least preferred set* in $\mathcal{X}$ if and only if $\mathcal{X}_i$ is the left-most set in $\mathcal{X}$ and $\mathcal{X}_i \not\subseteq FT_j$ where $\mathcal{F}_{\preceq}(\mathcal{T}) = \langle FT_1, \ldots, FT_k \rangle$. Insisting that $\mathcal{X}_i$ is not a subset of FT_i guarantees that it contains at least one graded proposition that could be given up (recall that the logical consequences of $\mathcal{T}$ can not be given up).

Definition 12. *Let* $\mathcal{X} = \langle \mathcal{X}_1, \ldots, \mathcal{X}_k \rangle$ *be a* $\perp$*-kernel of* $\mathcal{Q}$*,* $\mathfrak{T} = \langle \mathcal{T}, \mathfrak{O}, \mathfrak{s} \rangle$ *be a telescoping structure, and* $\mathcal{F}_{\preceq}(\mathcal{T}) = \langle FT_1, \ldots, FT_k \rangle$*. A proposition* p *does not survive* $\mathcal{X}$ *in* i *given* $\mathfrak{T}$ *if and only if* $p \notin FT_i$ *and all of the following are true.*

1. $p \in \mathcal{X}_i$ *where* $\mathcal{X}_i$ *is the least preferred set in* $\mathcal{X}$*.*
2. $p \in \mathfrak{s}_1(MinGrades)$ *where* $MinGrades$ *is the set of propositions with the minimum grades in the difference of* $\mathcal{X}_i$ *and* FT_i*.*

The question now is which $\perp$-kernels do we select from $\mathcal{Q}_{\perp}$ in order to give propositions up from them? We can just consider all of them, this will surely resolve all the inconsistencies. However, in doing this, we might end up giving up beliefs and motivations unnecessarily. Alternatively, we will only consider a subset of $\mathcal{Q}_{\perp}$. To define this subset, we need to define first what will be referred to as a *next-best* $\perp$-kernel. The idea is that a next-best $\perp$-kernel must be examined first to give up a proposition from according to Definition 12.

Definition 13. *A next-best $\bot$-kernel $\mathcal{X}^*$ is a $\bot$-kernel in $\mathcal{Q}_\bot$ such that:*

1. *$\mathcal{X}^*$ has the longest prefix of sets that are subsets of their corresponding sets in $\mathcal{F}_\preceq(\mathcal{T})$; and*
2. *if $\mathcal{X}_j$ is the least preferred set in $\mathcal{X}^*$, then $\mathcal{X}_j$ contains at least one graded proposition that has the maximum grade among the propositions with the minimum grades in the least preferred sets of all the $\bot$-kernels in $\mathcal{Q}_\bot$.*

The set of next-best kernels in $\mathcal{Q}_\bot$ will be referred to as $\mathcal{N}(\mathcal{Q}_\bot)$.

The motivation behind the first condition is to tend to the $\bot$-kernels that we are forced to remove a proposition from a more preferred attitude according to the agent's character. In Example 5, for instance, even if Doc's character prefers to give up its beliefs the least, we are forced to give up from the first kernel a belief since both the desires and obligations sets are empty. In this case, we consider the first kernel first in hope that resolving the inconsistency in it will resolve the inconsistency in other kernels so we end up considering less kernels and giving up less propositions. The second condition has a similar motivation. We want to tend to the $\bot$-kernels that contain the proposition with the maximum grade among the minimas in the least preferred set of all the kernels in $\mathcal{Q}_\bot$ in hope that if we give up this proposition up other inconsistencies will be resolved as well. We will come back to these two conditions when we present how the $\bot$-kernels in Example 5 will be treated to resolve all the inconsistencies.

We now define the subset of $\mathcal{Q}_\bot$ to be considered by utilizing $\mathcal{N}(\mathcal{Q}_\bot)$. We first pick using $\mathfrak{s}_2$ a next-best $\bot$-kernel $\mathcal{X}^*$ from $\mathcal{N}(\mathcal{Q}_\bot)$. Then, the proposition that does not survive $\mathcal{X}^*$ in i is removed from $\mathcal{Q}_i$, and we get the next-best kernels from the new tuple. This process is repeated until no more $\bot$-kernels remain. That is, all inconsistencies are resolved.

Definition 14. *Let $\mathfrak{T} = \langle \mathcal{T}, \mathfrak{O}, \mathfrak{s} \rangle$ be a telescoping structure. The set of $\bot$-kernels to be considered in $\mathcal{Q}$, denoted $\mathcal{N}^*(\mathcal{Q})$, is defined as follows.*

1. *$\mathcal{N}^*(\mathcal{Q}) = \varnothing$ if $\mathcal{Q}_\bot = \varnothing$.*
2. *Otherwise, $\mathcal{N}^*(\mathcal{Q}) = \{\mathfrak{s}_2(\mathcal{N}(\mathcal{Q}_\bot))\} \cup \mathcal{N}^*(\mathcal{R})$ where $\mathcal{R} = \langle \mathcal{R}_1, \dots, \mathcal{R}_k \rangle$ such that $\mathcal{R}_i = \mathcal{Q}_i - \{p\}$ with p being the proposition that does not survive $\mathfrak{s}_2(\mathcal{N}(\mathcal{Q}_\bot))$ in i given $\mathfrak{T}$, and $\mathcal{R}_j = \mathcal{Q}_j$ for $j \neq i$.*

Employing $\mathcal{N}^*(\mathcal{Q})$, we now present the construction of the *tuple of kernel survivors*. The tuple of kernel survivors constitutes the joint revision of beliefs and intentions as it is removes from the set of beliefs and sets of motivations just enough propositions to guarantee the joint consistency of its beliefs and filtered motivations making up the agents intentions.

Definition 15. *The tuple of kernel survivors of $\mathcal{Q}$ given $\mathfrak{T}$ is $\kappa(\mathcal{Q}, \mathfrak{T})$ where $\kappa(\mathcal{Q}, \mathfrak{T}) = \langle \mathcal{Q}_1 - \mathcal{O}_1, \dots, \mathcal{Q}_k - \mathcal{O}_k \rangle$ where $p \in \mathcal{O}_i$ if and only if p does not survive any $\bot$-kernel $\mathcal{X}$ in i given $\mathfrak{T}$ where $\mathcal{X} \in \mathcal{N}^*(\mathcal{Q})$.*

We finally define graded multifilters as the multifilter of the telescoping of $\mathcal{T}$.

Definition 16. *Let $\mathfrak{T} = \langle \mathcal{T}, \mathfrak{O}, \mathfrak{s} \rangle$ be a telescoping structure. The graded multifilter with respect to S, denoted $\mathfrak{F}_\preceq(\mathfrak{T})$, is $\mathcal{F}_\preceq(\tau_\mathfrak{T}(\mathcal{T}))$ where $\tau_\mathfrak{T}(\mathcal{T}) = \kappa(E(\mathcal{F}_\preceq(\mathcal{T})), \mathfrak{T})$.*

The following theorem states that if the union of the sets in the multifilter of $\mathcal{T}$ is consistent, then the union of the sets in the multifilter after getting $\mathcal{T}$-induced graded multifilter is consistent. This basically means that the process of telescoping is consistency preserving. Accordingly, we can do the revision of the agent's beliefs and intentions while maintaining consistency amongst all the beliefs and intentions.

Theorem 2. *Let $\mathfrak{T}$ be a telescoping structure, $\mathcal{F}_{\preceq}(\mathcal{T}) = \langle FT_1, \ldots, FT_k \rangle$, and $\mathfrak{F}_{\preceq}(\mathfrak{T}) = \langle FT'_1, \ldots, FT'_k \rangle$. If $\bigcup_{i=1}^{k} FT_i$ is consistent, then $\bigcup_{i=1}^{k} FT'_i$ is consistent.*

Just like we used multifilters to define logical consequence in Sect. 3, we now use graded multifilters to define *graded consequence* as follows. We start by defining what will be referred to as a *canon* which are the syntactic counterparts of the selection functions in the telescoping structure. In the sequel, let a canon be a pair $\mathcal{C} = \langle s_1, s_2 \rangle$ where $s_1 : 2^{\sigma_P} \longrightarrow \sigma_P$ and $s_2 : 2^{(2^{\sigma_P})^k} \longrightarrow (2^{\sigma_P})^k$. In what follows, we define a *relevant telescoping structure given a Log_APR theory, a valuation, and a canon.* This notion will be used in the definition of graded consequence.

Definition 17. *Let $\mathbb{T} = \langle \mathbb{A}, \mathbb{R}, \mathbb{C}, b \rangle$ be a Log_APR theory, $\mathcal{V}$ be a valuation, and $\mathcal{C} = \langle s_1, s_2 \rangle$ be a canon. A relevant telescoping structure given $\mathbb{T}$, $\mathcal{V}$, and $\mathcal{C}$ is $\mathfrak{T} = \langle \mathcal{T}, \mathfrak{O}, \mathfrak{s} \rangle$ such that:*

- *$\mathcal{T} = [\![\mathbb{A}]\!]^{\mathcal{V}}$;*
- *$\mathfrak{O}$ is an ultrafilter that extends $F([\![\mathbb{A}_b]\!]^{\mathcal{V}} \cap Range(\ll))^3$;*
- *$\mathfrak{s} = \langle \mathfrak{s}_1, \mathfrak{s}_2 \rangle$ for any a set of propositional terms Γ and propositional term ϕ, if $s_1(\Gamma) = \phi$, then $\mathfrak{s}_1([\![\Gamma]\!]^{\mathcal{V}}) = [\![\phi]\!]^{\mathcal{V}}$, and for any set of k-tuples of sets of propositional terms K and a k-tuple of sets of propositional terms X, if $s_2(K) = X$, then $\mathfrak{s}_2([\![K]\!]^{\mathcal{V}}) = [\![X]\!]^{\mathcal{V}}$.*

We are finally ready to define graded consequence. The intuition is that ϕ is a graded i-consequence if its denotation is in the ith set of the graded multifilter based on the $\mathbb{T}^{\mathcal{V}}$ induced order and a relevant telescoping structure $\mathfrak{T}$.

Definition 18. *Let $\mathbb{T} = \langle \mathbb{A}, \mathbb{R}, \mathbb{C}, b \rangle$ be a Log_APR theory, and canon $\mathcal{C} = \langle s_1, s_2 \rangle$. $\phi \in \sigma_P$ is a graded i-consequence of $\mathbb{T}$ with respect to canon $\mathcal{C}$, denoted $\mathbb{T} \models_i^{\mathcal{C}} \phi$, if, for every valuation $\mathcal{V}$, $\mathfrak{F}_{\preceq_{\mathbb{T}^{\mathcal{V}}}}(\mathfrak{T}) = \langle \mathcal{A}_1, \ldots, \mathcal{A}_k \rangle$ and $[\![\phi]\!]^{\mathcal{V}} \in \mathcal{A}_i$ where $\mathfrak{T}$ is a relevant telescoping structure given $\mathbb{T}$, $\mathcal{V}$, and $\mathcal{C}$. ϕ is a graded belief (intention) consequence of $\mathbb{T}$, denoted $\mathbb{T} \models_B^{\mathcal{C}} \phi$ ($\mathbb{T} \models_I^{\mathcal{C}} \phi$), if $\mathbb{T} \models_b^{\mathcal{C}} \phi$ ($\mathbb{T} \models_j^{\mathcal{C}} \phi$ for any $j \neq b$).*

Example 6. There are six possible characters for Doc reflected by the order of the sets in $\mathbb{A}$. We will indicate the order of the sets as a sequence of the initials of the attitudes in $\mathbb{A}$. We illustrate the construction of graded multifilters for all the possible orders. In each case, we illustrate which propositions Doc ends up believing and intending. The following table shows the set of next-best kernels to be considered $\mathcal{N}^*([\![\mathbb{A}]\!]^{\mathcal{V}})$ according to Doc's character and the removed propositions to resolve the inconsistency. We indicate the kernels by using their numbers in the list shown in Example 5.

[3] An ultrafilter U extends a filter F if $F \subseteq U$ [35].

Agent character	$\mathcal{N}^*([\![\mathbb{A}]\!]^{\mathcal{V}})$	Removed proposition(s)
$\langle B, O, D\rangle$	$\{2\}$	belief: $\forall x[t(x) \Leftrightarrow \neg h(x)]$
$\langle B, D, O\rangle$	$\{2\}$	belief: $\forall x[t(x) \Leftrightarrow \neg h(x)]$
$\langle O, B, D\rangle$	$\{(1\ or\ 3), (2\ or\ 4)\}$	belief: $\forall x[o(x) \Rightarrow h(x)]$, obligation: $h(G)$
$\langle O, D, B\rangle$	$\{1, (2\ or\ 4)\}$	belief: $\forall x[o(x) \Rightarrow h(x)]$, obligation: $h(G)$
$\langle D, B, O\rangle$	$\{4, (2\ or\ 3)\}$	belief: $t(G)$, desire: $t(G)$
$\langle D, O, B\rangle$	$\{1, 4\}$	belief: $\forall x[o(x) \Rightarrow h(x)]$, obligation: $h(G)$

- If Doc's character is selfish idealistic ($\langle B, O, D\rangle$) or idealistic selfish ($\langle B, D, O\rangle$), $\mathcal{N}^*([\![\mathcal{A}]\!]^{\mathcal{V}}) = \{2\}$ since it has the longest prefix of sets that are not subsets of the top propositions and contains the proposition with the maximum grade among the minimas (according to Definition 13). Hence, Doc will give up the belief $\forall x[t(x) \Leftrightarrow \neg h(x)]$, and ends up intending to treat the girl and take her to the hospital. Notice here that if we did not choose the kernel that contains the proposition with the minimum grade among the minimum grade of the propositions in the beliefs set instead, one of the first or third kernels will be chosen giving up the belief with the minimum grade $\forall x[o(x) \Rightarrow h(x)]$, but the fourth kernel will have to be considered next giving up $t(G)$ from Doc's beliefs as well. This illustrates that choosing the kernel that contains the proposition with the maximum grade can end up in giving up less propositions.
- If Doc's character is selfish realistic ($\langle O, B, D\rangle$), the set of next-bests to be considered will comprise one of the first or third kernels and one of the second or fourth kernels based on the choice of the selection function $\mathfrak{s}_2$. In any case, the belief $\forall x[o(x) \Rightarrow h(x)]$ will then be given up from either the first or third kernels and the obligation $h(G)$ will be given up from one of the second or fourth kernels. Doc ends up following its desire and intending to treat the girl on the spot as Doc's character prefers its desires the most. If Doc's character is realistic selfish ($\langle O, D, B\rangle$), the set of next-bests will comprise the first kernel as it has the longest prefix and one of the second or fourth kernels. In this case, again, the belief $\forall x[o(x) \Rightarrow h(x)]$ will be given up from the first kernel and the obligation $h(G)$ from one of the second or fourth kernels. Doc ends up following its desire to treat the girl. Note that here even though Doc's character prefers its beliefs the most, Doc ends up giving up a belief since its beliefs are inconsistent as indicated by the first kernel. Note here that if we did not choose the kernel with the longest prefix (Condition 1 of Definition 13), we would have chosen the second or fourth kernels first giving up the obligation $h(G)$, then the third kernel giving up the desire $t(A)$, then finally the first kernel giving up the belief $\forall x[o(x) \Rightarrow h(x)]$. In this case, Doc would have not have intended to do anything. This illustrates that choosing the kernel with the longest prefix of sets consisting of propositions that follow from the top propositions can end up in giving up less propositions.
- If Doc's character is idealistic realistic ($\langle D, B, O\rangle$), the set of next-bests will comprise the fourth kernel and one of the second or third kernels. In this case, the belief $t(G)$ will be given up from the fourth kernel and the desire $t(G)$ from the second or third kernels. Doc ends up following its obligation by intending to take the girl to the

hospital. If Doc's character is realistic idealistic ($\langle\langle D, O, B\rangle\rangle$), the set of next-bests will comprise the first and fourth kernels. In this case, the belief $\forall x[o(x) \Rightarrow h(x)]$ will be given up from the first kernel and the obligation $h(G)$ will be given up from the fourth kernel. Doc ends up following its desires and intending to treat the girl. Note that even though we expect this character to give up its desires before its obligations, an obligation was given up since the fourth kernel shows an inconsistency between beliefs and obligations and Doc prefers to give up obligations before beliefs. □

6 Explanations

Throughout this section, we assume a $Log_A\mathbf{PR}$ theory $\mathbb{T} = \langle\langle \mathbb{A}_1, \ldots, \mathbb{A}_k\rangle, \mathbb{R}, \mathbb{C}, b\rangle$ and a canon $\mathcal{C} = \langle s_1, s_2\rangle$. We start by presenting a simplistic form of explanations for a belief or an intention ϕ.

Definition 19. *A support for ϕ as a graded belief (intention) consequence of $\mathbb{T}$ given $\mathcal{C}$ is a $Log_A\mathbf{PR}$ theory $\mathbb{T}' = \langle(\mathbb{A}'_1, \ldots, \mathbb{A}'_k), \mathbb{R}', \mathbb{C}, b\rangle$ where $\mathbb{A}'_1, \ldots, \mathbb{A}'_k, \mathbb{R}'$ are the smallest subsets*[4] *of $\mathbb{A}_1, \ldots, \mathbb{A}_k, \mathbb{R}$ respectively where $\mathbb{T}' \approx\!\!\models^{\mathcal{C}}_{B} \phi$ ($\mathbb{T}' \approx\!\!\models^{\mathcal{C}}_{I} \phi$).*

Since $Log_A\mathbf{PR}$ is a non-monotonic logic capturing the joint revision of beliefs and intentions, it makes sense to augment our explanations with reasons why ϕ survived the revision. Moreover, in case ϕ does not survive the revision, the explanation should also include the reason why it did not. Recall that, according to Definition 15, in order for ϕ to be a belief/intention graded consequence, it has to survive all the $\perp$-kernels in the set of next-bests. Similarly, for ϕ not to be a belief/intention graded consequence it has to not survive one of the kernels in the set of next-bests. Hence, we will use the set of next-bests to explain why ϕ is or is not a belief/intention consequence. Since we will provide our explanations in syntactic terms, the syntactic representation of the set of next-bests is defined as follows.

Definition 20. *Let $Q = \langle Q_1, \ldots, Q_k\rangle$ be a tuple of sets of propositional terms. $X = \langle X_1, \ldots, X_k\rangle$ is a syntactic $\perp$-kernel of Q if and only if, for every valuation $\mathcal{V}$, $[\![X]\!]^{\mathcal{V}}$ is a $\perp$-kernel of $[\![Q]\!]^{\mathcal{V}}$ given $\mathfrak{T}$ as per Definition 11 where $\mathfrak{T}$ is a relevant telescoping structure given $\mathbb{T}, \mathcal{V}$, and $\mathcal{C}$. The syntactic representation of the set of next-bests $N^*(Q)$ is the set of syntactic representations of the $\perp$-kernels in $\mathcal{N}^*([\![Q]\!]^{\mathcal{V}})$ where $\mathcal{N}^*([\![Q]\!]^{\mathcal{V}})$ is as per Definition 14.*

We now define an explanation for any belief/intention graded consequence ϕ as a set of supports in $\mathbb{T}$ given $\mathcal{C}$ in addition to a set of syntactic $\perp$-kernels that ϕ survives. In what follows, for any tuple of sets of propositional terms $Q = \langle Q_1, \ldots, Q_k\rangle$, let $E(Q) = \langle EQ_1, \ldots, EQ_k\rangle$ where, $EQ_i = Q_i \cup \{\phi \mid \mathbf{A_i}(\phi, g) \in Q_i\}$ for all $1 \le i \le k$. Further, let $Q_\perp$ be the set of syntactic $\perp$-kernels of Q and $Cn(\mathbb{T}) = \langle \mathbb{A}'_1, \ldots, \mathbb{A}'_k\rangle$ with $\mathbb{A}'_i = \{\phi \mid \mathbb{T} \models_i \phi\}$ for all $1 \le i \le k$.

[4] By smallest subsets, we mean that if we remove any proposition from any of $\mathbb{A}'_1, \ldots, \mathbb{A}'_k$ or any rule from $\mathbb{R}'$, ϕ ceases to be an i-consequence.

Definition 21. *Let* $\mathbb{T} \models^{\mathcal{C}}_{B} \phi$ $(\mathbb{T} \models^{\mathcal{C}}_{I} \phi)$ *and* $Q = E(Cn(\mathbb{T})))$. *An explanation for accepting* ϕ *as a belief (intention) consequence is a pair* $(\mathcal{E}_S, \mathcal{E}_K)$ *where:*

- $\mathcal{E}_S$ *is the set of all supports for* ϕ *as a graded belief (intention) in* $\mathbb{T}$ *given* $\mathcal{C}$*; and*
- $\mathcal{E}_K = \{\langle X_1, \ldots, X_k\rangle \mid \langle X_1, \ldots, X_k\rangle \in N^*(Q)$ *and* $\phi \in X_b$ $(\phi \in X_j\, for\, some$ $j \neq b)\}$.

On the other hand, an explanation for rejecting ϕ as a belief/intention consequence is just one kernel K in $N^*(Q)$ that ϕ does not survive as a belief/intention. Recall that the only reason why ϕ would not survive is that it is in the least preferred set according to the agent's character with the least grade in this set.

Definition 22. *Let* $\mathbb{T} \not\models^{\mathcal{C}}_{B} \phi$ $(\mathbb{T} \not\models^{\mathcal{C}}_{I} \phi)$. *An explanation for rejecting* ϕ *as a graded belief (intention) consequence is* X *where* $X = \langle X_1, \ldots, X_k\rangle \in N^*(Q)$ *with* $Q = E(Cn(\mathbb{T}))$ *and* $\phi \in Q_b$ $(\phi \in Q_j$ *for some* $j \neq b)$.

Example 7. We go back one last time to our running example. Suppose that Doc is selfish realistic (the order of the sets is obligations, beliefs, then desires), and $\mathcal{C} = \langle s_1, s_2\rangle$. We show below some belief and intention graded consequences with their explanations:

- $\mathbb{T} \models^{\mathcal{C}}_{B} w(G)$. The explanation for $w(G)$ is $(\mathcal{E}_S, \mathcal{E}_K)$ where:
  $\mathcal{E}_S = \{\langle\langle\{\}, \{w(G)\}, \{\}\rangle, \{\}, \{2\}, 2\rangle\}$; and $\mathcal{E}_K = \{\}$

  Since $w(G) \in \mathbb{A}_1$, its explanation is just made up of its support. It was not involved in any $\bot$-kernels in $N^*(Q)$, hence, $\mathcal{E}_K$ is empty.

- $\mathbb{T} \models^{\mathcal{C}}_{B} o(G)$. The explanation for $o(G)$ is $(\mathcal{E}_S, \mathcal{E}_K)$ where:
  $\mathcal{E}_S = \{\langle\langle\{\}, \{\mathbf{B}(o(G), 6)\}, \{\},\rangle, \{\}, \{2\}, 2\rangle\}$;
  and $\mathcal{E}_K = \{\langle\{\}, \{\forall x[t(x) \Leftrightarrow \neg h(x)], o(G), \forall x[o(x) \Rightarrow h(x)]\}, \{t(G)\}\rangle\}$.

  The belief $o(G)$ was graded in $\mathbb{A}_1$. This is why its support contains its grading proposition. The $\bot$-kernel in $N^*(Q)$ where $o(G)$ appeared but survived appears in $\mathcal{E}_K$.

- $\mathbb{T} \models^{\mathcal{C}}_{I} t(G)$. The explanation for $t(G)$ is $(\mathcal{E}_S, \mathcal{E}_K)$ where:
  $\mathcal{E}_S = \{\langle\langle\{\}, \{\}, \{\mathbf{D}(p(D), 10)\}\rangle, \{\mathbf{r1}\}, \{2\}, 2\rangle\}$;
  and $\mathcal{E}_K = \{\langle\{h(G)\}, \{\forall x[t(x) \Leftrightarrow \neg h(x)]\}, \{t(G)\}\rangle\}$

  The intention $t(G)$ follows using a bridge rule and a desire. For this reason, both the supporting desire and the bridge rule $\mathbf{r1}$ appear in its support. The $\bot$-kernel in $N^*(Q)$ where $t(G)$ appeared but survived appears in $\mathcal{E}_K$.

We show below the explanations for why a belief and an intention were rejected by specifying the kernel they do not survive.

- $\mathbb{T} \not\models^{\mathcal{C}}_{B} \forall x[o(x) \Rightarrow h(x)]$.
  Explained by: $\langle\{\}, \{\forall x[t(x) \Leftrightarrow \neg h(x)], t(G), o(G), \forall x[o(x) \Rightarrow h(x)]\}, \{\}\rangle$.
- $\mathbb{T} \not\models^{\mathcal{C}}_{I} h(G)$.
  Explained by: $\langle\{h(G)\}, \{\forall x[t(x) \Leftrightarrow \neg h(x)]\}, \{t(G)\}\rangle$. □

7 Related Work

To the best of our knowledge, all of the existing approaches for practical reasoning in the literature fail to account for at least one of: representing the process of intention formation, capturing the joint belief and intention revision, or providing explanations for the agent's beliefs and intentions, which can all be accounted for in Log_A**PR** as we demonstrated in the previous section. In this section, we review the most prominent approaches to practical reasoning pointing out what each approach fail to account for.

The most widely studied approach to modelling practical reasoning is the BDI architecture [34] and its extensions to include other mental attitudes such as obligations [5]. However, these models do not accomodate modelling preferences among the agent's beliefs or motivations or the joint belief and intention revision [22]. The existing logical approaches to modelling preferences within the BDI architecture are the graded-BDI (g-BDI) model [8] and TEAMLOG [16]. While both approaches propose frameworks for joint reasoning with graded beliefs, desires, and intentions; neither has an account for the joint revision of the three mental attitudes. Moreover, the g-BDI model lacks precise semantics and TEAMLOG is based on a normal modal logic providing only a third-person account of reasoning about the mental attitudes unlike the first-person account offered by Log_A**PR**. On the other hand, the joint revision of beliefs and intentions has been attempted in [23]. These theories, however, do not account for desires or preferences over beliefs and intentions. Other approaches for goal generation like [9,14,38] also exist in the literature, yet, none of them provide any account for explanation just like all the previously mentioned approaches. A very recent approach to explaining BDI agents exists presented in [12], however, it assumes a library of prioritized plans and a selection function that selects intentions from the plan library. The process of intention formation is not thoroughly studied unlike Log_A**PR** and it is explicitly mentioned that the revision process is outside their scope.

According to a recent literature review on explainable agency [1], most of the currently existing approaches focus on providing explanations to humans, only tackle simple pre-set scenarios [4,21, for example], and are argumentation-based. The approaches proposed in [7,37] focus only on explaining beliefs without regarding the process of intention formation or joint revision. On the other hand, the argumentation framework presented in [30] provides an account for explaining the revision process, but it can only represent preferences amongst goals not beliefs, all the beliefs and goals are represented as ground literals, and the only allowed beliefs in their framework are beliefs about goals. It is worth noting here that Log_A**PR** can be used as the underlying logic on which any argumentation framework can be built. Notions employed by Log_A**PR** like supports and $\perp$-kernels can be readily utilized to define arguments, counter arguments, attacks, and defeaters. In this way, such argumentation framework built on top of Log_A**PR** will be able to account for the aspired trio of intention formation, joint revision, and explanation.

8 Concluding Remarks

As we become increasingly relying on autonomous agents to perform critical tasks, it becomes an impelling priority to endow artificial agents with explainable practical reasoning capabilities. An adequate framework for explainable practical reasoning must be able to capture the process of intention formation and revision while providing explanations for the chosen intentions after the revision. The problems of intention revision and intention explanation after the revision have been scarcely studied. To this end, we presented in this paper a powerful logical language for explainable practical reasoning we refer to as $Log_A\mathbf{PR}$. We demonstrated how $Log_A\mathbf{PR}$ provides an adequate framework for practical reasoning. Future work include studying the non-monotonic properties of our graded consequence relations, modelling the process of means-end reasoning within $Log_A\mathbf{PR}$ building on approaches to automated planning, and enhancing our explanations by translating them to a natural language to facilitate their communication to humans. We also plan on defining an argumentation framework build on $Log_A\mathbf{PR}$ as an underling logic to provide a novel argumentation based framework inheriting the capabilities of $Log_A\mathbf{PR}$.

References

1. Anjomshoae, S., Najjar, A., Calvaresi, D., Främling, K.: Explainable agents and robots: results from a systematic literature review. In: 18th International Conference on Autonomous Agents and Multiagent Systems (AAMAS 2019), Montreal, Canada, 13–17 May 2019, pp. 1078–1088. International Foundation for Autonomous Agents and Multiagent Systems (2019)
2. Aucher, G.: An internal version of epistemic logic. Stud. Logica. **94**(1), 1–22 (2010)
3. Bealer, G.: Theories of properties, relations, and propositions. J. Philos. **76**(11), 634–648 (1979)
4. Broekens, J., Harbers, M., Hindriks, K., van den Bosch, K., Jonker, C., Meyer, J.-J.: Do you get it? User-evaluated explainable BDI agents. In: Dix, J., Witteveen, C. (eds.) MATES 2010. LNCS (LNAI), vol. 6251, pp. 28–39. Springer, Heidelberg (2010). https://doi.org/10.1007/978-3-642-16178-0_5
5. Broersen, J., Dastani, M., Hulstijn, J., van der Torre, L.: Goal generation in the BOID architecture. Cogn. Sci. Q. **2**(3–4), 428–447 (2002)
6. Broome, J.: Practical reasoning. In: Bermúdez, J.L., Millar, A. (eds.) Reason and Nature: Essays in the Theory of Rationality, pp. 85–111. Clarendon Press, Oxford (2002)
7. Capobianco, M., Chesnevar, C.I., Simari, G.R.: Argumentation and the dynamics of warranted beliefs in changing environments. Auton. Agent. Multi-Agent Syst. **11**(2), 127–151 (2005)
8. Casali, A., Godo, L., Sierra, C.: A graded BDI agent model to represent and reason about preferences. Artif. Intell. **175**(7–8), 1468–1478 (2011)
9. Castelfranchi, C., Paglieri, F.: The role of beliefs in goal dynamics: prolegomena to a constructive theory of intentions. Synthese **155**(2), 237–263 (2007)
10. Church, A.: On Carnap's analysis of statements of assertion and belief. Analysis **10**(5), 97–99 (1950). https://doi.org/10.1093/analys/10.5.97
11. Cohen, P.R., Levesque, H.J.: Intention is choice with commitment. Artif. Intell. **42**(2–3), 213–261 (1990)
12. Dennis, L.A., Oren, N.: Explaining BDI agent behaviour through dialogue. In: Proceedings of the 20th International Conference on Autonomous Agents and MultiAgent Systems, pp. 429–437 (2021)
13. Dietrich, F., Staras, A., sugden, R.: Beyond Belief: Logic in Multiple Attitudes, January 2020. https://halshs.archives-ouvertes.fr/halshs-02431917. Working paper or preprint

14. Dignum, F., Kinny, D., Sonenberg, L.: From desires, obligations and norms to goals. Cogn. Sci. Q. **2**(3–4), 407–430 (2002)
15. van Ditmarsch, H., Halpern, J., Kooi, B.: An introduction to logics of knowledge and belief. In: van Ditmarsch, H., Halpern, J., van der Hoek, W., Kooi, B. (eds.) Handbook of Epistemic Logic, pp. 1–51. College Publications, London (2015)
16. Dunin-Keplicz, B., Nguyen, L.A., Szalas, A.: A framework for graded beliefs, goals and intentions. Fund. Inform. **100**(1–4), 53–76 (2010)
17. Ehab, N., Ismail, H.O.: Algebraic foundations for non-monotonic practical reasoning. In: Varzinczak, I.J., Martínez, M.V. (eds.) Proceedings of the 18th International Workshop on Non-Monotonic Reasoning (NMR2020), pp. 160–169 (2020)
18. Ehab, N., Ismail, H.O.: Appendix for towards explainable practical agency: a logical perspective. Technical report, German University in Cairo (2021). https://met.guc.edu.eg/Repository/Faculty/Publications/972/EXTRAAMAS-2021-Proofs.pdf
19. Ehab, N., Ismail, H.O.: $Log_A\mathbf{G}$: an algebraic non-monotonic logic for reasoning with graded propositions. Ann. Math. Artif. Intell. **89**(1), 103–158 (2021)
20. Hansson, S.O.: Kernel contraction. J. Symbolic Logic **59**(03), 845–859 (1994)
21. Harbers, M., van den Bosch, K., Meyer, J.J.: Design and evaluation of explainable BDI agents. In: 2010 IEEE/WIC/ACM International Conference on Web Intelligence and Intelligent Agent Technology, vol. 2, pp. 125–132. IEEE (2010)
22. Herzig, A., Lorini, E., Perrussel, L., Xiao, Z.: BDI logics for BDI architectures: old problems, new perspectives. KI-Künstliche Intelligenz **31**(1), 73–83 (2017)
23. Icard, T., Pacuit, E., Shoham, Y.: Joint revision of belief and intention. In: Proceedings of the 12th International Conference on Knowledge Representation, pp. 572–574 (2010)
24. Ismail, H.O.: $Log_A\mathbf{B}$: a first-order, non-paradoxical, algebraic logic of belief. Logic J. IGPL **20**(5), 774–795 (2012)
25. Ismail, H.O.: The good, the bad, and the rational: aspects of character in logical agents. In: El Bolock, A., Abdelrahman, Y., Abdennadher, S. (eds.) Character Computing. HIS, pp. 139–164. Springer, Cham (2020). https://doi.org/10.1007/978-3-030-15954-2_9
26. Jago, M.: Epistemic logic for rule-based agents. J. Logic Lang. Inform. **18**, 131–158 (2009)
27. Langley, P., Meadows, B., Sridharan, M., Choi, D.: Explainable agency for intelligent autonomous systems. In: Proceedings of the Thirty-First AAAI Conference on Artificial Intelligence, AAAI 2017, pp. 4762–4763. AAAI Press (2017)
28. McNamara, P.: Deontic logic. In: Zalta, E.N. (ed.) The Stanford Encyclopedia of Philosophy. Metaphysics Research Lab, Stanford University, Fall 2018 edn. (2018)
29. Morgan, P.R., Cohen, J.L., Pollack, M.E.: Intentions in Communication. MIT Press, Cambridge (1990)
30. Morveli-Espinoza, M., Tacla, C.A., Jasinski, H.M.R.: An argumentation-based approach for explaining goals selection in intelligent agents. In: Cerri, R., Prati, R.C. (eds.) BRACIS 2020. LNCS (LNAI), vol. 12320, pp. 47–62. Springer, Cham (2020). https://doi.org/10.1007/978-3-030-61380-8_4
31. Parsons, T.: On denoting propositions and facts. Philos. Perspect. **7**, 441–460 (1993)
32. Perlis, D., Brody, J., Kraus, S., Miller, M.J.: The internal reasoning of robots. In: Gordon, A.S., Miller, R., Turán, G. (eds.) Proceedings of the Thirteenth International Symposium on Commonsense Reasoning, COMMONSENSE 2017, London, UK, 6–8 November 2017. CEUR Workshop Proceedings, vol. 2052. CEUR-WS.org (2017)
33. Perlis, D.: Languages with self-reference II: knowledge, belief, and modality. Artif. Intell. **34**(2), 179–212 (1988)
34. Rao, A.S., Georgeff, M.P.: BDI agents: from theory to practice. In: In Proceedings of the First International Conference on Multi-Agent Systems (ICMAS 1995), pp. 312–319 (1995)
35. Sankappanavar, H., Burris, S.: A course in universal algebra. Graduate Texts Mathematics, vol. 78, New York (1981)

36. Shapiro, S.C.: Belief spaces as sets of propositions. J. Exp. Theoret. Artif. Intell. **5**(2–3), 225–235 (1993)
37. Sklar, E.I., Azhar, M.Q.: Explanation through argumentation. In: Proceedings of the 6th International Conference on Human-Agent Interaction, pp. 277–285 (2018)
38. Thomason, R.H.: Desires and defaults: a framework for planning with inferred goals. In: Proceedings of the Seventh International Conference on Principles of Knowledge Representation and Reasoning, KR 2000, pp. 702–713. Morgan Kaufmann Publishers Inc., San Francisco (2000)
39. Thomason, R.H., Gabbay, D.M., Guenthner, F.: The formalization of practical reasoning: problems and prospects. In: Gabbay, D.M., Guenthner, F. (eds.) Handbook of Philosophical Logic. HPL, vol. 18, pp. 105–132. Springer, Cham (2018). https://doi.org/10.1007/978-3-319-97755-3_3
40. Wooldridge, M.: An Introduction to Multiagent Systems. Wiley, Chichester (2009)
41. Zhang, Q.s., Zhu, S.C.: Visual interpretability for deep learning: a survey. Front. Inf. Technol. Electr. Eng. **19**(1), 27–39 (2018)

Explainable Reasoning in Face of Contradictions: From Humans to Machines

Timotheus Kampik[1(✉)] and Dov Gabbay[2,3,4]

[1] Umeå University, Umeå, Sweden
tkampik@cs.umu.se
[2] University of Luxembourg, Esch-sur-Alzette, Luxembourg
[3] King's College London, London, UK
dov.gabbay@kcl.ac.uk
[4] Bar Ilan University, Ramat Gan, Israel

Abstract. A well-studied trait of human reasoning and decision-making is the ability to not only make decisions in the presence of contradictions, but also to explain *why* a decision was made, in particular if a decision deviates from what is expected by an inquirer who requests the explanation. In this paper, we examine this phenomenon, which has been extensively explored by behavioral economics research, from the perspective of symbolic artificial intelligence. In particular, we introduce four levels of intelligent reasoning in face of contradictions, which we motivate from a microeconomics and behavioral economics perspective. We relate these principles to symbolic reasoning approaches, using abstract argumentation as an exemplary method. This allows us to ground the four levels in a body of related previous and ongoing research, which we use as a point of departure for outlining future research directions.

Keywords: Symbolic artificial intelligence · Explainable artificial intelligence · Non-monotonic reasoning

1 Introduction

Over the last decades, the public perception of what artificial intelligence is (and is not) has dramatically shifted. For example, in 1996 and 1997, when the reigning chess champion Gary Kasparov played against IBM's chess computer *Deep Blue*, the ability of playing chess well was considered a key characteristic of human intelligence. Today, as technically literate consumers can easily install a world champion-beating program on their mobile phones, the focus has shifted to other problems, which range from different games like Starcraft and Go to real-world challenges like fully autonomous driving in inner cities. Even the Turing test [31], which roughly speaking requires a machine to be able to deceive a human into thinking it is human, seems to fail the test of time; given current socio-technical information systems, distinguishing men from machines is increasingly challenging, even in contexts where the machine behavior is determined by simple scripts, for example when social media bots spread misinformation [28].

Hence, to define characteristics of intelligent behavior, more abstract approaches are required. Such approaches have, indeed, been introduced as principles of non-monotonic reasoning; most notable are relaxed forms of monotony, such as restricted

D. Calvaresi et al. (Eds.): EXTRAAMAS 2021, LNAI 12688, pp. 280–295, 2021.
https://doi.org/10.1007/978-3-030-82017-6_17

monotony [14] (also known as cautious monotony) and rational monotony [24][1]. From a symbolic artificial intelligence perspective, these properties are very useful because they can be formally verified. Still, the properties have some obvious limitations:

- The properties are merely *indicators* of intelligence; certainly, fairly "unintelligent agents" can also satisfy restricted monotony and rational monotony, simply by never inferring anything from any knowledge base.
- It is not clear how these properties relate to human intuitions of intelligence.

In this paper, we explore ways to address these limitations by *i)* building a conceptual bridge between formal principles of non-monotonic reasoning and empirical, as well as formal perspectives on human reasoning and decision-making and *ii)* illustrating how different formal approaches to non-monotonic reasoning reflect different levels of sophistication of human reasoning.

2 Human Intelligence: Bounded Rationality and Reasoning Backwards

As a preliminary for a bridge between human reasoning and formal methods of automated reasoning, let us provide a brief overview of the development of models of human reasoning and decision-making at the intersection of microeconomic theory and behavioral psychology. At least since the middle of the 20th century, studies in the fields of micro-economic theory and behavioral psychology attempt to identify patterns and *formal* models of human decision-making and reasoning, both for *descriptive* ("How do humans reason and make decisions?") and *prescriptive* ("How should humans reason and make decisions?") purposes[2]. An early theory that is still very influential is the formal model of *rational economic man*. According to the model (in its simplest variant), when faced with a choice, which is modeled as the selection from a set of items S, a rational decision-maker acts according to *clear preferences*, which are modeled as a partial order $\succeq$ on S. The partial order is established such that $\exists a^* \in S, \forall a \in S, a^* \succeq a$, *i.e.* a^* is preferred over all other elements in S. a^* is the decision-maker's choice. Given another set S', such that $S \subseteq S'$, for the decision-maker's choice $a'^* \in S'$ it must hold true that $a'^* \notin S$ or $a'^* = a^*$; *i.e.*, the preference relation on S must be consistent with the preference relation on S' (see, *e.g.* Osbourne and Rubinstein [25]). Consequently, a rational economic decision-maker can make a decision in any situation and the preferences this decision implies are consistent with the preferences implied by all previous decisions.

[1] Let us highlight that we do not introduce the so-called AGM postulates [2] here, because the *success* postulate stipulates (colloquially speaking) that "new" logical formulas are always added to the belief base and never rejected; however, we assume that, intuitively, an intelligent agent should be able to reject new beliefs under some circumstances.

[2] Less formal models of human decision-making and reasoning have been, of course, subject of in-depth study for much longer. Indeed, the management of contradictions that is at the center of this paper is also the subject of the *Shev Shema'tata*, a book on the treatment of doubt in Rabbinic law, written at the turn from the 18th to the 19th century [18].

While the model of rational economic man remains influential and is a common foundation of micro-economic curricula, its shortcomings have been criticized in high-profile scientific venues since the 1950s, notably in Herbert Simon's seminal paper *A Behavioral Model of Rational Choice* [29]. A key argument made by Simon is that the model is too simplistic in that it does not account for the information an agent has (from our perspective: the agent's *beliefs*) and hence the model can neither describe nor prescribe the real-life decision-making processes of agents or organizations. A simple example of economically irrational behavior is as follows: an agent chooses b from a set $\{b, c\}$ which establishes the preference $b \succeq c$, but chooses c from a set $\{b, c, d\}$, which establishes the preference $c \succeq b$. For instance, let us assume a choice from a set of beverages: $b := coffee, c := tea, d := juice$. After choosing coffee from the set of "tea and coffee", a rational decision-maker must not choose tea from the set of "tea, coffee, and juice", given all other things remain the same. From a knowledge representation perspective, one can of course argue that the presence of d allows us to infer something about b and/or c that makes us reverse $b \succeq c$ to $c \succeq b$[3].

Building on top of these initial insights, Tversky and Kahneman conducted a series of behavioral psychology experiments to systematically identify shortcomings of models of economic rationality that led to refined models of rational decision-making, like *prospect theory* [20], eventually winning Kahneman the Nobel Memorial Prize in Economic Sciences [19]. While a broad range of other formal models has been developed to address the aforementioned and similar shortcomings [27], further ground-breaking empirical research has emerged about other aspects of human reasoning. Most notably in the context of this paper is a line of research conducted by Jonathan Haidt (and others), showing that humans are prone to first make an intuition-based decision and, if required, then search for a "rational" (colloquially speaking) explanation [17].

To summarize, this brief overview of selected microeconomic and behavioral economics research history gives us the following insights on perspectives on human reasoning and decision-making:

1. Traditionally, humans are considered intelligent, *rational* decision-makers that act, at least roughly, according to formal model of clear and consistent preferences.
2. Empirical research about human behavior has systematically debunked assumptions about economic rationality in human decision-making, leading to a refinement of formal models of decision-making to *models of bounded rationality*.
3. More recently, additional empirical research has provided evidence for the hypothesis that humans are prone to make intuition-based decisions and then *reason backwards* to generate convincing, "rational" explanations if required.

3 Levels of Intelligent Reasoning in Face of Contradictions

From the overview of perspectives on human decision-making, we can generate three levels of intelligent reasoning in face of contradictions, which we outline in this section.

[3] Indeed, empirical studies (conducted decades after the publication of Simon's paper) show that humans sometimes do exactly this [6].

In addition, we describe a fourth level that prescribes desirable behavior that – by combining principle-based reasoning and learning perspectives – goes beyond existing perspectives on human decision-making and reasoning.

3.1 Clear Preferences

At the most primitive level, the only property one expects from a decision-maker is to be *decisive*. In microeconomic theory, this intuition is ingrained in the assumption that when observing a decision-maker who chooses one option from a set of options A, a partial order $\succeq$ on A that describes the decision-maker's preferences can be inferred, such that given the choice $a^* \in A$, it holds true that $\forall a \in A, a^* \succeq a$, *i.e.* the decision-maker strictly prefers the choice over all other possible alternatives that could have been chosen. In its most primitive form, this model can be considered to merely cover a one-shot observation: as long as an agent is decisive, clear preferences can be inferred from a single decision and no consistency check with regard to previous decisions is performed. From a reasoning perspective, this means that an inference method must always come to a conclusion when drawing inferences from a belief base; no further conditions need to be satisfied. This one-shot approach can be compared to the behavior of a populist politician, who makes his decisions based on gut-feeling, notwithstanding that he is aware of contradicting evidence, and does not care about the long-term consistency of his actions (and speech acts).

3.2 Consistent Preferences

As an obvious next step, economists assess whether a decision-maker's preferences are consistent over a sequence of decisions; *i.e.*, given a new choice $a'^* \in A'$, such that $A \subseteq A'$, if $a'^* \in A$ then $a'^* = a^*$; this property follows from the model of clear preferences as introduced in the previous subsection (see *e.g.* Rubinstein [27, p. 11] for a proof). Again, from a reasoning perspective, the analogy is obvious: when drawing inferences $concl(A)$ from a belief base A, for the inferences $concl(A')$ that are drawn from a belief base A', such that $A \subseteq A'$, it must hold true that $concl(A) = concl(A')$ unless a belief in $A' \setminus A$ is accepted as an element of $concl(A')$. Consequently, we can see that the consistent preferences principle is in its motivation similar to notions of "relaxed" monotony, in particular to cautious monotony [14], which can semi-formally be described as *if* $C \subseteq concl(A)$ *and* $B \subseteq concl(A)$ *then* $C \subseteq concl(A \cup B)$.

3.3 Explainable "Backwards Reasoning"

As summarized in the previous section, behavioral psychology research suggests that humans typically make intuition-based decisions and then find a "rational" explanation if necessary. This *reasoning backwards* approach has traditionally been favored by neoclassical economics, to the extent that the economist Steven Landsburg colloquializes it as follows in his best-selling popular science book *The Armchair Economist*:

"[We] stubbornly maintain the fiction that all people are rational at all times, and [...] insist on finding rational explanations, no matter how outlandish, for all of this apparently irrational behavior[4] [23]."

Landsburg does *not* describe observations of common human decision-making fallacies, but instead refers to – albeit with some overstatement to underline his point – a key aspect of the approach that he and some other economists use to build their models. However, in the real-world, reasoning backwards is not considered a "reasonable" approach to explain a decision or line of reasoning, which the following anecdote illustrates.

Example 1. In 2019, world-renowned association football coach José Mourinho, who at that time recently had joined Tottenham Hotspur F.C. ("the Spurs"), had the following exchange with a journalist during a press conference[5]:

- *Journalist*: "When you were at Chelsea, you were asked whether you would ever come to the Spurs and you said: 'Never, I love the Chelsea fans too much.' What has changed?"
- *Mourinho*: "[That was] before I was sacked [at Chelsea]."

From a reasoning perspective, one can say that when asked about the inconsistency between two conclusions, Mourinho produces a new belief that explains why the latter conclusion does not entail the initial conclusion. Technically, one could argue that Mourinho has successfully assured that his decision to join Tottenham is indeed consistent, because he has provided a new belief (an argument) that supports his change of mind, and when considering the adoption of a belief as a part of a choice process, his preferences are consistent (*economically rational*). From a logics perspective, the existence of a *conflict* between the new belief and the previous beliefs can explain why monotony of entailment is violated. Practically, it is – however – obvious that his stated commitment to Chelsea was implied to last beyond his tenure as a coach at the club. Indeed, both Mourinho and the journalists that are present laugh about the answer; they are aware of how ridiculous the explanation that Mourinho has provided must look from the perspective of a Chelsea fan (in particular when considering that Chelsea and Tottenham are London city rivals).

3.4 Evidence-Based Principle Revision

Similarly to Tversky and Kahneman, who started off by taking formal models of economic rationality and then systematically refined them as they observed diverging human behavior in the real world, an intelligent agent should be able to start off with an *explainable* model of reasoning and decision-making and then refine it based on the observations it makes; *i.e.*, the agent should make decisions/draw inferences as follows:

1. It should employ an *explainable* formal model that prescribes and describes its behavior and satisfies some formal principles.

[4] Note that this statement precedes a defense of the approach it describes.

[5] See: http://s.cs.umu.se/hlzdqf.

2. It should be able to refine the model and adjust its principles if it observes that changes are beneficial (based on feedback from its environment).

This hybrid approach requires a combination of symbolic (logic-based) and sub-symbolic (machine learning-based) approaches to artificial intelligence. Considering the example in the previous subsection, Mourinho could, for example, revise his reasoning principles after being subjected to the scorn of the Chelsea fans, and in the future be more conservative when discarding previously drawn conclusions, at least the ones he has publicly announced to be committed to.

4 Examples: Abstract Argumentation

Let us further illustrate the intuitions we have introduced in the previous section by providing precise formal examples. As our reasoning method, we employ abstract argumentation because it *a)* is a simple model that can be introduced without a lot of formal preliminaries and *b)* has a clear focus on managing conflicts/contradictions.

Definition 1 (Argumentation Framework [13]**).**
An abstract argumentation framework is a tuple $AF = (AR, AT)$*, where* AR *is a set of elements (*arguments*) and* $AT \subseteq AR \times AR$ *is a set of* attacks *between arguments in* AR*.*

Given an argumentation framework $AF = (AR, AT)$ and two arguments $a, b \in AR$, we say that "a attacks b" iff $(a, b) \in AT$. An argument $a \in AR$ is *acceptable* with regard to a set $S \subseteq AR$ iff for each $b \in AR$ it holds true that if b attacks a, then b is attacked by S. In abstract argumentation, key concepts are the notions of conflict-free and admissible sets.

Definition 2 (Conflict-free and Admissible Sets [13]**).**
Let $AF = (AR, AT)$ *be an argumentation framework. A set* $S \subseteq AR$ *is:*

- conflict-free *iff* $\nexists a, b \in S$*, such that* a *attacks* b*;*
- admissible *iff* S *is conflict-free and each argument in* S *is acceptable with regard to* S*.*

Given an argumentation framework $AF = (AR, AT)$ and a set $S \subseteq AR$, we define $S^+ = \{a | a \in AR, \exists b \in S$, such that b attacks $a\}$. *Argumentation semantics* determine which sets of arguments in an argumentation framework can be considered valid conclusions. A set of such valid conclusions is called an *extension*. All argumentation semantics that have been introduced by Dung in the initial paper are based on the notion of an admissible set.

Definition 3 (Admissible Set-based Argumentation Semantics [13]**).**
Given an argumentation framework $AF = (AR, AT)$*, an admissible set* $S \subseteq AR$ *is:*

- *a* stable extension *of* AF *iff* S *attacks each argument that does not belong to* S*.* $\sigma_{stable}(AF)$ *denotes all stable extensions of* AF*.*
- *a* complete extension *iff each argument that is acceptable w.r.t.* S *belongs to* S*.* $\sigma_{complete}(AF)$ *denotes all complete extensions of* AF*.*

- *a* preferred extension *of AF iff S is a maximal (w.r.t. set inclusion) admissible subset of AR. $\sigma_{preferred}(AF)$ denotes all preferred extensions of AF;*
- *a* grounded extension *of AF iff S is the minimal (w.r.t. set inclusion) complete extension of AF. $\sigma_{grounded}(AF)$ denotes all grounded extensions of AF.*

Given an argumentation framework $AF = (AR, AT)$ and an argumentation semantics σ, a set $S \subseteq AR$ is called a σ-extension of AF iff $S \in \sigma(AF)$. Other semantics have been defined that start of with the assumption of a maximal conflict-free (*naive*) set[6].

Definition 4 (Naive Set-based Argumentation Semantics [32]**).**
A conflict-free set $S \subseteq AR$ is a:

- naive extension *iff S is maximal w.r.t. set inclusion among all conflict-free sets. $\sigma_{naive}(AF)$ denotes all naive extensions of AF.*
- stage extension, *iff $S \cup S^+$ is maximal w.r.t. set inclusion among all conflict-free sets,* i.e. *there is no conflict-free set $S' \subseteq AR$, such that $(S' \cup S'^+) \supset (S \cup S^+)$. $\sigma_{stage}(AF)$ denotes all stage extensions of AF.*

In the context of this paper, we are interested in how agents draw inferences from a belief base to which new beliefs are added over time. For this, we depend on the notion of argumentation framework expansion, and in particular on *normal* expansions.

Definition 5 (Argumentation Framework Expansions [7]**).**
An argumentation framework $AF' = (AR', AT')$ is:

- *an **expansion** of another argumentation framework $AF = (AR, AT)$ (denoted by $AF \preceq_E AF'$) iff $AR \subseteq AR'$ and $AT \subseteq AT'$.*
- *a **normal expansion** of an argumentation framework $AF = (AR, AT)$ (denoted by $AF \preceq_N AF'$) iff $AF \preceq_E AF'$ and $\nexists(a, b) \in AT' \setminus AT$, such that $a \in AR \wedge b \in AR$.*

Colloquially speaking, a normal expansion of an argumentation framework adds new arguments to the argumentation framework, but neither removes arguments nor changes attacks between existing arguments. To support the design and analysis of argumentation semantics, formal argumentation principles have been defined [4,30]. For example, the *uniqueness* principle stipulates that an argumentation semantics must return exactly one extension, given any argumentation framework.

4.1 Clear Preferences

From an argumentation perspective, an agent has clear preferences iff it can reach an unambiguous conclusion, given any argumentation framework and the argumentation semantics it employs. We can illustrate perspectives on this property given a *particular* argumentation framework, *e.g.* $AF = (AR, AT) = (\{a, b, c\}, \{(a, a), (b, c), (c, b)\})$. Figure 1 depicts the argumentation framework. Below are some examples of how different argumentation semantics resolve AF:

[6] More semantics exist, some of which address well-known issues with the semantics whose definitions we provide in this paper. However, we consider an in-depth overview of argumentation semantics out-of-scope.

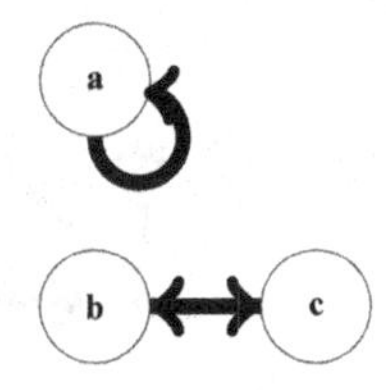

Fig. 1. Given stable semantics, self-contradicting arguments may lead to the inability to reach any conclusion, *e.g.*, $\sigma_{stable}(\{a, b, c\}, \{(a, a), (b, c), (c, b)\}) = \{\}$.

- Stable semantics: $\sigma_{stable}(AF) = \{\}$;
- Grounded semantics: $\sigma_{grounded}(AF) = \{\{\}\}$;
- Preferred semantics: $\sigma_{preferred}(AF) = \{\{b\}, \{c\}\}$.

It is obvious that stable semantics does not satisfy the notion of *clear preferences*: it does not return any extension for our argumentation framework. Conversely, preferred semantics returns the extensions $\{a\}$ and $\{b\}$. This does not reflect the clear preferences principle, either, because several extensions are returned. However, an intelligent agent that employs the semantics can certainly come to a decisive conclusion, for example by considering use case-specific meta-data (like a time-stamp or the source of an argument), or by simply *breaking the tie* with an arbitrary method that considers language-specific properties, like identifiers of the arguments[7]. Consequently, we argue that it depends on the exact application scenario whether one wants an argumentation semantics to be uniquely defined or not. For example, in one legal reasoning scenario, it can make sense to dismiss conflicting statements of two witnesses as mutually inconsistent, while in another scenario, it can be better to consider both statements and then select a preferred statement based on situational context or meta-data (which is aligned with the concept of *burden of persuasion*, see Prakken and Sartor [26]).

4.2 Consistent Preferences

To align with the *consistent preferences* property of economic rationality, we can create a straight-forward argumentation principle (see our ongoing line of work [21,22][8]): we assume that an agent, given an argumentation semantics σ, resolves an argumentation framework $AF = (AR, AT)$ by selecting any σ-extension E of AF ($E \in \sigma(AF)$). This selection establishes the preferences $\forall S \in 2^{AR}, E \succeq S$. When continuing the interaction with its environment, the agent adopts new, and potentially conflicting beliefs, *i.e.* it normally expands AF and creates $AF' = (AR', AT'), AF \preceq_N AF'$. When determining the σ-extensions of AF', the agent must find at least on extension ($\exists E' \in \sigma(AF)$), such that the preferences established by inferring E' from AF' ($\forall S' \subseteq 2^{AR'}, E' \succeq S$) are consistent with the preferences established by inferring E from AF. Figure 2 illustrates the concept of consistent preferences in abstract argumentation. For example, let us assume argument a denotes that a new business strategy should

[7] Note that this would be a violation of the *language independence* principle.

[8] In these works, we name the principle *weak reference independence*.

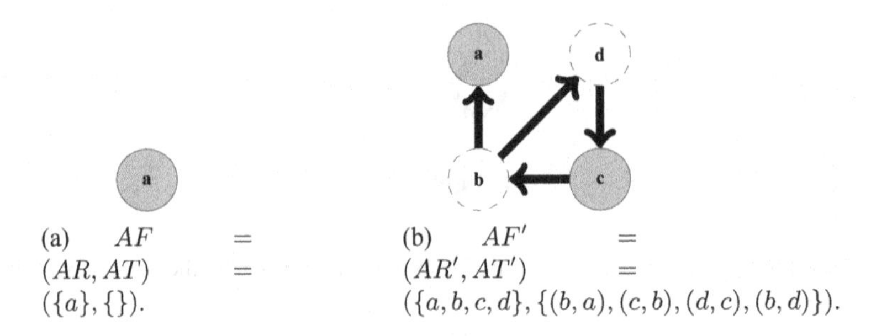

(a) $AF = (AR, AT) = (\{a\}, \{\})$. (b) $AF' = (AR', AT') = (\{a, b, c, d\}, \{(b, a), (c, b), (d, c), (b, d)\})$.

Fig. 2. Consistent preferences. Assuming stage semantics, we have $\sigma_{stage}(AF) = \{\{a\}\}$ and $\sigma_{stage}(AF') = \{\{a, c\}, \{a, d\}, \{b\}\}$. All σ_{stage}-extensions of AF' establish consistent preferences with regard to the only σ_{stage}-extension of AF. In contrast, assuming preferred semantics, we have $\sigma_{preferred}(AF) = \{\{a\}\}$ and $\sigma_{preferred}(AF') = \{\{\}\}$; the only $\sigma_{preferred}$-extension of AF establishes the preferences $\forall S \in 2^{AR}, \{a\} \succeq S$, which is inconsistent with the preferences established by the only $\sigma_{preferred}$-extension of AF: $\forall S' \in 2^{AR'}, \{\} \succeq S'$.

be executed, to which an agent first commits: $AF = (AR, AT) = (\{a\}, \{\})$, from which we obviously conclude $\{a\}$. However, by consulting multiple stakeholders, the agent collects the additional arguments b, c, and d that directly or indirectly argue for or against the strategy: $AF' = (AR', AT') = (\{a, b, c, d\}, \{(b, a), (c, b), (d, c), (b, d)\})$. Now, considering some argumentation semantics, for example preferred semantics, the only conclusion we can draw from AF' is $\{\}$ (the only extension/valid conclusion does not contain any arguments); this establishes the preference $\{\} \succeq \{a\}$, which is inconsistent with the preference $\{a\} \succeq \{\}$ as established by the previous decision. In contrast, some other semantics, such as stage semantics, do not establish inconsistent preferences *in this scenario*[9]. $\sigma_{stage}(AF) = \{\{a\}\}$ and $\sigma_{stage}(AF') = \sigma_{stage}(AF') = \{\{a, c\}, \{a, d\}, \{b\}\}$: because all σ_{stage}-extensions of AF' include an argument that is not in AR, the preferences established by selecting any of the extensions are obviously consistent with the preferences established by inferring $\{a\}$ from AF.

Let us note that an open question – which we touch upon in Sect. 5 – is how to adjust the consistent preferences principle to account for "undecided" arguments, *i.e.* arguments that are, given an extension, neither part of the extension nor attacked by any argument in the extension. Also, similar argumentation principles that are based on other well-known properties can be and have been introduced, for example an abstract argumentation equivalent of restricted (cautious) monotony [21,22].

4.3 Explainable "Backwards Reasoning"

An important feature of an intelligent agent is the ability to explain its inferences and the resulting actions. Indeed, economists who build formal models of human decision-making typically do not claim that their models are accurate representations of what

[9] Let us note that stage semantics does not generally establish consistent preferences, given any argumentation framework and any of its normal expansions, see [22].

goes on in a human's mind, but instead argue that when observing a human decision-maker, their models are sufficiently precise to describe the decision-maker's behavior in an explainable (that is: formally analyzable) manner. In the artificial intelligence community, the design and analysis of *explainable* agents is a research direction that has gained tremendous traction over the past years [3]. Agents that employ symbolic approaches to automated reasoning – such as formal argumentation – are often considered *explainable*, because each inference and action can be linked to the formal model that generated it (see *e.g.* Zhong *et al.* [33]). However, when considering the iterative argumentation approach we take in the context of this paper, it is clear that merely pointing out general semantics behavior is not always sufficient to explain why exactly the inferences drawn from an argumentation framework are fundamentally different than the inferences that are subsequently drawn from one of its (normal) expansions. To some extent, merely explaining an inference process by pointing to the entire formal model that has been used to infer it resembles the *reasoning backwards* approach as introduced as a description of human reasoning in behavioral economics. From an argumentation perspective, we argue that an agent can take two approaches to reasoning backwards:

1. It can take a principle that happens to be satisfied to explain the result of its inference process.
2. If asked why a specific principle is violated, it can generate arguments and add them to the argumentation framework, so that the principle is no longer violated.

As mentioned above, the first approach is obvious, and reflected in the way *explainable* argumentation is typically presented. The second approach reflects the *Mourinho* example (Example 1), which is illustrated as a sequence of argumentation frameworks by Fig. 3:

1. We start with an initial argumentation framework $AF = (\{a, b\}, \{(a, b), (b, a)\})$. a denotes the obligation of maintaining the respect of the Chelsea fans while b denotes taking a job at Tottenham; a and b attack each other. Our agent (Mourinho) infers a, deciding to stay committed to Chelsea.
2. Later, our agent has a change of mind, and instead infers $\{b\}$ from $AF' = AF$ and takes a job at Tottenham.
3. Another agent (the journalist) scrutinizes the Mourinho agent by highlighting that the inference process implies inconsistent preferences.
4. The Mourinho agent responds to the scrutiny by producing an argument c (the relief of the loyalty obligation because Chelsea has sacked him), which is in mutual conflict with argument a. Note that inferring $\{b, c\}$ from AF'' does not imply preferences that are inconsistent with the preferences implied by inferring $\{a\}$ from AF.

4.4 Evidence-Based Principle Revision

Let us go back to the previous example (Fig. 3). However, we now assume the *Mourinho* agent is using the relational principle we have semi-formally introduced in Subsect. 4.2 to ensure consistent preferences. In the example, this means that the agent must not infer

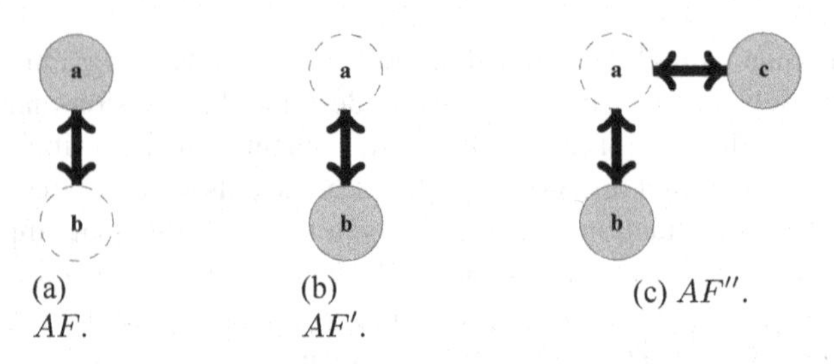

Fig. 3. Reasoning backwards. Given the argumentation framework $(\{a, b\}, \{(a, b), (b, a)\})$, an agent first concludes $\{a\}$ and at a later stage concludes $\{b\}$. When asked about the reason for the inconsistency (preference reversal), the agent produces argument c (generating AF'') that restores consistent preferences.

$\{b\}$ from AF' after having inferred $\{a\}$ from AF; the expansion to AF'' is required to then infer $\{b, c\}$, which is a principle-compliant conclusion. Let us assume that after drawing this inference (and joining Tottenham), our agent's reputation is severely damaged, which makes the agent reflect about its inference process. Satisfying the consistent preferences principle may have been a reasonable starting point, but we want to be able to further evolve from there. Ideally, the agent analyzes its own inference process and searches for principle-based improvements it can make. In our example, the agent can, for instance assuming that it is using stage semantics, observe that the semantics also supports inferring $\{a\}$ from AF'': *i.e.* $\sigma_{stage}(AF'') = \{\{a\}, \{b, c\}\}$. Consequently, the agent can "learn" a new principle that stipulates the following: given two argumentation frameworks AF^* and AF^{**} and a conclusion E^* of AF^* ($E^* \in \sigma_{stage}(AF^*)$), if inferring a conclusion E' from AF^{**} ($E^{**} \in \sigma_{stage}(AF^{**})$) is possible such that $E^* \subseteq E^{**}$, do not infer a conclusion D^{**} from AF^{**} such that $E^* \not\subseteq D^{**}$. The agent can apply this principle and draw inferences in future scenarios accordingly (depicted by Fig. 4). However, first the agent would need to (formally) verify whether enforcing this new principle implies a violation of any other principle that the agent has already adopted (in our example, the agent may still want to satisfy the consistent preferences

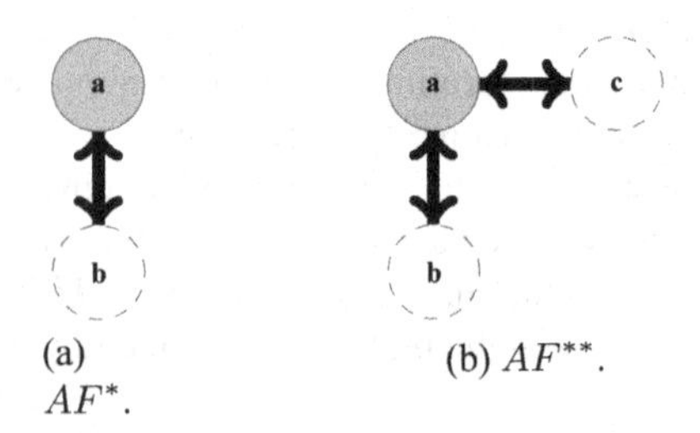

Fig. 4. Evidence-based principle revision. Let us assume our agent has received negative feedback from its actions that were based on the inferences drawn in Fig. 3. To learn from this experience, the agent adjusts its reasoning principles and now always keeps previously inferred conclusions (arguments) to the extent its semantics supports this.

principle), and if so, whether previously adopted principles should be relaxed or entirely discarded.

5 Research Directions

Based on the position we establish in the previous sections, we provide an overview of relevant ongoing research directions and highlight open challenges. Again, our focus is on formal argumentation as an exemplary method for automated non-monotonic reasoning.

5.1 Consistent Preferences and Undecided Beliefs

Some of the argumentation examples we present in this paper draw from ongoing research on economic rationality and formal argumentation [21,22]. An open question in this line of research is how to adjust the model of consistent preferences in abstract argumentation to support the notion of *undecided* arguments[10]. Let us highlight that this question cannot be addressed by straight-forward tweaks of the economic rationality-based argumentation principle, in particular because an agent must eventually commit to a course of action; *i.e.*, some arguments must not remain undecided. This can be illustrated with the help of a simple example. We have two weather report sources: one reports that it will rain (argument r) and the other reports it will not rain (argument $\neg r$). Obviously, r and $\neg r$ attack each other. We want to decide whether to take an umbrella with us (argument u). If we think it does not rain, we do not take an umbrella with us ($\neg r$ attacks u). Figure 5 depicts the corresponding argumentation framework.

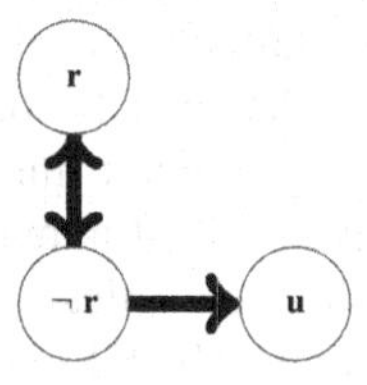

Fig. 5. $AF = (\{r, \neg r, u\}, \{(r, \neg r), (\neg r, r), (\neg r, u)\})$. How can we manage undecided arguments if we cannot be undecided about actions?

For example, given grounded semantics $\sigma_{grounded}$, all arguments in the argumentation framework are undecided. However, we must eventually make a decision on whether or not to take the umbrella with us; *i.e.*, to support undecided arguments, we need to define two argument types: *belief arguments* that may be undecided and *action arguments* that must never be undecided.

[10] Given an argumentation framework and a semantics' extension of this framework, the undecided arguments are all arguments that are neither in the extension, nor attacked by any of the arguments in the extension.

5.2 Burdens of Persuasion

When analyzing consistency and monotony properties of inference methods like formal argumentation approaches, it can be useful to apply intuitions that are provided by well-established practical research domains. In this regard, a particularly interesting concept is the notion of the *burden of persuasion* in legal research and practice. In case of two conflicting statements, the burden of persuasion can be placed on one of the statements, which implies that this statement requires additional justification; otherwise, it will be automatically defeated. For example, given two contradicting witness statements, of which one provides an alibi for the defendant, whereas the other one claims the defendant was at the crime scene at the time of the crime, the burden of persuasion could be laid on the latter argument to reflect the notion of *in dubio pro reo*[11]. Models of burdens of persuasion have already been introduced to formal argumentation approaches [9,26]. In these approaches, the burden of persuasion is explicitly modeled. In contrast, from the perspective of *consistent* inference, the burden of persuasion can automatically be placed on new arguments when expanding an argumentation framework; *i.e.*, if considering a new argument as part of the conclusion violates a consistency/monotony property (because the new argument is, directly or indirectly, in conflict with an argument that is part of a previous conclusion), the burden of persuasion is placed on this argument; additional conditions must be satisfied to allow for this argument to "kick out" the previously inferred argument[12]. Formally integrating this intuition with models of burdens of persuasion and consistency/monotony properties of formal argumentation can be considered promising future research.

5.3 Intuitive Rationality

Independently of the research on formal models of economic rationality and formal argumentation, recent research has started to shed light on what humans intuitively think are "reasonable" conclusions that can be drawn from argumentation frameworks [10,11]. The results suggest that while there is not necessarily one semantics whose behavior is more intuitive to most humans than all other semantics, some semantics (notably grounded and CF2 semantics[13]) seem to exhibit particularly intuitive behavior. As a result of these studies, SCF2 semantics has been introduced, which addresses some issues CF2 semantics has with regard to the handling of self-attacking arguments and even cycles that exceed a certain length [12]. The studies shed some light on human evaluations of argumentation principles, which can, however, be investigated more comprehensively. In particular, it is worth examining how well intuitive human assessments align with the consistent preference argumentation principle that is based on economic rationality (see Subsect. 4.2), as well as with other principles that can emerge from cross-disciplinary perspectives on "rational" and "consistent" reasoning and decision-making.

[11] This is a constructed example that does not fully reflect real-world legal reasoning.

[12] This notion is reflected by *loop-busting* approaches that have been proposed in the context of formal argumentation and that are based on Talmudic logic [1].

[13] For the sake of conciseness we do not introduce CF2 semantics in this paper; the semantics is introduced by Baroni *et al.* in [5].

5.4 Neuro-Symbolic Artificial Intelligence

Recently, combining machine learning and symbolic reasoning approaches has re-emerged as a hot topic in artificial intelligence research [16]. This trend is possibly accelerated because the machine learning break-throughs of the last decade have created the initial expectation of rapid and continuous progress, which machine learning alone cannot live up to. However, the integration of machine learning approaches and symbolic methods (which is sometimes referred to as *neural-symbolic AI*) has been a well-established research direction since several decades [15]. In Subsect. 4.4, we illustrate by example that a neuro-symbolic AI approach can be considered promising to allow for the evidence-based revision of reasoning (argumentation) principles. While formal argumentation has been integrated with machine learning methods, in particular in the context of argument mining [8], to our knowledge no research combines these hybrid approaches with a principle-based perspective.

To realize our proposal of an agent that can learn reasoning principles as it observes and interacts with its environment, we need create formal models and implementations at the intersection of non-monotonic symbolic reasoning and reinforcement learning, to find answers to the following questions. *i)* Which principles should an agent inhibit statically by design and which principles should be learnable? *ii)* How can we design principles that allow for a parameterization that facilitates learning? *iii)* To what extent is principle revision use-case agnostic, to what extent is it use-case-dependent? *iv)* When an agent learns new principles and hence updates its inference method, how does it trade-off consistency with regard to previously drawn inferences and compliance with the newly learned principles?

6 Conclusion

In this paper, we have introduced a formal perspective that takes inspirations from models of human models of decision-making reasoning to define levels of intelligent reasoning, *i.e.* the ability of an agent to:

1. reason in face of contradictions;
2. reason according to well-established principles, like the *clear and consistent preferences* principle that follows from economic rationality;
3. explain the resolution of contradictions according to whatever reasoning principles that are satisfied in a given scenario;
4. dynamically revise a principle-based inference process based on feedback the agent perceives as the result of interactions with its environment.

This perspective integrates well with a long-running line of research on non-monotonic reasoning approaches, which we have illustrated for formal (abstract) argumentation. In particular, *dynamic* models of formal argumentation that cover the expansion and iterative resolution of argumentation frameworks, considering fundamental properties of non-monotonic reasoning. However, as outlined in this paper, these models need further refinement to fully reflect the idea of explainable intelligent reasoning in face of contradictions.

Acknowledgments. The authors thank Amro Najjar, Michele Persiani, and the anonymous reviewers for their useful feedback. This work was partially supported by the Wallenberg AI, Autonomous Systems and Software Program (WASP) funded by the Knut and Alice Wallenberg Foundation.

References

1. Abraham, M., Gabbay, D.M., Schild, U.J.: The handling of loops in talmudic logic, with application to odd and even loops in argumentation. HOWARD-60: a Festschrift on the Occasion of Howard Barringer's 60th Birthday (2014)
2. Alchourrón, C.E., Gärdenfors, P., Makinson, D.: On the logic of theory change: partial meet contraction and revision functions. J. Symbolic Logic **50**(2), 510–530 (1985)
3. Anjomshoae, S., Najjar, A., Calvaresi, D., Främling, K.: Explainable agents and robots: results from a systematic literature review. In: Proceedings of the 18th International Conference on Autonomous Agents and MultiAgent Systems. AAMAS 2019, pp. 1078–1088. International Foundation for Autonomous Agents and Multiagent Systems, Richland, SC (2019)
4. Baroni, P., Giacomin, M.: On principle-based evaluation of extension-based argumentation semantics. Artif. Intell. **171**(10), 675–700 (2007). Argumentation in Artificial Intelligence. https://doi.org/10.1016/j.artint.2007.04.004, http://www.sciencedirect.com/science/article/pii/S0004370207000744
5. Baroni, P., Giacomin, M., Guida, G.: SCC-recursiveness: a general schema for argumentation semantics. Artif. Intell. **168**(1), 162–210 (2005). https://doi.org/10.1016/j.artint.2005.05.006
6. Bateman, I., Munro, A., Rhodes, B., Starmer, C., Sugden, R.: A test of the theory of reference-dependent preferences. Q. J. Econ. **112**(2), 479–505 (1997)
7. Baumann, R., Brewka, G.: Expanding argumentation frameworks: enforcing and monotonicity results. COMMA **10**, 75–86 (2010)
8. Cabrio, E., Villata, S.: Five years of argument mining: a data-driven analysis. In: Proceedings of the 27th International Joint Conference on Artificial Intelligence. IJCAI 2018, pp. 5427–5433. AAAI Press (2018)
9. Calegari, R., Riveret, R., Sartor, G.: The burden of persuasion in structured argumentation. In: Proceedings of the Nineteenth International Conference on Artificial Intelligence and Law. ICAIL 2021, Association for Computing Machinery, New York, NY, USA (2021)
10. Cramer, M., Guillaume, M.: Empirical cognitive study on abstract argumentation semantics. Front. Artif. Intell. Appl. **305**, 413–424 (2018). https://ebooks.iospress.nl/volume/computational-models-of-argument-proceedings-of-comma-2018
11. Cramer, M., Guillaume, M.: Empirical study on human evaluation of complex argumentation frameworks. In: Calimeri, F., Leone, N., Manna, M. (eds.) JELIA 2019. LNCS (LNAI), vol. 11468, pp. 102–115. Springer, Cham (2019). https://doi.org/10.1007/978-3-030-19570-0_7
12. Cramer, M., van der Torre, L.: SCF2-an argumentation semantics for rational human judgments on argument acceptability. In: Proceedings of the 8th Workshop on Dynamics of Knowledge and Belief (DKB-2019) and the 7th Workshop KI\& Kognition (KIK-2019) co-located with 44nd German Conference on Artificial Intelligence (KI 2019), Kassel, Germany, pp. 24–35 (2019)
13. Dung, P.M.: On the acceptability of arguments and its fundamental role in nonmonotonic reasoning, logic programming and n-person games. Artif. Intell. **77**(2), 321–357 (1995)
14. Gabbay, D.M.: Theoretical Foundations for Non-Monotonic Reasoning in Expert Systems. In: Apt, K.R. (ed.) Logics and Models of Concurrent Systems. NATO ASI Series (Series F: Computer and Systems Sciences), vol. 13, pp. 439–457. Springer, Heidelberg (1985). https://doi.org/10.1007/978-3-642-82453-1_15

15. Garcez, A.S.D., Lamb, L.C., Gabbay, D.M.: Neural-symbolic learning systems. In: Lamb, L.C. (ed.) Neural-Symbolic Cognitive Reasoning. COGTECH, Springer, Heidelberg (2009). https://doi.org/10.1007/978-3-540-73246-4_4
16. Geffner, H.: Model-free, model-based, and general intelligence. In: Proceedings of the 27th International Joint Conference on Artificial Intelligence. IJCAI 2018, pp. 10–17. AAAI Press (2018)
17. Haidt, J.: The emotional dog and its rational tail: a social intuitionist approach to moral judgment. Psychol. Rev. **108**(4), 814 (2001)
18. Jacobs, L.: Rabbi aryeh laib heller's theological introduction to his "shev shema'tata". Modern Judaism **1**(2), 184–216 (1981). http://www.jstor.org/stable/1396060
19. Kahneman, D.: Maps of bounded rationality: psychology for behavioral economics. Am. Econ. Rev. **93**(5), 1449–1475 (2003)
20. Kahneman, D., Tversky, A.: Prospect theory: an analysis of decision under risk. Econometrica **47**(2), 263–291 (1979)
21. Kampik, T., Gabbay, D.: Towards DIARG: an argumentation-based dialogue reasoning engine. In: SAFA@ COMMA, pp. 14–21 (2020)
22. Kampik, T., Nieves, J.C.: Abstract argumentation and the rational man. J. Logic Comput. **31**(2), 654–699 (2021). https://doi.org/10.1093/logcom/exab003
23. Landsburg, S.: The Armchair Economist (revised and updated May 2012): Economics and Everyday Life. Free Press (2007)
24. Lehmann, D., Magidor, M.: What does a conditional knowledge base entail? Artif. Intell. **55**(1), 1–60 (1992). http://www.sciencedirect.com/science/article/pii/000437029290041U
25. Osborne, M.J., Rubinstein, A.: Models in Microeconomic Theory. Open Book Publishers, Cambridge (2020). https://doi.org/10.11647/OBP.0204
26. Prakken, H., Sartor, G.: A logical analysis of burdens of proof. In: Legal Evidence and Proof: Statistics, Stories, Logic, pp. 223–253 (2009)
27. Rubinstein, A.: Modeling Bounded Rationality. MIT Press, Cambridge (1998)
28. Shao, C., Ciampaglia, G.L., Varol, O., Yang, K.C., Flammini, A., Menczer, F.: The spread of low-credibility content by social bots. Nat. Commun. **9**(1), 1–9 (2018)
29. Simon, H.A.: A behavioral model of rational choice. Q. J. Econ. **69**(1), 99–118 (1955). https://doi.org/10.2307/1884852
30. van der Torre, L., Vesic, S.: The principle-based approach to abstract argumentation semantics. IfCoLog J. Logics Appl. **4**(8), 34 (2017)
31. Turing, A.M.: Computing machinery and intelligence. In: Epstein, R., Roberts, G., Beber, G. (eds.) Parsing the Turing Test, pp. 23–65. Springer, Dordrecht (2009). https://doi.org/10.1007/978-1-4020-6710-5_3
32. Verheij, B.: Two approaches to dialectical argumentation: admissible sets and argumentation stages. Proc. NAIC **96**, 357–368 (1996)
33. Zhong, Q., Fan, X., Luo, X., Toni, F.: An explainable multi-attribute decision model based on argumentation. Expert Syst. Appl. **117**, 42–61 (2019). https://doi.org/10.1016/j.eswa.2018.09.038, http://www.sciencedirect.com/science/article/pii/S0957417418306158

Towards Transparent Legal Formalization

Tomer Libal[1,2] and Tereza Novotná[3(✉)]

[1] American University of Paris, Paris, France
tomer.libal@uni.lu
[2] University of Luxembourg, Esch-sur-Alzette, Luxembourg
[3] Masaryk University, Brno, Czech Republic
tereza.novotna@law.muni.cz

Abstract. A key challenge in making a transparent formalization of a legal text is the dependency on two domain experts. While a legal expert is needed in order to interpret the legal text, a logician or a programmer is needed for encoding it into a program or a formula. Various existing methods are trying to solve this challenge by improving or automating the communication between the two experts. In this paper, we follow a different direction and attempt to eliminate the dependency on the target domain expert. This is achieved by inverting the translation back into the original text. By skipping over the logical translation, a legal expert can now both interpret and evaluate a translation.

Keywords: Legal knowledge base · Annotation editor · Formal representation

1 Introduction

Machine legal reasoning has been around for more than 30 years and various implementations have been developed [3,11,14]. Nevertheless, the authors consider it safe to claim that legal machine reasoning is still not widely used, despite the effort invested.

One of the most challenging tasks towards automated legal reasoning is the ability to encode legislation in a machine readable form [13]. In this step, a legislation written by law and policy makers needs to be converted into a formal representation, which can then be read and analyzed by computer programs.

Among the difficulties in obtaining a perfect translation, one can mention:

1. The need to translate a possibly ambiguous legal text into an unambiguous formal representation.
2. The need to decide on, usually, a specific interpretation of various legal terms, which are left open in the legislation.
3. The fact that two separate domain experts are needed, one for the target, formal, domain, and a legal expert capable of interpreting the legislation.

The first two difficulties are hard to overcome, although there is some work in that direction [15]. On the other hand, there are more than a few approaches and solutions to the third problem.

D. Calvaresi et al. (Eds.): EXTRAAMAS 2021, LNAI 12688, pp. 296–313, 2021.
https://doi.org/10.1007/978-3-030-82017-6_18

One way to bridge the gap between the two domain experts is to facilitate communications between them. In [2], the problem of bridging between two domain experts is likened to the software engineering problem of bridging between client requirements and software. The agile methodology in software engineering suggests that short iterations and frequent consultation with the client can help to close the gap. The methodology described in this paper aims at translating an intermediate legislation back into text, so the legal expert can communicate problems back to the logician.

A second way to bridge the gap is described in [9] and is based on the same legislation editor used in this paper. This methodology, which is based on another Agile technique called Behavior-driven development[1], supports dividing the work between the two domain experts. The legal expert task is to write legal "cases" (users stories in Agile), while the logician is still in charge of the legal formalization process. The interaction between the two is achieved by executing the cases against the formalized legislation. If a case fails, then the logician must fix the formalization.

Another way is to introduce a formalization language which is closely related to the legal text. A popular such language is SBVR [1]. While such languages greatly facilitate the formalization effort [4], they are usually not expressive enough to capture the semantics of legal texts, as is attested by the contiguous search for more expressive languages. Several tools for multi-agent normative reasoning exists, for example Operetta for constructing normative specifications[2], however our tool is different mainly because of the interaction with the user.

Our solution is different and is based on taking out from the equation one of the experts. By building on top of a previous work [5], in which a one-to-one mapping between a legislation and its translation was established, we hope that a legal expert can execute both the formalization and the evaluation steps.

The work in [5] was concerned with the question of "How similar is the translation to the original legislation?. Existing translations, such as [12] and [14], are normally very different from the legislation. For example, to obtain the translation of a sentence of the form (part of the Regulation (EU) 2016/679 - General Data Protection Regulation, hereinafter as "GDPR", Article 13 par. 1):

> the controller shall, at the time when personal data are obtained, provide the data subject with all of the following information:
> a) the identity and the contact details of the controller and, where applicable, of the controller's representative;
> b) the contact details of the data protection officer, where applicable;"

Current approaches will generate two or three different statements, each containing parts of the general conditions and adding new conditions and conclusion. On the other hand, the extension in [5] to the annotation language described in [8] enables a one-to-one mapping between the original legislation and its translation.

[1] https://www.agilealliance.org/glossary/bdd/.

[2] http://www.cs.uu.nl/research/projects/opera/.

In this paper, we build on the fact that there is a one-to-one mapping between the legislation and its translation and translate it back into a version of the original legislation. This last translation also enjoys a one-to-one mapping, which provides us with the main advantage of the method. The new methodology requires only a legal expert for asserting the quality of the translation. The expert is provided with the input and output of each paragraph and can determine if the translated formalization faithfully represents the specific legal interpretation of the input text, thus enhancing the transparency of the legal knowledge base.

This paper is organized as follows. In the next section, we present the editor and describe the additional features which were implemented in order to support the results of this paper. The third section is dedicated to describing the methodology and its application to article 7 of the GDPR. We conclude with an overview of future work and improvements.

2 The Legislation Editor

This section is adapted from [8].

The legislation editor integrates theorem proving technology into a usable graphical user interface (GUI) for the computer-assisted formalization of legal texts and applying automated normative reasoning procedures on these artifacts. In particular, the system includes

1. a legislation editor that graphically supports the formalization of legal texts via the use of annotations,
2. means of assessing the quality of entered formalizations, e.g., by automatically conducting consistency checks and assessing logical independence,
3. ready-to-use theorem prover technology for evaluating user-specified queries wrt. a given formalization, and
4. the possibility to share and collaborate, and to experiment with different formalizations and underlying logics.

The system is realized using a web-based Software-as-a-service and is available using a browser. It comprises a GUI that is implemented as a Javascript browser application, and a NodeJS application on the back-end side which connects to theorem provers, data storage services and relevant middleware. Using this architectural layout, no further software is required from the user perspective for using the editor and its reasoning procedures, as all necessary software is made available on the back end and the computationally heavy tasks are executed on the remote servers only. The results of the different reasoning procedures are sent back to the GUI and displayed to the user.

2.1 The Annotation Editor

The annotation editor allows users to create formalizations of legal documents that can be subsequently used for formal legal reasoning. The general functionality of the editor is described in the following.

One of the main ideas of the editor is to hide the underlying logical details and technical reasoning input and outputs from the user. We consider this essential, as the primary target audience of the tool are legal experts who are not necessarily logicians. It could greatly decrease the usability of the tool if a solid knowledge about formal logic was required. This is realized by letting the user annotate legal texts and queries graphically and by allowing the user to access the different reasoning functionalities by simply clicking buttons that are integrated into the GUI.

Another main idea of the editor is to enable a direct annotation of the original text, without the need to modify it. This property is essential to the core result of this paper - the ability to regenerate the original text from the annotations, which can then be compared. It should be noted that many times, legal texts contain implicit information. For example, the GDPR often talks about the processing of data, but without always specifying that this is done by a processor. In such cases, the users need to add the implicit text explicitly in the editor, so it can be annotated. We encourage the user to put such text in square brackets, in order to simplify the comparison between the original and the generated text at the end of the process.

These two properties raise the need for an expressive annotation language, which can capture not only language properties but also meta-language properties, such as exceptions and other complex relations. In order to support such a language the annotations are organized into three layers.

In order to demonstrate the different elements of the editor, we will use GDPR article 7, paragraph 2 as an example. The full annotation of this paragraph can be seen in Fig. 1.

The editor employs a hierarchical approach to annotations. At the base, there are "term" annotations, which are used to mark all relevant entities and relations. These annotations are assigned specific colors, corresponding to their entities. Entities normally denote a class of items or people while relations denote actions or relations between different entities. An example of an entity which can be seen in Fig. 1 is "give_consent", which corresponds to a relation between a data subject, a controller and a specific data, processing and time.

On top of the term annotations, there are logical and normative properties, such as obligations, conditions and negations. These annotations place a specific context around term annotations, as well as other logical and normative ones. For example, a term annotation might imply another, in which case the two relate to each other as a condition and implication. If an annotation is an obligation, such an annotation should be applied. In Fig. 1, we can see that the relation between the first two terms and the following five is a condition in which the conclusion is an obligation. This is obtained by setting each group of the two groups of terms as a conjunction and setting the relationship between the two groups as a conditional obligation, where the first part consists of the conditions and the second consists of the obligation.

Lastly, the editor enables the user to annotate concepts which go beyond direct logical and normative properties. For example, a sentence might be an exception of another, in which case the two must be bound together, despite the fact that they may not appear together. Another example is when a sentence refers to another but requires the replacement of some concepts with others. The current text requires but a simple use of this feature. In Fig. 1, we can see that each paragraph is being associated with its specific numbering using the "Labeling" annotation. Such labels can be used later, for example in exceptions. More complex examples can be found in the formalization of article 13 of the GDPR, which is described in [6].

The formalization proceeds as follows: The user selects some text from the legal document and annotates it, either as a term or as a composite (complex) statement. In the first case, a name for that term is computed automatically, but it can also be chosen freely. Different terms are displayed as different colors in the text. In the latter case, the user needs to choose among the different possibilities, which correspond to either logical connectives or higher-level sentence structures called macros. The composite annotations are displayed as a box around the text and can be done recursively.

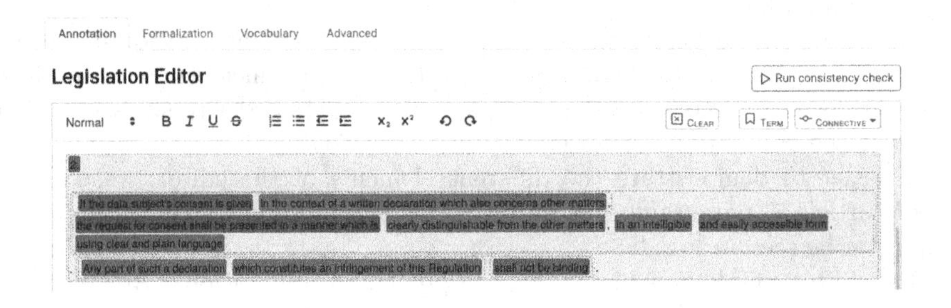

Fig. 1. Article 7, par 2: full annotation

The editor also features direct access to the consistency check and logical independence check procedures (as buttons). When such a button is clicked, the current state of the formalization will be translated and sent to the back-end provers, which determine whether it is consistent resp. logically independent.

User queries are also created using such an editor. In addition to the steps sketched above, users may declare a text passage as *goal* using a dedicated annotation button, whose contents are again annotated as usual. If the query is executed, the back-end provers will try to prove (or refute) that the goal logically follows from the remaining annotations and the underlying legislation. This way, the tool can answer YES/NO questions on whether the goal logically follows the facts and logical relations of formalized legislation.

A very important feature of the editor is the ability to check the formalization. Often times, formalizations cannot be checked for correctness and can contain, therefore, many errors and typos[3]. By clicking the "Save" button, the

[3] Please refer to [5] for more information and examples.

annotations are being converted into a formal structure, which is then being converted into a legal text, for comparison. This process fails if the structure of the annotations is incorrect and serves as a kind of "compiler".

In the next section, we are going to discuss in more detail the generated legal text, which can be found under the "Formalization" tab.

3 Legal Formalization

Formalization of legal text into a machine readable format is a rather complex process. In such a process, the user inevitably encounters a number of problems as described in Sect. 1. In previous research, the authors tackled a few such difficulties by proposing an approach of interactive legal text formalization, where a user or legal expert decides about some of the types of difficulties based on her expertise in certain legal field [8]. An interpretation of different statements and terms in a legislation or legal text is generally one of the biggest problems during automatic formalization. Authors deal with the interpretation issue in such a manner that legal experts as users solely interpret the legal text being formalized. Using this approach the editing tool becomes more of a support tool while formalizing legal text and only depends on the decisions of the user. This approach is further advantageous because it can flexibly react on the development of interpretation of a certain legal text in time.

In this article, the authors develop this approach further and tackle another relevant problem. As described in Sect. 1, legal formalization using an editing tool requires knowledge of both law and logic which in most cases requires the cooperation of a logician and a lawyer. The authors propose a solution in providing user-friendly output of a formalization of legal text which is easily comprehensible for lawyers with basic knowledge of logic. This approach aims to reduce the level of expertise in logic required and assumes that the legal expert is able to both formalize the legislation and assess the accuracy of the output.

3.1 Article 7 of GDPR

We use GDPR as a representative legislation for the editing tool functioning presentation. GDPR is one of the most discussed and in many aspects controversial European legislations, which makes it an optimal use case for the purpose of this study. The GDPR defines the principles of personal data processing and the conditions of lawfulness of their processing. It also regulates the conditions for expressing the consent given to the processing of data and the provision of information and access to personal data. The GDPR is universally binding and applicable in all Member States of EU. Regarding its universality, uniqueness and relative strictness in duties and sanctions, it is not surprising that a great deal of effort is being made to ensure compliance of personal data processing with the regulation in the private sector and to analyze and interpret the wording of the legislation in the public and academic sectors. This statement may be also supported by the rich case law concerning GDPR of both the general and European courts and other administrative institutions. The robustness of the GDPR application then creates the need to simplify and make the legal document ideally accessible to a wide range of citizens and institutions in a transparent and simple form.

In a current state, the user is capable of formalizing the whole legislation, which is the recommended way of using the editing tool. Regarding the scope of this article, we decided to present the formalization of the whole of Article 7 of GDPR - *Conditions for consent.* For the same reason we do not formalize the Article in context with the rest of the legislation, however this systematic formalization could be done following the same methodology.

Consent, in the context of GDPR, is one of the lawful ways of processing personal data in accordance with Article 6 par. 1 (a) of GDPR and it is one of the most common ways to lawfully process personal data as it is part of a majority of services a human can use or buy nowadays. Article 7 states necessary conditions for the consent of a subject of personal data to be lawful and in accordance with the European regulation.

The full wording of Article 7 GDPR is:

1. Where processing is based on consent, the controller shall be able to demonstrate that the data subject has consented to processing of his or her personal data.

2. If the data subject's consent is given in the context of a written declaration which also concerns other matters, the request for consent shall be presented in a manner which is clearly distinguishable from the other matters, in an intelligible and easily accessible form, using clear and plain language. Any part of such a declaration which constitutes an infringement of this Regulation shall not be binding.

3. The data subject shall have the right to withdraw his or her consent at any time. The withdrawal of consent shall not affect the lawfulness of processing based on consent before its withdrawal. Prior to giving consent, the data subject shall be informed thereof. It shall be as easy to withdraw as to give consent.

4. When assessing whether consent is freely given, utmost account shall be taken of whether, inter alia, the performance of a contract, including the provision of a service, is conditional on consent to the processing of personal data that is not necessary for the performance of that contract.

3.2 Formalization of Article 7 Using the Legislation Editor

The formalization of legal text means the transformation of the text, including its semantics, into the logical representation. The legislation editor uses the formalization performed by the user - legal expert - as an input. The formalization is performed in the editor using annotations of the text as described in Sect. 2.1 in the Annotation tab.

The formalization in the legislation editor is based on the annotations of three categories: terms, connectives and macro concepts. Terms are parts of the text that the tool recognizes as an entity, entities can appear in the formalization any number of times. Connectives are logical constructs aimed to catch the logical relationship among terms. Terms are marked as colored rectangles and connectives and macros as grey borders in Fig. 1, the same term is always colored with the same color. Annotations are performed by the expert who marks the text according to her interpretation of the text and regarding the goal of the formalization - what kind of questions about the legal text the user wants to answer.

As it was mentioned in Sect. 1, the annotations methodology follows hierarchical order. It is recommended to firstly determine the legal text or parts of the legal text which user wants to formalize and then to clarify the logical relationships between the parts of the text. Given the example of Article 7 GDPR, this article contains 4 paragraphs, all paragraphs relate to the conditions of consent, which is gradually refined, therefore it is possible to start the formalization of individual paragraphs separately.

Subsequent formalization shall follow a similar trend - firstly determine all the terms (entities) in the text which the user considers important and mark them as terms in the Annotation tab. After this step, the user shall mark the logical relationships among marked terms. During this step, it is again necessary to annotate the relations hierarchically, i.e. to start with external relationships and gradually move on to internal ones.

Given Article 7 par. 1: *1. Where processing is based on consent, the controller shall be able to demonstrate that the data subject has consented to processing of his or her personal data.*

This paragraph contains an obligation of a controller to demonstrate the consent of a specific subject for the processing of his or her personal data if the processing is based on consent. The essential entities of this paragraph are 6 following: processing, based on consent, controller, to demonstrate, data subject and consent to processing of personal data. Therefore, we accept these 6 parts of the sentence as terms bearing the meaning of the sentence. We use the legislation editor annotations scheme to mark all these terms and choose the term names and variables. Variables follow first-order logic methodology and they are contained in the brackets behind the term names. They need to start with upper case letter to be recognized as variables. Variables allow the user to address specific real-life subjects and objects and reason about them. Let us take the first term *processing*. While processing, there is a specific subject and her personal data to be processed by a specific controller in a specific situation in time. All these characteristics of processing are changeable depending on the specific situation, therefore we accept them as variables. When annotating in the Annotation tab of the tool, we mark first part of the sentence "*Where processing*" as a term using the "Term" button as in Fig. 1 and choose the name **processing(Data, Subject, Time, Controller)**, where the order of variables is arbitrary, as in Fig. 2. We annotate the other 5 terms in the same manner with the variables as following:

"*is based on consent*" as term **based_on_consent(Data, Subject, Time, Controller, Consent, Processing)**

"*the controller*" as term **controller(Controller)**
"*shall be able to demonstrate*" as term **to_demonstrate(Controller, Consent)**
"*that the data subject*" as term **data_subject(Subject)** "*has consented to processing of his or her personal data*" as term **give_consent(Data, Subject, Time, Controller, Consent, Processing)**

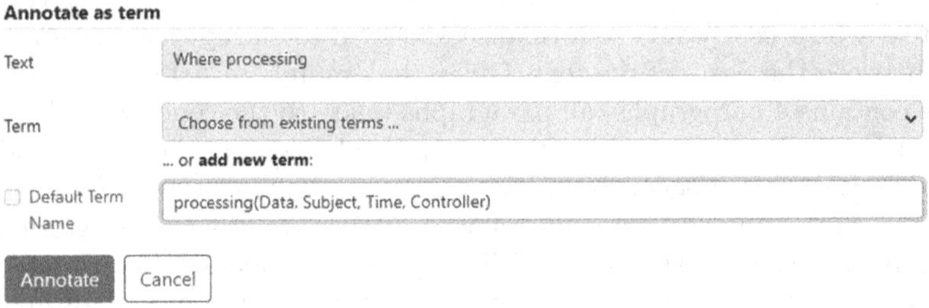

Fig. 2. Annotation of a "Term"

Additionally, we annotate the number of paragraph "1." as a label for this paragraph using the "Term" button and choosing the name **paragraph1**. We do this step in order to be able to use the sentence as a whole later in the situation, where it is needed to state a specific logical relationship between different paragraphs. Therefore, there is no need to use variables for this term, since we are choosing just a name for the paragraph.
"*1.*" as term **paragraph1**

This way we divided the first paragraph of Article 7 into different parts marked as terms in the Annotation tab, all of the marked terms bearing different meaning and having different roles in the sentence. The second step in formalization using the legislation editor is to define the logical relationships between the terms in legal text.

As it was mentioned beforehand, during the annotation of logical relationships, it is necessary to proceed from external logical relationships to internal ones. Practically, annotation of logical relationships is similar to annotation of the terms. We have to mark the whole text to which we intend to assign a logical relationship and choose the correct logical relationship from the "Connectives" button, which provides several options as we can see in Fig. 3.

First, we have to use the "Labelling" of the paragraph, which is a macro concept listed in "Connectives" as "Labelling sentences". This macro concept accepts the first term in the marked text as a label and the rest as a sentence to be labelled. In this case, the first marked term is the term **paragraph1**. This means that if we need to refer to this paragraph in the future while formalizing of the regulation, we can use the term **paragraph1** as a reference to this paragraph. Labelling is a usual first step as it delimits the part of a text we will formalize as a whole.

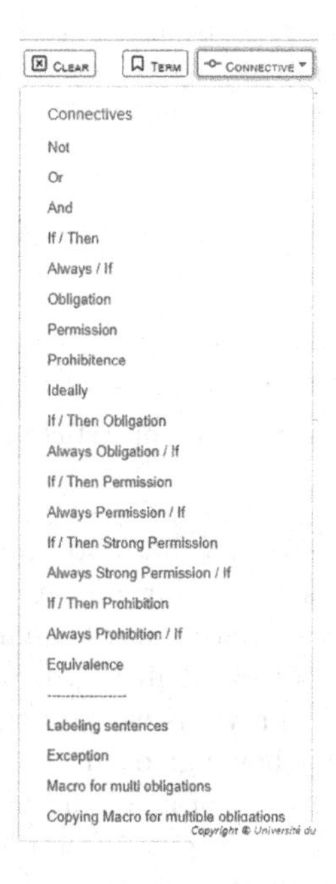

Fig. 3. The list of connectives

Secondly, we have to decide on the logical relationship of a paragraph (sentence). This task requires knowledge of legal notions and legal language used for expressing different legal statements. In the case of the first paragraph of Article 7, this sentence expresses an obligation of the controller to demonstrate the consent, however this obligation is conditioned by the first part of sentence, according to which the processing must be based on a consent for the following obligation to apply. We can deduce that the obligation of the controller in the second part of the paragraph is conditioned by the first part of the paragraph. This situation clearly leads to an implication of obligation, in the "Connectives" list in Fig. 3 marked as "If/Then Obligation". Therefore, we mark the whole sentence, except the label **paragraph1** as "If/Then Obligation".

The first two terms - **processing(Data, Subject, Time, Controller)** and **based_on_consent(Data, Subject, Time, Controller, Consent, Processing)** are conditions for the obligation of the controller to apply. Thus, we mark these two terms with another internal logical connective "And" (as a conjunction).

The obligation of the controller is expressed in rest of the terms in this paragraph - **controller(Controller)**, **to_demonstrate(Controller, Consent)**, and **data_subject(Subject)** and **give_consent(Data, Subject, Time, Controller, Consent, Processing)**. Similarly, we mark these terms again with the internal logical connective "And".

To summarize this process, we can say that the controller is legally obligated to demonstrate a consent expressed by the data subject for specific processing of her personal data, if such processing is based on consent. Full obligation of Article 7 paragraph 1 is in Fig. 4.

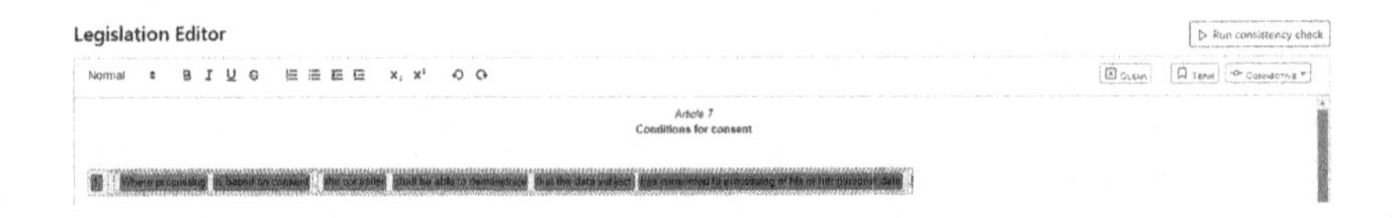

Fig. 4. Annotation of Article 7 par. 1

The legislation editor contains another two important buttons - "Save" and "Run consistency check". We save the formalization of the first paragraph with the "Save" button and we check whether our formalization is correct by pressing the "Run consistency check" button as in Fig. 4. This feature of the legislation editor was already described in previous work of the authors, however the issue tackled in this article is the following: even though the consistency check is correct, is the user (lawyer) capable of assessing her formalization and deciding whether it is correct and meaningful from both a legal and logical perspective?

To tackle this issue and simplify this task, we propose a work-in-progress solution in the Formalization tab in the Legislation Editor.

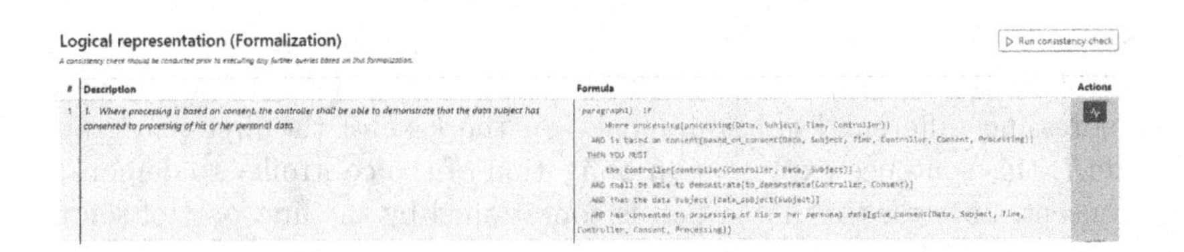

Fig. 5. Formalization of Article 7 par. 1

The Formalization tab in Fig. 5 offers comparison of the original text and logical formulae applied to the original text. Furthermore, logical formulae are provided in such a way that the user can easily understand the logical relationships and evaluate their correctness. The current presentation displays the internal logic formulae next to the relevant text. These formulae are denoted in first-order Deontic logic [7] and variables are implicitly quantified over each of the whole formulae displayed on the formalization tab. In future versions, this logical terms will be replaced by their matching entities in a relevant ontology (see for example, [10]).

It should also be noted that we make a distinction between the formula used for representing the knowledge, and the formula used for actual reasoning over the knowledge. Once a theorem prover is being called, the representation formula is being translated into the prover input logic. Issues such as conditional obligations, negation-as-failure and others are resolved at this point, depending on the target theorem prover.

The goal of the methodology of translating the original text into a formula and then back into text, is that the output is as comprehensible and as similar as the input, i.e. the original legal text in the left part of the Formalization tab. At the same time, it is necessary to preserve the logical formulae used while formalizating the text and present them in the output in a compatible way with the text. To find the right balance of these two goals - to present both the original text and logical formulae comprehensibly together - is the future work for the authors of this article, the current setting is just the first attempt on this path.

As we can see in Fig. 5, first the legislation editor shows the label of the paragraph. Subsequently, it shows the specific part of the original legal text and the term allocated to this specific part of the text behind together with selected variables. To clearly distinguish different semantic parts of texts, it lists the terms on separate rows in the order as in the original text, therefore it preserves the sequence of the original legal text. Additionally, different terms (different parts of original text) are connected with logical words representing different logical relationships.

Specifically, we can first observe the word "IF" which means that the following terms are conditions. These following terms are connected with "AND". The obligation dependent on previous conditions is marked as "THEN YOU MUST" and the following terms are the consequences that need to happen in order to meet the obligation of this part of the regulation. The legal consequences are again connected with "AND" to show that all of them need to apply. Detailed formalization of the first paragraph is in Fig. 6.

Formula **Actions**

```
paragraph1)  IF
     Where processing[processing(Data, Subject, Time, Controller)]
   AND is based on consent[based_on_consent(Data, Subject, Time, Controller, Consent, Processing)]
  THEN YOU MUST
     the controller[controller(Controller, Data, Subject)]
   AND shall be able to demonstrate[to_demonstrate(Controller, Consent)]
   AND that the data subject [data_subject(Subject)]
   AND has consented to processing of his or her personal data[give_consent(Data, Subject, Time,
Controller, Consent, Processing)]
```

Fig. 6. Formalization of Article 7 par. 1 - formulae

In this case, the comprehensibility is achieved through the indentation, separation of the original text and allocated terms and differentiation of connectives in upper case and terms or original text in lower case. The visual distinction of logical parts of legal text is another way of presenting the output, although we are aware that in this direction, the legislation editor is limited and we accept this as a future work issue. Nevertheless, we believe that our approach is the right first step in the comprehensibility of formalization of legal text.

Regarding paragraph 1 of Article 7, the output is easily comparable with the input in the left part of the Formalization tab. The check of formalization is then easily performed by simply reading through the right formalized output and assessing its legal meaning when comparing to the meaning of the original text.

Given this formalization methodology, we can proceed with the rest of the paragraphs in the same manner. In paragraph 2 as we can see in Fig. 1 we used following terms:

"*2.*" as term **paragraph2**

"*If the data subject's consent is given*" as term **give_consent(Data, Subject, Time, Controller, Consent, Processing)**

"*in the context of a written declaration which also concerns other matters*" as term **written_declaration_other_matters(Data, Subject, Time, Controller, Consent)**

"*the request for consent shall be presented in a manner which is*" as term **request_for_consent(Data, Subject, Time, Controller, Consent)**

"*clearly distinguishable from the other matters*" as term **distinguishable(Data, Subject, Time, Controller, Consent)**

"*in an intelligible*" as term **intelligible(Data, Subject, Time, Controller, Consent)**

"*and easily accessible form*" as term **accessible(Data, Subject, Time, Controller, Consent)**

"*using clear and plain language*" as term **language(Data, Subject, Time, Controller, Consent)**

"*Any part of such a declaration*" as term **written_declaration_other_matters (Data, Subject, Time, Controller, Consent)**

"*which constitutes an infringement of this Regulation*" as term **infringement**

"*shall not be binding*" as term **binding**

Subsequently, after labelling the paragraph, we chose two obligations based on conditions as connectives for the two separate sentences in paragraph 2. Therefore, we annotated them separately as "IF/THEN OBLIGATION" connectives, where **give_consent(Data, Subject, Time, Controller, Consent, Processing)** and **written_declaration_other_matters(Data, Subject, Time, Controller, Consent)** are conditions for first implication and the following terms are consequences. **written_declaration_other_matters(Data, Subject, Time, Controller, Consent)** and **infringement** are conditions for second implication and **binding** (annotated as a "NEGATION" first) is a consequence.

The resulting formalization is in Fig. 7.

```
paragraph2)     IF
        If the data subject's consent is given[give_consent(Data, Subject, Time, Controller, Consent,
Processing)]
    AND in the context of a written declaration which also concerns other
matters[written_declaration_other_matters(Data, Subject, Time, Controller, Consent)]
   THEN YOU MUST
        the request for consent shall be presented in a manner which is[request_for_consent(Data,
Subject, Time, Controller, Consent)]
    AND clearly distinguishable from the other matters[distinguishable(Data, Subject, Time, Controller,
Consent)]
    AND in an intelligible[intelligible(Data, Subject, Time, Controller, Consent)]
    AND and easily accessible form[accessible(Data, Subject, Time, Controller, Consent)]
    AND using clear and plain language[language(Data, Subject, Time, Controller, Consent)]

  AND    IF
        Any part of such a declaration[written_declaration_other_matters(Data, Subject, Time, Controller,
Consent)]
    AND which constitutes an infringement of this Regulation[infringement]
   THEN YOU MUST
        NOT
     shall not be binding[binding]
```

Fig. 7. Formalization of Article 7 par. 2 - formulae

Again, we can observe two separated implications with obligations starting with "IF" statement and continuing with "THEN YOU MUST". These two implications (sentences) are connected with the connective "AND" as two parts of one paragraph labelled as "paragraph2". Given the scope of this article, we are not able to go through the formalization of all the paragraphs step by step, however we are providing the outputs in Fig. 8 and Fig. 9.

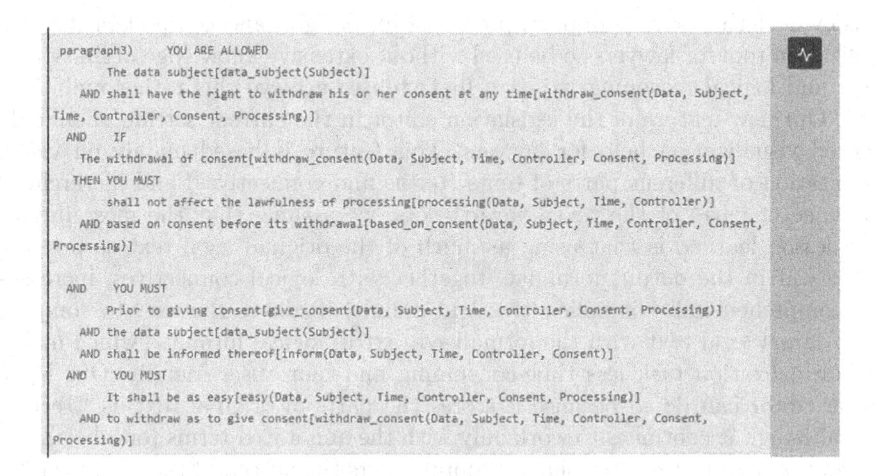

paragraph3)   YOU ARE ALLOWED
The data subject[data_subject(Subject)]
AND shall have the right to withdraw his or her consent at any time[withdraw_consent(Data, Subject,
Time, Controller, Consent, Processing)]
AND   IF
The withdrawal of consent[withdraw_consent(Data, Subject, Time, Controller, Consent, Processing)]
THEN YOU MUST
shall not affect the lawfulness of processing[processing(Data, Subject, Time, Controller)]
AND based on consent before its withdrawal[based_on_consent(Data, Subject, Time, Controller, Consent,
Processing)]
AND   YOU MUST
Prior to giving consent[give_consent(Data, Subject, Time, Controller, Consent, Processing)]
AND the data subject[data_subject(Subject)]
AND shall be informed thereof[inform(Data, Subject, Time, Controller, Consent)]
AND   YOU MUST
It shall be as easy[easy(Data, Subject, Time, Controller, Consent, Processing)]
AND to withdraw as to give consent[withdraw_consent(Data, Subject, Time, Controller, Consent,
Processing)]

Fig. 8. Formalization of Article 7 par. 3 - formulae

We used a permission to formalize the first sentence of paragraph 3 as a right of subject of personal data to withdraw the consent. This permission is

```
paragraph4)  IF
     When assessing whether consent[assessing(Data, Subject, Time, Controller, Consent)]
   AND is freely given[freely_given(Data, Subject, Time, Controller, Consent)]
  THEN YOU MUST
     utmost account shall be taken of [take_account]
   AND whether, inter alia, the performance of a contract, including the provision of a
service[performace_of_contract(Data, Subject, Time, Controller, Consent)]
   AND is conditional[conditional(Data, Subject, Time, Controller, Consent, Processing)]
   AND on consent to the processing of personal data[give_consent(Data, Subject, Time, Controller,
Consent, Processing)]
   AND     NOT
     that is not necessary[necessary(Data, Subject, Time, Controller, Consent, Processing)]
   AND for the performance of that contract[performace_of_contract(Data, Subject, Time, Controller,
Consent)]
```

Fig. 9. Formalization of Article 7 par. 4 - formulae

simple, it is not conditioned by any other facts or actions. Similarly, we used simple obligations without implications for the third and fourth sentences in this paragraph given that these sentences mean simple obligations non-conditioned by any precedent facts or actions to apply.

Regarding paragraph 4, we annotated the whole sentence as an obligation with the implication "IF/THEN OBLIGATION" given that "*whether assessing if the consent is freely given*" are conditions and considering the necessity of processing of personal data for the performance of a contract are consequences that must be met to comply with the GDPR regulation.

3.3 The Comprehensibility of Formalization - Future Work

It is evident that our attempt to provide a meaningful and comprehensible formalization tool for lawyers to be used without extensive knowledge of first-order logic and logical reasoning is just a first step to achieve this rather ambitious task. Our new feature of the legislation editor in the current setting offers only limited visualization help for lawyers. This feature is based mainly on visual indentation of different parts of texts (terms and connectives) and hierarchical sequence of parts of the text. Nevertheless, we assume that the most important lesson learned is that using as much of the original legal text as possible to present in the output formulae, together with logical connectives, increases the comprehensibility significantly. And mainly, it allows the user to compare the original legal text with the formalized text in logical formulae which makes the formalization task less time-consuming and more user-friendly. Our legislation editor can do so because it stores the parts of original text as "Terms" and although it continues to work only with the annotated terms (or labels), the stored text is very helpful when the tool presents the output of the formalization to the user.

As a future work, we intend to achieve higher comprehensibility with better visualization of the output formulae using graphical features such as colours or diagrams and connecting the formalized terms to the ontological dictionary. We believe that enhancing the comprehensibility of output formulae is the main task to be solved on the way to usable formalization tool for lawyers without further knowledge of advanced logic.

4 Conclusion

One of the major obstacles for the creation of a formal legal knowledge base, is the dependency on both target and source domain experts. The target domain experts, who are usually logicians or programmers, are most suitable for encoding knowledge into formulae and programs, but they usually lack a deep understanding in the legal domain. On the other hand, most legal experts would find it difficult to follow the legal meaning of a legislation in a completely formal form. This dependency on two domain experts means that the trust legal experts can place in a knowledge base is limited.

In this article, we build on our previous research ideas and applications in legal reasoning and formalization of legal texts in order to make the formalization process transparent. This is achieved by increasing the confidence of the legal expert in the formalized knowledge.

Our approach is based on an assumption that if the formalization of legal text is user-friendly enough and if the output of formalization is comprehensible enough, the formalization of legal text can be performed by an expert in specific legal domain without a deeper knowledge of first-order logic. We are trying to achieve this goal by extending an existing legislation editor with a visualized output of a formalization.

Using our approach, a lawyer with limited knowledge of first-order logic is capable of translating certain legal text into formal language using a simple annotation editor according to his expertise in a legal domain related to this legal text. The formalization is only dependent on his expert interpretation of the text. Furthermore, the lawyer is capable of properly checking the formalized formulae of legal text and eventually, to correct the formalization according to his best knowledge of interpretation and legal domain.

To demonstrate this approach, we formalize Article 7 of GDPR. We show the practical use of the legislation editor in the formalization of Article 7 of GDPR in a step by step manner. Subsequently, we show the visualized output of the logical formulae as a result of legal text formalization. We argue that the proposed design of the output formulae is sufficient and comprehensible enough for a lawyer to assess the logical formulae and compare it to the original plain legal text and potentially use it to correct the formalization.

This paper describes a very basic prototype with the sole purpose of demonstrating the approach. In the future, we plan on enhancing the translation back into legal text so it will be as similar to the original text as possible. For example, a legal text might use various words to denote an obligation, such as "must", "should" and "obliged". All these verbs are currently uniformly being translated into "It is an obligation that". By parsing the original text and extracting the specific "obligation" verb, we can increase the similarity of the generated text to the original.

Another enhancement to the current version would be to create a two-dimensional display of the legal text. Currently, both the original text and the generated formula which denote a certain term are appearing in the translation. In a two-dimensional approach, only the original text would be displayed and the user would have the possibility to see which ontological entity is mapped to it by clicking on or hovering over it.

Acknowledgment. Tereza Novotn acknowledges the support of the ERDF project "Internal grant agency of Masaryk University" (No. CZ.02.2.69/0.0/0.0/19_\0073/0016943).

References

1. Bajwa, I.S., Lee, M.G., Bordbar, B.: SBVR business rules generation from natural language specification. In: 2011 AAAI Spring Symposium Series. Citeseer (2011)
2. Bartolini, C., Lenzini, G., Santos, C.: An agile approach to validate a formal representation of the GDPR. In: Kojima, K., Sakamoto, M., Mineshima, K., Satoh, K. (eds.) JSAI-isAI 2018. LNCS (LNAI), vol. 11717, pp. 160–176. Springer, Cham (2019). https://doi.org/10.1007/978-3-030-31605-1_13
3. Governatori, G., Shek, S.: Regorous: a business process compliance checker. In: Proceedings of the Fourteenth International Conference on Artificial Intelligence and Law, pp. 245–246 (2013)
4. Johnsen, Å., Berre, A.: A bridge between legislator and technologist-formalization in SBVR for improved quality and understanding of legal rules. In: International Workshop on Business Models, Business Rules and Ontologies, Bressanone, Brixen, Italy. Citeseer (2010)
5. Libal, T.: A meta-level annotation language for legal texts. In: Dastani, M., Dong, H., van der Torre, L. (eds.) CLAR 2020. LNCS (LNAI), vol. 12061, pp. 131–150. Springer, Cham (2020). https://doi.org/10.1007/978-3-030-44638-3_9
6. Libal, T.: Towards automated GDPR compliance checking. In: Proceedings of the Workshop on the Scientific Foundations of Trustworthy AI, Integrating Learning, Optimisation and Reasoning (to appear)
7. Libal, T., Pascucci, M.: Automated reasoning in normative detachment structures with ideal conditions. In: Proceedings of the Seventeenth International Conference on Artificial Intelligence and Law, pp. 63–72 (2019)
8. Libal, T., Steen, A.: NAI: the normative reasoner. In: Proceedings of the Seventeenth International Conference on Artificial Intelligence and Law, pp. 262–263 (2019)
9. Libal, T., Steen, A.: Towards an executable methodology for the formalization of legal texts. In: Dastani, M., Dong, H., van der Torre, L. (eds.) CLAR 2020. LNCS (LNAI), vol. 12061, pp. 151–165. Springer, Cham (2020). https://doi.org/10.1007/978-3-030-44638-3_10
10. Zeiler, M.D., Fergus, R.: Visualizing and understanding convolutional networks. In: Fleet, D., Pajdla, T., Schiele, B., Tuytelaars, T. (eds.) ECCV 2014. LNCS, vol. 8689, pp. 818–833. Springer, Cham (2014). https://doi.org/10.1007/978-3-319-10590-1_53
11. Prakken, H., Wyner, A., Bench-Capon, T., Atkinson, K.: A formalization of argumentation schemes for legal case-based reasoning in ASPIC+. J. Logic Comput. **25**(5), 1141–1166 (2015)

12. Robaldo, L., Bartolini, C., Palmirani, M., Rossi, A., Martoni, M., Lenzini, G.: Formalizing GDPR provisions in reified I/O logic: the DAPRECO knowledge base. J. Logic Lang. Inf. **29**(4), 401–449 (2020)
13. Routen, T., Bench-Capon, T.: Hierarchical formalizations. Int. J. Man Mach. Stud. **35**(1), 69–93 (1991)
14. Sergot, M.J., Sadri, F., Kowalski, R.A., Kriwaczek, F., Hammond, P., Cory, H.T.: The British nationality act as a logic program. Commun. ACM **29**(5), 370–386 (1986)
15. Vitali, F., Zeni, F.: Towards a country-independent data format: the Akoma Ntoso experience. In: Proceedings of the V Legislative XML Workshop, pp. 67–86 (2007)

Applying Abstract Argumentation to Normal-Form Games

You Cheng, Beishui Liao, and Jieting Luo(✉)

Institute of Logic and Cognition, Zhejiang University, Hangzhou 310028, China
{ichyou,baiseliao,luojieting}@zju.edu.cn

Abstract. Game theory is the most common approach to studying strategic interactions between agents, but it provides little explanation for game-theoretical solution concepts. In this paper, we use a game-based argumentation framework to solve normal-form games. The result is that solution concepts in game theory can be interpreted by extensions of a game-based argumentation framework. We can use our framework to solve normal-form games, providing explanation for solution concepts.

Keywords: Game theory · Argumentation framework · Explanation

1 Introduction

In a game, agents can influence one another's payoffs by choosing different strategies. Game theory provides mathematical models to deal with strategic interactions between agents [9,12,21]. Argumentation copes with inconsistency and support relation among arguments put forward by agents [2]. Therefore, it is obvious that both game theory and argumentation address problems concerning interaction among agents.

The interplay between games and argumentation can be treated in two dimensions, one is to use games to analyze agents' behavior in argumentation, and the other is to apply argumentation to games. Lots of work has been done in the former dimension [14–16,18]. For the later dimension, Dung has showed that argumentation can be used to solve cooperative n-person games and the stable marriage problem in his seminal paper about abstract argumentation framework (AF) [5]. Dung's work about applying argumentation to games is advanced by [3,22], which respectively focus on cooperative n-person games and the stable marriage problem (SMP). Fan and Toni extend the later dimension by applying argumentation to normal-form games [7], which shows how assumption-based argumentation can be used to solve normal-form games via dialogue. Fan and Toni's pioneering work is of great originality, but their approach is not efficient and does not provide strict correspondence between argumentation frameworks and game theory. This is the problem we try to solve.

Argumentation deals with support and attack among arguments in the process of reasoning, and is widely used for providing explanation. Moulin et al.

D. Calvaresi et al. (Eds.): EXTRAAMAS 2021, LNAI 12688, pp. 314–328, 2021.
https://doi.org/10.1007/978-3-030-82017-6_19

propose research perspectives for the integration of explanation and argumentation capabilities [11]. Argumentation has been used to give explanation for decision-making of a value-driven agent [10], scientific debates [20] and case-based reasoning [4].

The field of applying argumentation to normal-form games largely remains unexplored and may bring interesting research opportunities. An opportunity is that normal-form games can benefit from argumentation in terms of explanation. By mapping a normal-form game into an abstract argumentation framework, strategic interactions between players can be explained by defenses and attacks among arguments. In this article, we study how to represent normal-form games using abstract argumentation frameworks and how to compute game-theoretical solution concepts by argumentation semantics. In particular, game-theoretical solution concepts are explained by how the corresponding arguments are defended.

This article is written as follows: In Sect. 2, the theoretical background of the article is presented. In Sect. 3, we construct a game-based argumentation framework (GBA framework for short) to solve normal-form games with principled techniques from the field of argumentation. Therefore, in Sect. 4, we prove the correspondence between concepts in a GBA framework and game-theoretical solution concepts. For example, stable extensions of a GBA framework correspond to strict Nash equilibria of a game, and the grounded extension of a GBA framework corresponds to the equilibrium in strictly dominant strategies of a game. We compare the process of solving normal-form games using game theory with that of our framework, and discuss why our framework provides better explanation for solution concepts.

2 Backgrounds

In this section, we briefly review some important solution concepts in game theory and some concepts in argumentation, which set up the theoretical basis of this paper.

2.1 Normal-Form Games

The normal form is the most familiar representation of strategic interactions in game theory. In a normal-form game, each player can select a single action and play it. We call such a strategy a *pure strategy*. In this paper, scope of our research is limited to pure strategy games, in which a strategy represents taking an action with probability of 1. So according to [21], a pure strategy game can be defined as follows:

Definition 1. ***Normal-form game.*** *A (finite, n-person) normal-form game is a tuple (N, S, u), where:*

- N *is a finite set of* n *players, indexed by* i*;*
- $S = S_1 \times \cdots \times S_n$*, where* S_i *is a finite set of strategies available to player* i*. Each vector* $s = (s_1, \ldots, s_n)$ *is called a strategy profile;*
- $u = (u_1, ..., u_n)$*, where* $u_i : S \to \mathbb{R}$ *is a* real-valued *utility (or payoff) function for player* i*.*

We use s_i to denote strategies available to player i, and use S_i to denote the set of pure strategies for player i. We call a choice of strategy for each player a *pure-strategy profile*, denoted by s. Formally, define $s_{-i} = (s_1, s_2, s_3, \ldots s_{i-1}, s_{i+1}, \ldots s_n)$, a strategy profile s without player i's strategy. Thus we write $s = (s_i, s_{-i})$. S_{-i} is the set of all strategy profiles of players without player i.

There are many solution concepts in game theory [21], some of which are defined as follows.

Definition 2. ***Strictly dominant strategy.*** *Let* s_i *and* s'_i *be two strategies of player* i*, then we say that* s_i *strictly dominates* s'_i *if* $\forall s_{-i} \in S_{-i}$*,* $u_i(s_i, s_{-i}) > u_i(s'_i, s_{-i})$*. A strategy is strictly dominant for an agent if it strictly dominates every other strategy for that agent.*

Definition 3. ***Equilibrium in strictly dominant strategies.*** *A strategy profile* $s = (s_1, ..., s_n)$ *in which every* s_i *is strictly dominant for player* i *is an equilibrium in strictly dominant strategies.*

Definition 4. ***Best response.*** *Player* i*'s best response to the strategy profile* s_{-i} *is a strategy* $s_i^* \in S_i$ *such that* $u_i(s_i^*, s_{-i}) \geq u_i(s_i, s_{-i})$ *for all strategies* $s_i \in S_i$*.*

Definition 5. ***Nash equilibrium.*** *A strategy profile* $s = (s_1, ...s_n)$ *is a Nash equilibrium if, for all agents* i*,* s_i *is a best response to* s_{-i}*.*

An equilibrium in strictly dominant strategies is necessarily the unique Nash equilibrium. Nash equilibria can be divided into two categories, **strict and weak**, depending on whether or not every agent's strategy constitutes a unique best response to the other agents' strategies.

Definition 6. ***Strict Nash.*** *A strategy profile* $s = (s_1, ...s_n)$ *is a strict Nash equilibrium if, for all agents* i *and for all strategies* $s'_i \neq s_i$*,* $u_i(s_i, s_{-i}) > u_i(s'_i, s_{-i})$*.*

Definition 7. ***Weak Nash.*** *A strategy profile* $s = (s_1, ...s_n)$ *is a weak Nash equilibrium if, for all agents* i *and for all strategies* $s'_i \neq s_i$*,* $u_i(s_i, s_{-i}) \geq u_i(s'_i, s_{-i})$*.*

Intuitively, weak Nash equilibria are less stable than strict Nash equilibria, because in the former case at least one player can deviate from Nash equilibria. There can be only one strict Nash equilibrium in a game (but there is not necessarily one).

Example 1. **The Prisoners' Dilemma** [9]. There are two prisoners, each of which has two strategies available: confess (F) or not confess (M). The game can be represented as $G = (N, S, u)$ with $N = \{1, 2\}$ and $S_1 = \{F_1, M_1\}$,

		Prisoner 2	
		M_2	F_2
Prisoner 1	M_1	−1, −1	−9, 0
	F_1	0, −9	−6, −6

$S_2 = \{F_2, M_2\}$. The payoff to the two prisoners when a particular pair of strategies is chosen is given in the table:
In this example, for prisoner 1, F_1 is the best response to M_2 and F_2, for prisoner 2, F_2 is the best response to M_1 and F_1. Therefore, strategy profile (F_1, F_2) is a unique Nash equilibrium.

2.2 Abstract Argumentation Framework

The notions of an argumentation framework (AF) were first introduced in [5]. An AF can be viewed as a directed graph in which arguments are represented by the nodes and the attack relation is represented as edges.

Definition 8. *An **argumentation framework** is a pair $AF = (\mathcal{A}, \mathcal{R})$, where $\mathcal{A}$ is a set of arguments, and $\mathcal{R}$ is a binary relation over $\mathcal{A}$, i.e., $\mathcal{R} \subseteq \mathcal{A} \times \mathcal{A}$.*

We use $(\alpha, \beta) \in \mathcal{R}$ to denote that α attacks β. Given an AF, statuses of arguments is evaluated, producing sets of arguments that are acceptable together, which are based on the following three notions [10]:

Definition 9. *Given $AF = (\mathcal{A}, \mathcal{R})$ and $\mathcal{B} \subseteq \mathcal{A}$*

- *A set $\mathcal{B}$ of arguments is conflict-free iff $\nexists \alpha, \beta \in \mathcal{B}$ such that $(\alpha, \beta) \in R$.*
- *An argument $\alpha \in \mathcal{A}$ is acceptable w.r.t. a set $\mathcal{B}$ (α is defended by $\mathcal{B}$), iff $\forall (\beta, \alpha) \in \mathcal{R}$ $(\beta \notin B, \beta \neq \alpha)$, $\exists \gamma \in \mathcal{B}$ such that $(\gamma, \beta) \in \mathcal{R}$.*
- *A conflict-free set of arguments $\mathcal{B}$ is admissible iff each argument in $\mathcal{B}$ is acceptable w.r.t. $\mathcal{B}$.*

An extension-based argumentation semantics can be viewed as a pre-defined criterion, according to which the acceptability of arguments in an AF can be determined.

Definition 10. *Let $AF = (\mathcal{A}, \mathcal{R})$ be an argument framework, and $\mathcal{E} \subseteq \mathcal{A}$ be an admissible set of arguments.*

- *$\mathcal{E}$ is a complete extension of AF iff each argument in A that is acceptable w.r.t. E is in $\mathcal{E}$.*
- *$\mathcal{E}$ is the grounded extension of AF iff $\mathcal{E}$ is a minimal (w.r.t. set inclusion) complete extension.*
- *$\mathcal{E}$ is a stable extension of AF iff $\mathcal{E}$ is conflict-free and $\forall \beta \in \mathcal{A} \backslash \mathcal{E}$, $\exists \alpha \in \mathcal{E}$ such that $(\alpha, \beta) \in \mathcal{R}$.*

We use $\mathcal{E}_{\mathcal{CO}}(AF), \mathcal{E}_{\mathcal{GR}}(AF), \mathcal{E}_{\mathcal{ST}}(AF)$ to denote the set of complete extensions, grounded extension, and stable extensions of AF respectively.

3 Game-Based Argumentation Framework

3.1 Applying Argumentation to Games

Dung introduced basic ideas of applying argumentation frameworks to n-person cooperative games and the SMP in [5]. When dealing with n-person cooperative games, Dung interprets $< IMP, \rightarrow >$ as an abstract AF, where argument is an argument for a imputation (given payoff distribution among the agents) and each attack denotes domination between imputations. Building on Dung's results, Young et al. further the correspondence between Dung's argumentation semantics and solution concepts in cooperative game theory [22]. When dealing with the SMP, an argument is denoted with (m, w) that represents a man m marries woman w. (m', w') attacks (m, w) iff: ① $m' = m$ and m prefers w' to w; ② $w' = w$ and w prefers m' to m. Bistarelli and Santini take Dung's approach and advanced the research of applying argumentation to the SMP [3].

Fan and Toni extend Dung's work by mapping games in normal form into assumption-based argumentation frameworks. To find the Nash equilibrium, they translate each strategy profile into an assumption $d(\sigma_\alpha, \sigma_\beta)$ that represent this strategy profile is a Nash equilibrium, and a conclusion $nD(\sigma_\alpha, \sigma_\beta)$ that represent this strategy profile is not a Nash equilibrium. An argument $\{d(\sigma_\alpha, \sigma_\beta)\} \vdash d(\sigma_\alpha, \sigma_\beta)$ is attacked by an argument with a conclusion $nD(\sigma_\alpha, \sigma_\beta)$. Dung considers an imputation as an argument. Fan and Toni consider a strategy profile as an argument. Both imputation and strategy profiles involve all players. From this perspective, the approach Fan and Toni take is similar to Dung's. Dung's approach proves to be rather successful in cooperative games, since a strict correspondence can be established between solutions in cooperative games and semantics of abstract AF [22]. But in Fan and Toni's work about normal-form games, the correspondence between solution concepts of normal-form games and semantics of ABA is weak.

Dung uses different definition of argument and attack in different games. Normal-form games are different from cooperative games. In cooperative games, joint actions of groups of players are the primitives; in Non-cooperative games, actions of individual players are the primitives [13].

3.2 Game-Based Argumentation Framework

In this paper, we propose a Game-based Argumentation (GBA) framework to solve normal-form games through argumentation. We directly transform a strategy available to a player into an argument for the player. The best response relationship is transformed into attack relationship.

Definition 11 (Game-based argument). *Given a normal-form game $G = (N, S, u)$, a game-based argument a_i is interpreted as*

"player i should choose strategy s_i". $\mathcal{A}_i$ is a set of game-based arguments available to player i. $\mathcal{A}_G$ denotes the set of game-based arguments for all players, that is to say $\mathcal{A}_G = \cup_{i=1}^{n} \mathcal{A}_i$.

Example 2. In the Prisoners' Dilemma, the game-based arguments available to player 1 is $\mathcal{A}_1 = \{F_1, M_1\}$. The game-based arguments available to player 2 is $\mathcal{A}_2 = \{F_2, M_2\}$. So $\mathcal{A}_G = \mathcal{A}_1 \cup \mathcal{A}_2$ (Fig. 1).

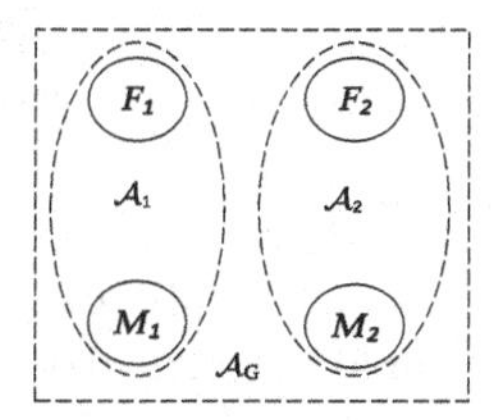

Fig. 1. Game-based arguments and sets of game-based arguments in the Prisoners' Dilemma.

If player i's strategy s_i is not the best response facing player j's strategy s_j, it is obvious that when player j choose strategy s_j, player i will not choose strategy s_i as a reaction. Similarly, in an argumentation framework, given that a_i is attacked by a_j, a_i will be unacceptable if a_j has been decided as acceptable. According to such a similarity, we have a definition as follows:

Definition 12. *Given a normal-form game $G = (N, S, u)$ and the set of game-based arguments $\mathcal{A}_G$, $\mathcal{R}_G \subseteq \mathcal{A}_i \times \mathcal{A}_j$ (i and j are two different players) is a set of game-based attacks. $(a_j, a_i) \in \mathcal{R}_G$ iff s_i is not the best response to all s_{-i} that contain s_j.*

$\mathcal{R}_G$ represents attack relation between game-based arguments put forward by different players. Given a_i a_i' and a_j (the corresponding strategies are s_i s_i' and s_j), a_i is attacked by a_j if s_i' brings higher payoff than s_i w.r.t. s_j.

Example 3. In the Prisoners' Dilemma, if prisoner 2 chooses to play the strategy M_2, then prisoner 1 will prefer to choose the strategy F_1 rather than M_1, since the payoff from playing strategy F_1 is higher than M_1. Using the above definition, we can say that M_1 is attacked by M_2, as there exists F_1 for player 1 such that $u_1(F_1, M_2) > u_1(M_1, M_2)$. Similarly, M_1 is attacked by F_2. The strategy attacks are depicted in Fig. 2.

Figure 2 can be regarded as an AF and F_1, F_2 are acceptable. The corresponding strategies are chosen in the Prisoners' Dilemma.

However, the above approach to translating a game to a game-based argumentation framework is problematic. In the Battle of the Sexes [9], Pat and Chris must choose to attend either the opera or a prize fight. The game is represented in the accompanying bimatrix.

The Battle of the Sexes can be represented in $\mathcal{A}_G$ and $\mathcal{R}_G$ through the Fig. 3:

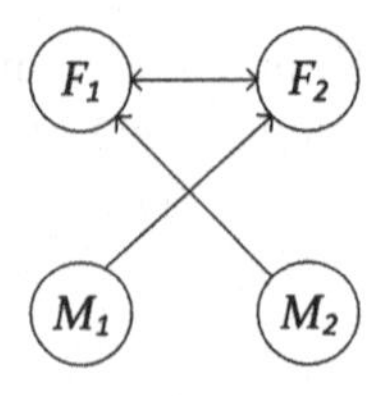

Fig. 2. Game-based arguments and strategy attacks in the Prisoners' Dilemma.

		Pat2	
		O_2	F_2
Chris1	O_1	2,1	0,0
	F_1	0,0	1,2

Figure 3 can be regarded as an AF, in which $\{F_1, O_1\}$ is acceptable if $\{F_2, O_2\}$ is unacceptable. But the corresponding strategies do not hold in the game, for Chris cannot take 2 strategies and Pat takes no strategy.So there should be some restrictions on the argument sets that represent solution concepts to a game.

Since a strategy profile in a normal-form game is in the form of $(s_1, s_2, s_3, \ldots, s_n)$, each player can choose only one strategy. A strategy profile arguments set that represents a strategy profile should be in the form of $\{a_1, a_2, a_3, \ldots, s_n\}$ and is defined as follows:

Definition 13. *Given a normal-form game $G = (N, S, u)$ and the corresponding game-based arguments set $\mathcal{A}_G$, a strategy profile arguments set (denoted as $\mathcal{A}_{sp}$) is a set of arguments, which represents a strategy profile such that $|\mathcal{A}_{sp}| = n$, for any $a, b \in \mathcal{A}_{sp}(a \neq b)$ it is not the case that there exists a player i such that $a \in \mathcal{A}_i$ and $b \in \mathcal{A}_i$. That is, every argument in $\mathcal{A}_{sp}$ belongs to a strategy of a distinct player.*

We use $\mathcal{A}_{SP}$ to denote the set containing all the possible $\mathcal{A}_{sp}$. Our game-based argumentation (GBA) framework is defined as follows:

Definition 14. *Given a normal-form game $G = (N, S, u)$, the corresponding GBA framework is defined as $AF_G = (\mathcal{A}_G, \mathcal{A}_{SP}, \mathcal{R}_G)$, where $\mathcal{A}_G$ is the set of game-based arguments, $\mathcal{A}_{SP}$ is the set of strategy profile arguments sets, and $\mathcal{R}_G$ is the set of game-based attacks.*

In a GBA framework, argument sets that is acceptable should satisfy semantics of AF and should contain arguments of each player. So the semantics of a GBA framework is defined as follows:

Definition 15. *Given a normal-form game $G = (N, S, u)$, the corresponding game-based framework $AF_G = (\mathcal{A}_G, \mathcal{A}_{SP}, \mathcal{R}_G)$ and a semantic σ, $\mathcal{E}_\sigma(AF_G) = \mathcal{E}_\sigma((\mathcal{A}_G, \mathcal{R}_G)) \cap \{\mathcal{A} \mid \mathcal{A} \in 2^{\mathcal{A}_G}$ and $\forall i \in N$, $\exists a_i \in \mathcal{A}\}$.*

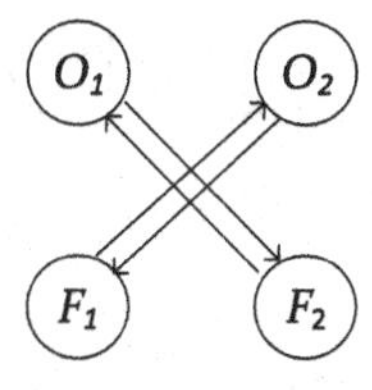

Fig. 3. Game-based arguments and strategy attacks in the Battle of the Sexes.

According to the Definition14 and Definition 15, argument sets correspond to solution concepts in normal-form games should be both in $\mathcal{A}_{SP}$ and an extension of AF_G.

4 Properties of a Game-Based Argumentation Framework

4.1 Correspondences Between Game-Theoretical Solution Concepts and Argument Extensions

The GBA framework is proposed to solve pure-strategy normal-form games. When a two-person normal-form game is transformed into a GBA framework, payoff functions are separately transformed into attacks. So there are interesting correspondences between game-theoretical solution concepts computed from payoff functions, such as dominant strategy and Nash equilibrium, and extensions of a GBA framework. More precisely, an stable extensions of a GBA framework that is $\mathcal{A}_{sp}$ corresponds to a strict Nash equilibria of the underlying game, and the grounded extension of a GBA framework that is in $\mathcal{A}_{SP}$ corresponds to the equilibrium in strictly dominant strategies of the underlying game.

In a n-person game, according to Definition 12, we have the following proposition:

Proposition 1. *Given a normal-form game $G = (N, S, u)$ and the corresponding AF_G, if there does not exist a set $\mathcal{A}_{sp}$ in the AF_G which is conflict-free, then there does not exist $s = (s_1, \ldots, s_n)$ which is an equilibrium.*

Proof. Because there does not exist a set $\mathcal{A}_{sp} = \{a_1, \ldots, a_n\}$ in the AF_G which is conflict-free, we have that given any a_j there exists $i \in N$, such that $(a_i, a_j) \in \mathcal{R}_G$. According to the Definition 12, s_j is not the best response to s_{-j} which includes s_i. That is to say, for any s_j, there exists s_{-j} to which s_j is not the best response, so there does not exist $s = (s_1, \ldots, s_n)$ which is an equilibrium. ■

In a two-person game, we have the following proposition.

Proposition 2. *Given a normal-form game $G = (N, S, u)$ where $N = \{1, 2\}$ and the corresponding AF_G, a_1 is attacked by a_2 iff s_1 is not the best response to s_2.*

Proof. In a two-person game where $N = \{1, 2\}$, s_2 is s_{-1}. Since there are no other players, according to **Definition** 12, it is obvious that a_1 is attacked by a_2 iff s_1 is not the best response to s_2. ■

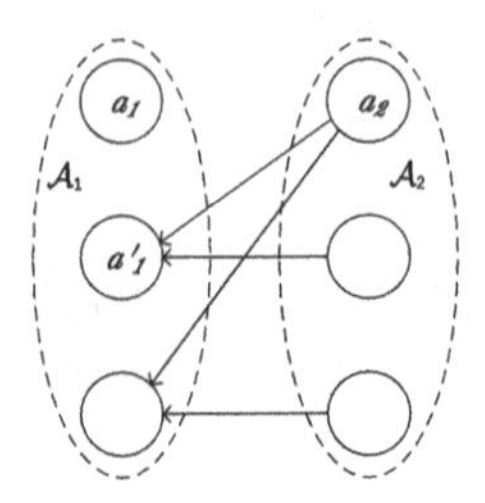

Fig. 4. Game-based argument corresponding to dominant strategy is unattacked.

A dominant strategy is the best response to every strategy from opponents. As shown in Fig. 4, an argument represent a dominant strategy is unattacked. We have the following proposition:

Proposition 3. *Given a normal-form game $G = (N, S, u)$ where $N = \{1, 2\}$ and the corresponding AF_G, a strategy s_1 is dominant iff a_1 is unattacked, i.e., $\nexists a_2 \in \mathcal{A}_2$, such that $(a_2, a_1) \in \mathcal{R}_G$.*

Proof. Forwards: If a strategy s_1 is dominant, according to the definition, $\forall s'_1 \in S_1$ where $s'_1 \neq s_1$, s_1 dominates s'_1. Thus $\forall s_2 \in S_2$, $u_i(s_1, s_2) \geq u_i(s'_1, s_2)$, so $\nexists s_2 \in S_2$ such that $u_1(s_1, s_2) > u_1(s'_1, s_2)$, hence $\nexists a_2 \in \mathcal{A}_2$ such that $(a_2, a_1) \in \mathcal{R}_G$.

Backwards: Given an AF_G transferred from G, if a game-based argument a_1 is unattacked, we have: $\forall s_2 \in S_2$ and $\forall s'_1 \in S_1$, $u_1(s_1, s_2) \geq u_1(s_1, s_2)$. Thus s'_1 is dominant. ■

As shown in Fig. 5, arguments set correspond to an equilibrium in strictly dominant strategies is not attacked and attack arguments represent dominated strategies. For an equilibrium in strictly dominant strategies, we have the theorem:

Theorem 1. *Given a normal-form game $G = (N, S, u)$ where $N = \{1, 2\}$ and the corresponding AF_G, a strategy profile $s = (s_1, s_2)$ is an equilibrium in strictly dominant strategies, iff $\{a_1, a_2\} = \mathcal{A}_{SP} \cap \mathcal{E}_{\mathcal{GR}}(AF_G)$.*

Proof. Forwards: Let $G = (N, S, u)$ be a two-person game where $s = (s_1, s_2)$ is an equilibrium in strictly dominant strategies. According to Definition 2, both s_1 and s_2 are strictly dominant, and according to Definition 2, $\forall s'_1 \in S_1$ where $s'_1 \neq s_1$, s'_1 is strictly dominated.

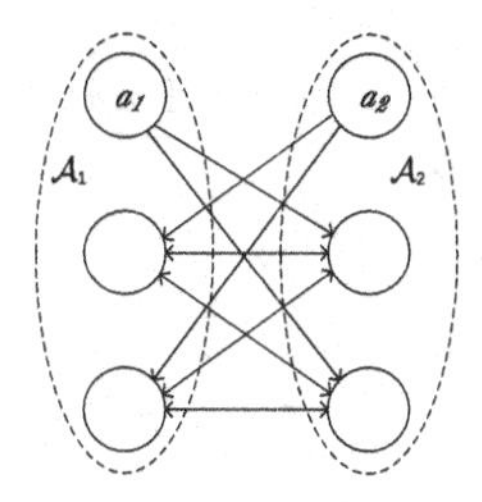

Fig. 5. Arguments set correspond to an equilibrium in strictly dominant strategies is not attacked and attack arguments outside the set.

We now transfer G to AF_G. According to Proposition 3, a_1, a_2 is unattacked. According to Definition 16 in [5], F_{AF} is a characteristic function that maps a set S to a set A that is acceptable w.r.t. S. In AF_G, there is $F_{AF}(\emptyset)=\{a_1, a_2\}$. According to Definition 2, $\forall s_1' \in S_1$ where $s_1' \neq s_1$, $u_1(s_1, s_2) > u_1(s_1', s_2)$. According to Definition 12, we have that a_2 attacks every a_1' in $\mathcal{A}_1$ except a_1. Similarly, a_1 attacks every a_2' in $\mathcal{A}_2$ except a_2. Then we have that $F_{AF}(\{a_1, a_2\}) = \{a_1, a_2\}$, hence $\{a_1, a_2\}$ is a *complete extension*. Also $\{a_1, a_2\}$ is the least fixed point of F_{AF}, so $\{a_1, a_2\}$ is the grounded extension.

Backwards: Given the AF_G transferred from G. If $\mathcal{E}_{\mathcal{GR}}(AF_G) = \{\{a_1, a_2\}\}$, then $\{a_1, a_2\}$ is the minimal complete extension and is conflict-free. According to the Definition 12, there is no attack from a_2 to $\mathcal{A}_2$, and there is no attack from a_1 to $\mathcal{A}_1$. So, either a_1, a_2 are unattacked, or a_1 defends itself from $\mathcal{A}_2$ and a_2 defends itself from $\mathcal{A}_1$. In the later case, $F_{AF}(\emptyset) = \emptyset$ and contradicts the premise that $\{a_1, a_2\}$ is the minimal complete extension. So both a_1, a_2 are unattacked and there is no other unattacked arguments. According to Proposition 3, a_1 is the only dominant strategy and hence is a strictly dominant strategy. Similarly a_2 is a strictly dominant strategy. Then we have that strategy profile $s = (s_1, s_2)$ is an equilibrium in strictly dominant strategies. ■

Proposition 4. *Given a normal-form game $G = (N, S, u)$ where $N = \{1, 2\}$ and the corresponding AF_G, a strategy profile $s = (s_1, s_2)$ is a Nash equilibrium iff in the AF_G transferred from G, the corresponding $\mathcal{A}_{sp} = \{a_1, a_2\}$ is conflict-free.*

Proof. Forwards: Given a normal-form game $G = (N, S, u)$ where $N = \{1, 2\}$, if a strategy profile $s = (s_1, s_2)$ is a Nash equilibrium, then s_1 is the best response to s_2 and vice versa. According to Proposition 2, we have a_2 does not attack a_1. Similarly, a_1 does not attack a_2, so the set $\mathcal{A}_{sp} = \{a_1, a_2\}$ is conflict-free.

Backwards: Given an AF_G transferred from G, if the set $\mathcal{A}_{sp} = \{a_1, a_2\}$ is conflict-free, then a_1 does not attack a_2 and vice versa. According to Proposition 2, s_1 is the best response to s_2 and vice versa. So $s = (s_1, s_2)$ is a Nash equilibrium. ■

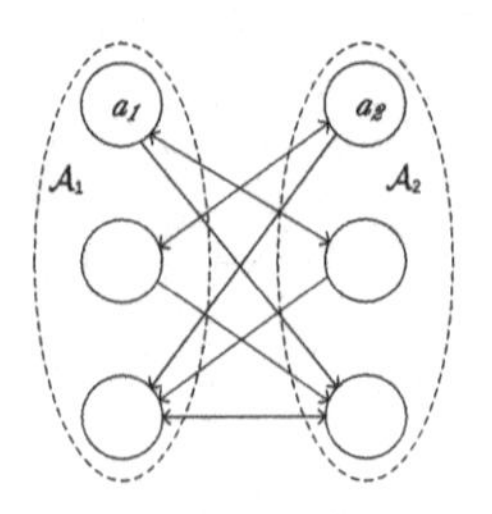

Fig. 6. Arguments set correspond to a strict Nash equilibrium attack every argument outside the set.

A Nash equilibrium is a stable strategy profile: no agent wants to change its strategy if it knows what strategies other players has chosen. A weak Nash equilibria are less stable than a strict Nash equilibria [17]. As shown in Fig. 6, arguments set correspond to a strict Nash equilibrium attack every argument outside it. For a strict Nash equilibrium, we have the following theorem:

Theorem 2. *Given a normal-form game $G = (N, S, u)$ where $N = \{1, 2\}$ and the corresponding AF_G, a strategy profile $s = (s_1, s_2)$ is a strict Nash equilibrium iff $\{a_1, a_2\} \in \mathcal{A}_{SP} \cap \mathcal{E}_{ST}(AF_G)$.*

Proof. Forwards:

Given a normal-form game $G = (N, S, u)$ where $N = \{1, 2\}$, if a strategy profile $s = (s_1, s_2)$ is a strict Nash equilibrium, we have: s_1 is the only best response to s_2 and vice versa, according to the Proposition 2, a_1 does not attack a_2 and vice versa. So $\{a_1, a_2\}$ is *conflict-free.*

Since s_1 is the only best response to s_2, $\forall s'_1 \in S_1$ where $s'_1 \neq s_1$, s'_1 is not the best response to s_2. According to the Proposition 2, $\forall a'_1 \in \mathcal{A}_1$ where $a'_1 \neq a_1$, a_2 attacks a'_1. Similarly, $\forall a'_2 \in \mathcal{A}_2$ where $a'_2 \neq a_2$, a_1 attacks a'_2. So in AF_G, $\{a_1, a_2\}$ attacks every argument that is not in $\{a_1, a_2\}$, hence $\mathcal{A}_{sp} = \{a_1, a_2\}$ is a stable extension.

Backwards:

Given an AF_G transferred from G, if $\mathcal{A}_{sp} = \{a_1, a_2\}$ is a stable extension, we have: a_1 attack every $a'_2 \in \mathcal{A}_2$ where $a'_2 \neq a_2$. Similarly, a_2 attack every $a'_1 \in \mathcal{A}_1$ where $a'_1 \neq a_1$.

According to Definition 11, s_1 is the only best response to s_2 and vice versa. So the strategy profile $s = (s_1, s_2)$ is a strict Nash equilibrium. ■

4.2 Towards Explanation for Normal-Form Games

In an AF, acceptability of an argument is easy to understand through analyzing whether it is defended. A set of arguments that contributes to defending an argument a is regarded as explanation for the argument a [6,23]. Argumentation frameworks are very interpretable and are widely used to provide explanations for other models [1,8,10].

The process of getting a solution provides an explanation for the solution. Traditionally, game theory solves a game by first enumerating all possible strategy profiles, and then eliminating dominated strategies and matching the best responses based on the corresponding payoffs. To judge whether a strategy profile is a solution in a normal-form game, players need to compute and compare the payoffs of relevant strategy profiles. This process is not intuitive enough and can be hard to understand. Our framework provides a graphic model to depict strategic interaction between players. To judge whether a strategy profile is a solution, players only need to judge whether corresponding arguments are conflict-free and defended. This process is transparent and easy to understand.

Our framework provides a qualitative analysis of solution concepts in Normal-form games, and draws some enlightening conclusions. For instance, we have proved the correspondence between strict Nash and stable semantics. According to [17], only strict Nash equilibria are asymptotically stable in the replicator dynamics. Our framework has the potential to provide explanation for stability of Nash equilibria.

5 Related Works

This paper focus on the interplay between games and argumentation. There is lots of work about applying extensive games to argumentation dialogues, both of which can be represented as trees. Games in extensive form have been proved to be useful to determine optimal strategies in dialogues [18]. Dialectical argumentation frameworks can be enhanced by using payoff functions that assign values to every possible valid dialogue, hence one can lucidly realize a dialectical argumentation framework as a game in extensive form [19].

There is several work about applying argumentation to games. Dung proposes the basic idea of applying abstract argumentation to n-person cooperative games and the SMP [5]. Young et al. further the correspondence between Dung's four argumentation semantics and solution concepts in cooperative game theory [22] Bistarelli and Santini advanced Dung's work on the SMP [3]. Fan and Toni propose that Normal-form games can be translated into an ABA framework, and correspondence is built between computing solution concepts and constructing successful dialogues [7]. The problem Fan and Toni deal with is different from Dung's, but the approach they take is similar to Dungs.

When dealing with normal-form games, we take an approach that is different from Fan and Toni's. In our approach, an argument is transferred from an action of an individual player instead of a strategy profile involving all players. Attack relationship among arguments is based on the best response. Compared with Fan and Toni's approach, our approach has the following advantages: First of all, the number of arguments is our approach is less. Since in a normal-form game strategy, the number of strategies is less than the number of strategy profiles. Secondly, we do not need to figure out whether a strategy profile is a solution before generating attacks. In Fan and Toni's approach, an argument $\{d(\sigma_\alpha, \sigma_\beta)\} \vdash d(\sigma_\alpha, \sigma_\beta)$ is to be attacked by $\{\} \vdash nD(\sigma_\alpha, \sigma_\beta)$ only after strategy profile $(\sigma_\alpha, \sigma_\beta)$ has been proved not a solution by game theory. Thirdly, in Fan and Toni's approach, an argument $\{d(\sigma_\alpha, \sigma_\beta)\} \vdash d(\sigma_\alpha, \sigma_\beta)$ can only be attacked by $\{\} \vdash nD(\sigma_\alpha, \sigma_\beta)$, which shows no interaction between players. Our approach shows strategic interactions between players through attacks and defenses among arguments, which can serve as an explanation for game solutions. Finally, we establish a strict correspondence between solution concepts in normal-form games and semantics of argumentation frameworks. In our work, Nash equilibrium and dominant strategy correspond to different semantics, while in Fan and Toni's they are similar.

6 Results and Future Work

In this paper, We propose a new approach of applying abstract argumentation to normal-form games. We put forward a GBA framework that can be used to compute game-theoretical solution concepts. We have proved the correspondence between concepts in a GBA framework and game-theoretical solution concepts. These results show that games can be solved through argumentation, and game-theoretical solution concepts can be explained through attacks and defences among arguments. Based on the GBA frameworks, we will try to provide a formal explanation for the stability of Nash equilibrium in our future work. We may also expand the field of applying argumentation to games. Some related questions include applying argumentation to extensive-form games and repeated games. This paper has considered pure strategies, and applying argumentation to mixed strategies games is an interesting and unstudied field.

Acknowledgment. The research reported in this paper was partially supported by the "2030 Megaproject"- New Generation Artificial Intelligence of China under Grant No. 2018AAA0100904, and the Major Project of National Social Science Foundation of China under Grant No. 20&ZD047.

References

1. Albini, E., Rago, A., Baroni, P., Toni, F.: Relation-based counterfactual explanations for Bayesian network classifiers. In: IJCAI International Joint Conference on Artificial Intelligence, January 2021, pp. 451–457. International Joint Conferences on Artificial Intelligence, July 2020. https://doi.org/10.24963/ijcai.2020/63. https://ec.europa.eu/digital-single-market/en/news/ethics-
2. Bench-Capon, T.J., Dunne, P.E.: Argumentation in artificial intelligence. Artif. Intell. **171**(10–15), 619–641 (2007). https://doi.org/10.1016/j.artint.2007.05.001
3. Bistarelli, S., Santini, F.: Abstract argumentation and (optimal) stable marriage problems. Argument Comput. **11**(1–2), 15–40 (2020). https://doi.org/10.3233/AAC-190474
4. Čyras, K., Satoh, K., Toni, F.: Abstract argumentation for case-based reasoning. In: Proceedings of the Fifteenth International Conference on Principles of Knowledge Representation and Reasoning, pp. 549–552 (2016). https://doi.org/10.5555/3032027.3032100

5. Dung, P.M.: On the acceptability of arguments and its fundamental role in non-monotonic reasoning, logic programming and n-person games. Artif. Intell. **77**(2), 321–357 (1995). https://doi.org/10.1016/0004-3702(94)00041-X
6. Fan, X., Toni, F.: On computing explanations in abstract argumentation. Front. Artif. Intell. Appl. **263**(Dung 1995), 1005–1006 (2014). https://doi.org/10.3233/978-1-61499-419-0-1005
7. Fan, X., Toni, F.: On the interplay between games, argumentation and dialogues. In: Proceedings of the 2016 International Conference on Autonomous Agents & Multiagent Systems, pp. 260–268 (2016). https://doi.org/10.5555/2936924.2936964
8. García, A.J., Chesñevar, C.I., Rotstein, N.D., Simari, G.R.: Formalizing dialectical explanation support for argument-based reasoning in knowledge-based systems. Expert Syst. Appl. **40**(8), 3233–3247 (2013). https://doi.org/10.1016/j.eswa.2012.12.036
9. Gibbons, R.S.: Game Theory for Applied Economists. Princeton University Press (1992). https://doi.org/10.1515/9781400835881
10. Liao, B., Anderson, M., Anderson, S.L.: Representation, justification, and explanation in a value-driven agent: an argumentation-based approach. AI Ethics **1**(1), 5–19 (2020). https://doi.org/10.1007/s43681-020-00001-8
11. Moulin, B., Irandoust, H., Bélanger, M., Desbordes, G.: Explanation and argumentation capabilities: towards the creation of more persuasive agents. Artif. Intell. Rev. **17**(3), 169–222 (2002). https://doi.org/10.1023/A:1015023512975
12. MYERSON, R.B.: Game Theory. Harvard University Press, March 2013. https://doi.org/10.2307/j.ctvjsf522. http://www.jstor.org/stable/10.2307/j.ctvjsf522
13. Narahari, Y.: Game Theory and Mechanism Design, IISc Lecture Notes Series, vol. 4. World Scientific, May 2014. https://doi.org/10.1142/8902. https://www.worldscientific.com/worldscibooks/10.1142/8902
14. Prakken, H.: Coherence and flexibility in dialogue games for argumentation. J. Logic Comput. **15**(6), 1009–1040 (2005). https://doi.org/10.1093/logcom/exi046
15. Rahwan, I., Larson, K.: Pareto optimality in abstract argumentation. In: Proceedings of the National Conference on Artificial Intelligence, vol. 1, pp. 150–155 (2008). https://www.aaai.org/Library/AAAI/2008/aaai08-024.php
16. Rahwan, I., Larson, K.: Argumentation and game theory. In: Simari, G., Rahwan, I. (eds.) Argumentation in Artificial Intelligence, pp. 321–339. Springer, Boston (2009). https://doi.org/10.1007/978-0-387-98197-0_16
17. Ritzberger, K., Weibull, J.W.: Evolutionary selection in normal-form games. Econometrica **63**(6), 1371 (1995). https://doi.org/10.2307/2171774
18. Riveret, R., Prakken, H., Rotolo, A., Sartor, G.: Heuristics in argumentation: a game-theoretical investigation. In: Frontiers in Artificial Intelligence and Applications, vol. 172, pp. 324–335. IOS Press, December 2008. https://doi.org/10.2139/ssrn.1317349. https://papers.ssrn.com/abstract=1317349
19. Rosenschein, J.S., Procaccia, A.D.: Extensive-form argumentation games. Technical report (2005). https://www.researchgate.net/publication/221613087
20. Šešelja, D., Straßer, C.: Abstract argumentation and explanation applied to scientific debates. Synthese **190**(12), 2195–2217 (2013). https://doi.org/10.1007/s11229-011-9964-y. https://link.springer.com/article/10.1007/s11229-011-9964-y
21. Shoham, Y., Leyton-Brown, K.: Multiagent Systems: Algorithmic, Game-Theoretic, and Logical Foundations, vol. 9780521899. Cambridge University Press, January 2008. https://doi.org/10.1017/CBO9780511811654. https://www.cambridge.org/core/books/multiagent-systems/B11B69E0CB9032D6EC0A254F59922360

22. Young, A.P., Marzagao, D.K., Murphy, J.: Applying abstract argumentation theory to cooperative game theory. arXiv preprint arXiv:1905.10922 (2019)
23. Zeng, Z., Miao, C., Leung, C., Shen, Z., Chin, J.J.: Computing argumentative explanations in bipolar argumentation frameworks. In: 33rd AAAI Conference on Artificial Intelligence, AAAI 2019, 31st Innovative Applications of Artificial Intelligence Conference, IAAI 2019 and the 9th AAAI Symposium on Educational Advances in Artificial Intelligence, EAAI 2019, pp. 10079–10080 (2019). https://doi.org/10.1609/aaai.v33i01.330110079

Decentralized and Heterogeneous XAI

EXPECTATION: Personalized Explainable Artificial Intelligence for Decentralized Agents with Heterogeneous Knowledge

Davide Calvaresi[4(✉)], Giovanni Ciatto[1], Amro Najjar[2], Reyhan Aydoğan[3], Leon Van der Torre[2], Andrea Omicini[1], and Michael Schumacher[4]

[1] Alma Mater Studiorum – Università di Bologna, Cesena, Italy
{giovanni.ciatto,andrea.omicini}@unibo.it
[2] University of Luxembourg, Esch-sur-Alzette, Luxembourg
{amro.najjar,leon.vandertorre}@uni.lu
[3] Özyeğin University, Istanbul, Turkey
reyhan.aydogan@ozyegin.edu.tr
[4] University of Applied Sciences and Arts Western Switzerland HES-SO, Sierre, Switzerland
{davide.calvaresi,michael.schumacher}@hevs.ch

Abstract. Explainable AI (XAI) has emerged in recent years as a set of techniques and methodologies to interpret and explain machine learning (ML) predictors. To date, many initiatives have been proposed. Nevertheless, current research efforts mainly focus on methods tailored to specific ML tasks and algorithms, such as image classification and sentiment analysis. However, explanation techniques are still embryotic, and they mainly target ML experts rather than heterogeneous end-users. Furthermore, existing solutions assume data to be centralised, homogeneous, and fully/continuously accessible—circumstances seldom found altogether in practice. Arguably, a system-wide perspective is currently missing.

The project named "Personalized Explainable Artificial Intelligence for Decentralized Agents with Heterogeneous Knowledge" (EXPECTATION) aims at overcoming such limitations. This manuscript presents the overall objectives and approach of the EXPECTATION project, focusing on the theoretical and practical advance of the state of the art of XAI towards the construction of *personalised* explanations in spite of *decentralisation* and *heterogeneity* of knowledge, agents, and explainees (both humans or virtual).

To tackle the challenges posed by personalisation, decentralisation, and heterogeneity, the project fruitfully combines abstractions, methods, and approaches from the multi-agent systems, knowledge extraction/injection, negotiation, argumentation, and symbolic reasoning communities.

Keywords: Multi-agent systems · eXplanable AI · CHIST-ERA IV · Personalisation · Decentralisation · EXPECTATION

D. Calvaresi et al. (Eds.): EXTRAAMAS 2021, LNAI 12688, pp. 331–343, 2021.
https://doi.org/10.1007/978-3-030-82017-6_20

1 Background and Motivations

In recent decades, data-driven decision-making processes have increasingly influenced strategic choices. This applies to both virtual and humans' decisional needs. The application domains of Machine learning (ML) algorithms are broadening [1,2]. Ranging from finance to healthcare, ML supports humans in making informed decisions based on the information buried within enormous amounts of data. However, most effective ML methods are inherently *opaque*, meaning that it is hard for humans (if possible at all) to grasp the reasoning *hidden* in their predictions (so-called black boxes). To mitigate the issues arising from such opaqueness, several techniques and methodologies aiming at inspecting ML models and predictors have been proposed under the eXplainable Artificial Intelligence (XAI) umbrella [3,4] (e.g., feature importance estimators, rule lists, and surrogate trees [5]). Such tools enable humans to understand, inspect, analyse – and therefore trust – the operation and outcomes of AI systems effectively.

The many XAI-related initiatives proposed so far constitute the building blocks for making tomorrow's intelligent systems explainable and trustable. However, to date, the ultimate goal of letting intelligent systems provide not only valuable recommendations but also *motivations* and *explanations* for their suggestions – possibly, interactively – is still unachieved. Indeed, current research efforts focus on specific methods and algorithms, often tailored to single ML tasks—e.g. classification and, in particular, image classification. For instance, virtually all approaches proposed so far target supervised learning, and in particular, classification tasks [3,4,6]—and many of them are tailored on neural networks [7]. In other words, there is still a long way to *generality* [8].

Moreover, while existing XAI solutions do an excellent job on inspecting ML algorithms, current interpretation/explanations provide valuable insights solely profitable by human *experts*, entirely neglecting the need for producing more broadly accessible or personalised explanations that everybody could understand. Recalling their social nature, explanations should rather be *interactive* and tailored on the explainee's cognitive capabilities and background knowledge to be effective [9,10].

To complicate this matter, existing XAI solutions assume data to be centralised, homogeneous, and fully/continuously available for operation [8]. Such circumstances rarely occur in real-world scenarios. For example, data is often scattered through many administrative domains. Thus, even when carrying similar information, datasets are commonly structured according to different schemas—when not lacking structure at all. Privacy and legal constraints complete the picture by making it unlikely for data to be fully available at any given moment. In other words, the availability of data is more frequently *partial* rather than total. Therefore, explainable intelligent systems should be able to deal with scattering, decentralisation, heterogeneity, and unavailability of data, rather than requiring data to be centralised and standardised before even starting to process it—which would impose heavy technical, administrative, and legal constraints on the production of both recommendations and explanations.

Summarising, further research is needed to push XAI towards the construction of *personalised* explanations, which can be built in spite of *decentralisation* and *heterogeneity* of information—possibly, out of the interaction among intelligent software systems and human or virtual explainees.

Clearly, tackling personalisation, decentralisation, and heterogeneity entails challenges from several perspectives. On the one hand, personalisation of explanations must cope with the need for providing human-intelligible (i.e., *symbolic*) explanations of incremental complexity, possibly *iteratively* adapting to the cognitive capabilities, and background knowledge of the users who are receiving the explanation. In turn, it requires enabling an *interactive* explanation process both within the intelligent systems themselves (i.e., agent to agent) and with the end-users. On the other hand, decentralisation of data opens to questioning how explanations can be produced or aggregated without letting data cross administrative borders. Therefore, the need for *collaboration* among multiple cross-domain software entities is imperative. Finally, the challenge of heterogeneity, of both data and ML techniques used to mine information out of it, dictates the detection of some *lingua franca* to present recommendations and explanations to the users in intelligible forms.

To address these challenges, the EXPECTATION project has been recently recommended for funding – along with other 11 projects – as part of the CHIST-ERA 2019 call[1] concerning "Explainable Machine Learning-based Artificial Intelligence". The project has started on April 1, 2021 and it will last up to the of March 2024. In the remainder of this paper, we discuss how the project plans to tackle the challenges posed by personalisation, decentralisation, and heterogeneity, by fruitfully combining abstractions, methods, and approaches from the multi-agent systems, knowledge extraction/injection, negotiation, argumentation, and symbolic reasoning research areas.

2 State of the Art

The generation of personalised explanation for decentralised and heterogeneous intelligent agents roots in several disciplines, including XAI, agreement technologies, personalisation, and AI ethics.

2.1 Explainable Agency

Neuro-symbolic integration [11,12] aims at bridging the gap between symbolic and sub-symbolic AI, reconciling the two key branches of AI (connectionist AI – relying on connectionist networks inspired from human neurons, and symbolic AI – relying on logic, symbols, and reasoning) [13]. Sub-symbolic techniques (e.g., pattern recognition and classification) can offer excellent performance. However, their outcomes can be biased and difficult to understand (if possible at all). Seeking trust, transparency, and the possibility to debug sub-symbolic predictors (so-called black boxes), the XAI community relies on reverse engineering

[1] https://www.chistera.eu/projects-call-2019.

models trained on unknown datasets generating plausible explanations fitting the outcome produced by the black box [14]. A typical practice is to train an interpretable machine learning model (e.g., decision trees, linear model, or rules) with the outcome of a black box [3,15,16].

Explainable agents go beyond the mere application of sub-symbolic ML mechanisms. Agents can leverage symbolic AI techniques (e.g., logic and planning languages), which are easier to trace, reason about, understand, debug, and explain [17]. However, they can still partially rely on ML predictors, thus deeming necessary to be explaining their overall behavior (relying on neuro-symbolic integration). Endowing virtual agents with explanatory abilities raises trust, acceptability, and reduces possible failures due to misunderstandings [14,18]. Yet, it necessary to consider user characterisation (e.g., age, background, and expertise), the context (e.g., why do the user need the explanation), and the agents' limits [14].

Built-in explainability is still rare in literature. Most of the works utterly provide indicators which "should serve" as an explanation for the human user [3]. To date, such approaches have been unable to produce satisfying human-understandable explanations. Nevertheless, more recent contributions employ neuro-symbolic integration to identifying factors influencing the human comprehension of representation formats and reasoning approaches [19].

2.2 Agreement Technologies

Understanding other parties' interests and preferences is crucial in human social interaction. It enables the proposal of reasonable bids to resolve conflicts effectively [20,21]. *Agreement technologies* (AT) [22] literature counts several techniques to automatically learn, reproduce, and possibly predict an opponent's preferences and bidding strategies in conflict resolution scenarios [23].

AT are mostly based on heuristics [24,25] and traditional ML methods (e.g., decision trees [26,27], Bayesian learning [28–30], and concept-based learning [31,32]) and rely on possibly numerous bid exchanges regulated by negotiation protocols [33]. By exploiting such techniques, machines can negotiate with humans seamlessly, resolving conflicts with a high degree of mutual understanding [34]. Nevertheless, in human-agent negotiation, the complexity skyrockets. Humans leverage on semantic and reasoning (e.g., employing similarities/differences) while learning about the competitors' preferences and generating well-targeted offers. Conversely to agent-agent, the number of exchanged bids between parties is limited due to the nature of human interactions, and may employ unstructured data. Therefore, classical opponent modeling techniques used in automated negotiation in which thousands of bids are exchanged may not be suitable, and additional reasoning to understand humans' intentions, interests, arguments, and explanations supporting their proposals is required [35,36]. To the best of our knowledge, there is no study incorporating exchanged arguments or *explanations* into opponent modeling in agent-based negotiation literature.

Without explanations, human users may attribute a wrong *state of mind* to agents/robots [18]. Thus, the creation of an effective agent-based explainable

AT for human-agent interactions and the realisation of a common understanding would require the integration of *(i)* ontology reasoning, *(ii)* understanding humans' preferences/interests by reasoning on any type of information provided during the negotiation, and *(iii)* generating well-targeted offers with their supportive explanations or motivations (i.e., why the offer can be acceptable for their human counterpart). To the best of our knowledge, the state of the art still needs concrete contributions concerning the three directions mentioned above. Moreover, albeit the need for personalised motivations and arguments (e.g., considering user expertise, personal attributes, and goals) is well known in literature [14], most of the existing works are rather conceptual and do not consider the overall big picture [37]. Furthermore, no work addresses explanation personalisation in the context of heterogeneous systems combining sub-symbolic (e.g., neural network) and symbolic (agents/robots) AI mechanisms.

2.3 AI Ethics

Due to the growing adoption of intelligent systems, machine ethics and AI ethics have received a deserved increasing attention from scientists working in various domains [38]. The growing safety, ethical, societal, and legal impacts of AI decisions are the main reason behind this surge of interest [39]. In literature, AI ethics includes implicitly- and explicitly-moral agents. In both cases, intelligent systems depend on human intervention to distinguish moral from immoral behaviour. However, on the one hand, implicitly-moral agents are ethically constrained from having immoral behaviour via rules set by the human designer [38]. On the other hand, explicitly-ethical agents (or agents with functional morality) presume to be able to morally judge themselves (having guidelines or examples of what is good and bad [38]).

Summarising, AI systems can have implicit and explicit ethical notions. The main advantage of implicit AI ethics is that they are simple to develop and control, being incapable of unethical behaviour. Nevertheless, this simplicity implies mirroring the ethic standing point and perception of the designer. Explicit-ethics systems affirm to autonomously evaluate the normative status of actions and reason independently about what they consider unethical, thus being able to solve normative conflicts. Furthermore, they could bend/violate some rules, resulting in better fulfilment of overarching ethical objectives. However, the main shortcoming of these systems is their complexity and possible unexpected behaviour.

3 The EXPECTATION Approach

This section elaborates on the limitations elicited from the state of art, the related challenges, and formalises the needed interventions. The six major limitations identified are:

(L1) Opaqueness of sub-symbolic predictors. Most ML algorithms leverage a sub-symbolic representation of knowledge that is hard to debug for experts and hard to interpret for common people. Thus, the compliance of internal mechanisms and results with ethical principles and regulations cannot be verified/ensured.

(L2) Heterogeneity of rule extraction techniques. Extracting general-purpose symbolic rules from any sort of sub-symbolic predictor can be a difficult task (if possible, at all). Indeed, the nature of the data and the particular predictor at hand significantly impact the quality (i.e., the intelligibility) of the extracted rules. Furthermore, existing techniques to extract rules to produce explanations mostly leverage structured, low-dimensional data, given the scarcity of methods supporting more complex data (i.e., images, videos, or audios). In particular, most of the existing works interpreting sub-symbolic mechanisms place interpretable mechanisms (i.e., decision-tree) on top of the predictors, thereby interpreting (e.g., reconstructing) from outside their outcomes without really mirroring their internal mechanisms.

(L3) Manual amending and integration of heterogeneous predictors. The update and integration of already pre-trained predictors are usually handcrafted and poorly automatable. Moreover, it heavily relies on datasets that might be available only for a limited period. Therefore, a sustainable, automatable, and seamless sharing/reusing/integrating of knowledge from diverse predictors is still unsatisfactory.

(L4) Lack of personalisation. Current XAI approaches are mostly one-way processes (e.g., interactive interactions are rarely involved) and do not consider the explainee's context and background. Thus, the customisation and personalisation of the explanations are still open challenges.

(L5) Tendency of centralisation in data-driven AI. The development of sub-symbolic predictors usually involves the centralisation of training data in a single point, which raises privacy concerns. Thus, letting a system composed of several distributed intelligent components learning without centralising data is still an open challenge.

(L6) Lack of explanation integration in Agreement Technologies. Current negotiation and argumentation frameworks mostly leverage well-structured interactions and clearly defined objectives, resources, and goals. Current AT are not suitable for providing interactive explanations nor for reconciling fragmented knowledge. Moreover, although a few works explored more sophisticated mechanisms (e.g., adopting semantic similarities via subsumption to relate alternative values within a single bid), the need for ontological reasoning to infer the relationship between several issues – possibly pivotal in negotiation and argumentation of explanations – is still unmet.

To overcome the limitation mention, Expectation formalises the following objectives:

(O1) To define an agent-based model embedding ML predictors relying on heterogeneous (though potentially similar/complementary) knowledge, as in training datasets, contextual assumptions & ontologies.

(O2) To design and implement a decentralised agent architecture capable of integrating symbolic knowledge and explanations produced by individual agents.

(O3) To define and implement agent strategies for cooperation, negotiation, and trust establishment for providing personalised explanations according to the user context.

(O4) To investigate, implement, and evaluate multi-modal explanation communication mechanisms (visual, auditory, cues, etc.), the role of the type of agent providing these explanations (e.g., robot, virtual agents), and their role in explanation personalisation.

(O5) To validate and evaluate the personalised explainability results, as well as the agent-based XAI approach for heterogeneous knowledge, within the context of a prototype, focused on food and nutrition recommendations.

(O6) To investigate the specific ethical challenges that XAI is able to meet and when and to what extent explicability is legally required in European regulations, considering the AI guidelines and evaluation protocols published by the national and European institutions (e.g., the Data Protection Impact Analysis thanks to the open-source software PIA, CNIL guidelines), as well as recent research on the ethics of recommender systems w.r.t. values such as transparency and fairness.

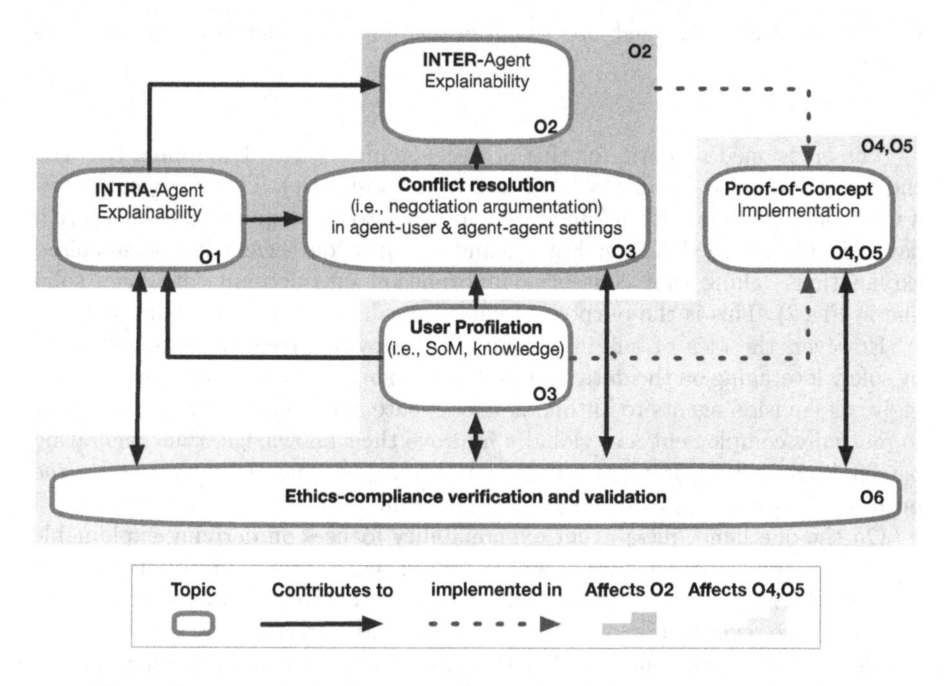

Fig. 1. EXPECTATION's objectives, topics, and respective interconnections.

The aforementioned objectives are clearly interdependent. In particular, Fig. 1 groups and organises the objectives per contribution, effect, and implementation among each other.

3.1 Research Method

Despite being still in its early stage, the project's roadmap has already been established. EXPECTATION's research and development activities will be carried out along two orthogonal dimensions – namely *intra-* and *inter*-agent ones –, as depicted in Fig. 2.

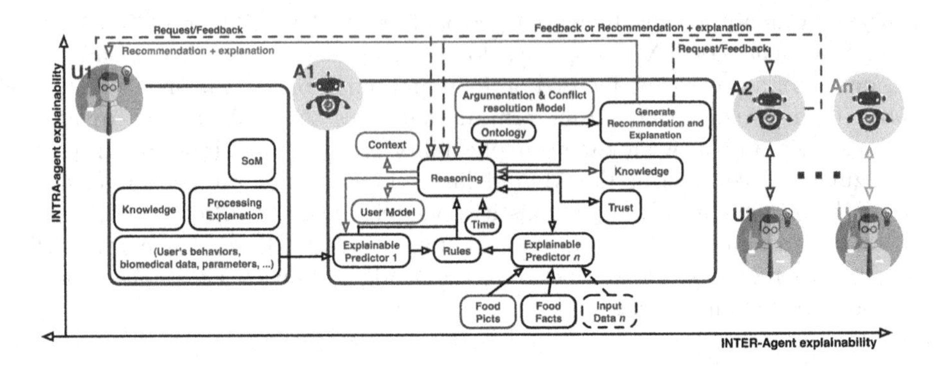

Fig. 2. Main components and interactions of the proposed architecture. (Color figure online)

The envisioned scenario for this project assumes a 1-to-1 mapping between end-users and software agents (cf. Fig. 2, rightmost part). Therefore, each software agent interacts with a single user in order to *(i)* acquire their contextual data (cf. blue dashed line in Fig. 2), and *(ii)* provide them with personalised explanations taking that contextual information into account (cf. green solid line in Fig. 2). This is the purpose of what we call *intra*-agent explainability.

However, the idea of building agents that provide precise recommendations by solely leveraging on the data acquired from a single user is unrealistic. Accordingly, we envision agents to autonomously debate and negotiate with each other to mutually complement and globally improve their knowledge, thus generating personalised and accurate recommendations. Addressing this challenge is the purpose of what we call *inter*-agent explainability.

On the one hand, intra-agent explainability focuses on deriving explainable information at the local level – where contextual information about the user is most likely available – and on presenting it to the user in a personalised way. To do so, symbolic knowledge extraction and injection play a crucial role. The former lets agents fully exploit the predictive performance of conventional ML-based black-box algorithms while still enabling the production of intelligible information to be used for building personalised explanations. Conversely, by injecting symbolic knowledge in ML-based systems, agents will be able to update, revise, and correct the functioning of ML-based predictors by taking into account users' contextual information and feedback.

On the other hand, inter-agent explainability focuses on enabling the agents to exploit negotiation and argumentations to mutually improve their predictive capabilities by exchanging the symbolic knowledge they have extracted from given black boxes. Even in this context, the role of symbolic knowledge extraction is of paramount importance as it enables exchanges of aggregated knowledge coming from different ML-predictors—which possibly offer different perspectives on the problem at hand. To this end, inter-agent explainability requires formalising interaction protocols specifying what actions are possible and how to represent this information so that both parties can understand and interpret it seamlessly. Moreover, inter-agent interactions will require reasoning mechanisms handling heterogeneous data received from other agents, including techniques to detect conflicts and adopt resolution or mitigation policies accordingly.

By combining intra- and inter-agent explainability, EXPECTATION will be able to tackle decentralisation (of both data and agents), heterogeneity (of both data and analysis techniques), and users' privacy simultaneously. Indeed, the proposed approach does not require data to be centralised to allow training and knowledge extraction. Therefore, each agent can autonomously take care of the local data it has access to by exploiting the ML-based analysis technique it prefers, while joint learning is delegated to decentralised negotiation protocols which only exchange aggregated knowledge. Users' personal data is expected to remain close to the user, while agents are in charge of blending the extracted symbolic knowledge with the general-purpose background knowledge jointly attained by the multi-agent systems via negotiation and argumentation. Heterogeneity is addressed indirectly via knowledge extraction, which provides a *lingua franca* for knowledge sharing in the form of logic facts and rules.

Notably, knowledge extraction is what enables bridging *intra*- and *inter*-agent explainability too, as it enables the exchange of the extracted knowledge via negotiation and argumentation protocols—which already rely on the exchange of symbolic information.

Knowledge injection closes the loop by letting the knowledge acquired via interaction to be used to improve the local data and analytic capabilities of each individual agent. Finally, the purposes of preserving privacy and complying with ethical implications are addressed by only allowing agents to share aggregated symbolic knowledge. Moreover, we envision to equip the agents with ethics reasoning engines combining techniques from both implicit and explicit ethics.

4 Discussion

To test the advancement produced by EXPECTATION, we envision combining the techniques mentioned above in a proof of concept cantered on a topic which nowadays is delicate more than ever: a nutrition recommender system, fostering a responsible and correct alimentation. Such a prototype will be tested and evaluated according to the user-subjective such as understandability, trust, acceptability, soundness, personalisation, perceived system autonomy, perceived user autonomy, and fairness. The envisioned agent-based recommender system is intended to operate as a virtual assistant equipped with personalised explanatory capabilities. This would make it possible to tackle two dimensions of the quest for a correct regime *(i)* trust and acceptance, and *(ii)* autonomous personalisation, education, and explicability. In particular, the user will be provided with transparent explanations about the recommendation received. The purpose of the explanations is multi-faceted: *(i)* educative (i.e., improve the user knowledge and raising his/her awareness about a given topic/suggestion), *(ii)* informative (i.e., indicate the user on how the system works), and *(iii)* motivational (i.e., it helps the user understanding how personal characteristics and decisions lead to favorable/adverse outcomes).

Overall, EXPECTATION is expected to impact beyond its lifespan. Such an impact encompasses several aspects and is four-folded.

Impact of theoretical outcomes. Production of mechanisms to extract, combine, explain, negotiate heterogeneous symbolic knowledge as well as cooperation and negotiation strategies.

Impact of technological outcomes. Fostering the adoption of intelligent systems in health and safety-critical domains and inspiring new technology leveraging innovative multi-modal explanation communication mechanisms.

Impact in application domains. We expect uptake of the project results in sectors (commercial/academic) such as eHealth, prevention, wellbeing applications, and distribution and restoration.

Impact of ethical aspects. Given the sensitive nature of personal data in the context of the project, the proposed XAI prototype will develop generalisable mechanisms to ensure compliance, fairness, transparency, and trust.

Acknowledgments. This work has been partially supported by the CHIST-ERA grant CHIST-ERA-19-XAI-005, and by *(i)* the Swiss National Science Foundation (G.A. 20CH21_195530), *(ii)* the Italian Ministry for Universities and Research, *(iii)* the Luxembourg National Research Fund (G.A. INTER/CHIST/19/14589586 and INTER/Mobility/19/13995684/DLAl/van), *(iv)* the Scientific and Research Council of Turkey (TÜBİTAK, G.A. 120N680).

References

1. Fadlullah, Z.M., et al.: State-of-the-art deep learning: evolving machine intelligence toward tomorrow's intelligent network traffic control systems. IEEE Commun. Surv. Tutor. **19**(4), 2432–2455 (2017)
2. Helbing, D.: Societal, economic, ethical and legal challenges of the digital revolution: from big data to deep learning, artificial intelligence, and manipulative technologies. In: Helbing, D. (ed.) Towards Digital Enlightenment, pp. 47–72. Springer, Cham (2019). https://doi.org/10.1007/978-3-319-90869-4_6
3. Guidotti, R., Monreale, A., Ruggieri, S., Turini, F., Giannotti, F., Pedreschi, D.: A survey of methods for explaining black box models. ACM Comput. Surv. **51**(5), 93:1–93:42 (2019)
4. Arrieta, A.B., et al.: Explainable explainable artificial intelligence (XAI): concepts, taxonomies, opportunities and challenges toward responsible AI. Inf. Fusion **58**, 82–115 (2020)

5. Calegari, R., Ciatto, G., Omicini, A.: On the integration of symbolic and sub-symbolic techniques for XAI: a survey. Intell. Artif. **14**(1), 7–32 (2020)
6. Dosilovic, F.K., Brcic, M., Hlupic, N.: Explainable artificial intelligence: a survey. In: Skala, K., et al. (eds.) 41st International Convention on Information and Communication Technology, Electronics and Microelectronics (MIPRO 2018), Opatija, Croatia, 21–25 May 2018, pp. 210–215. IEEE (2018)
7. Dağlarli, E.: Explainable artificial intelligence (xAI) approaches and deep meta-learning models. In: Aceves-Fernandez, M.A. (ed.) Advances and Applications in Deep Learning, chapter 5. IntechOpen, London, UK (2020)
8. Ciatto, G., Calegari, R., Omicini, A., Calvaresi, D.: Towards XMAS: eXplainability through multi-agent systems. In: Savaglio, C., Fortino, G., Ciatto, G., Omicini, A. (eds.) AI&IoT 2019 - Artificial Intelligence and Internet of Things 2019. CEUR Workshop Proceedings, vol. 2502, pp. 40–53. Sun SITE Central Europe, RWTH Aachen University (2019)
9. Ciatto, G., Schumacher, M.I., Omicini, A., Calvaresi, D.: Agent-based explanations in AI: towards an abstract framework. In: Calvaresi, D., Najjar, A., Winikoff, M., Främling, K. (eds.) EXTRAAMAS 2020. LNCS (LNAI), vol. 12175, pp. 3–20. Springer, Cham (2020). https://doi.org/10.1007/978-3-030-51924-7_1
10. Ciatto, G., Calvaresi, D., Schumacher, M.I., Omicini, A.: An abstract framework for agent-based explanations in AI. In: Seghrouchni, A.E.F., Sukthankar, G., An, B., Yorke-Smith, N. (eds.) 19th International Conference on Autonomous Agents and MultiAgent Systems, Auckland, New Zeland, May 2020, pp. 1816–1818. International Foundation for Autonomous Agents and Multiagent Systems. Extended Abstract (2020)
11. Pisano, G., Ciatto, G., Calegari, R., Omicini, A.: Neuro-symbolic computation for XAI: towards a unified model. In: Calegari, R., Ciatto, G., Denti, E., Omicini, A., Sartor, G. (eds.) WOA 2020–21th Workshop "From Objects to Agents", Volume 2706 of CEUR Workshop Proceedings, Aachen, Germany, October 2020, pp. 101–117. Sun SITE Central Europe, RWTH Aachen University, Bologna, 14–16 September 2020
12. Wagner, B., d'Avila Garcez, A.: Neural-symbolic integration for fairness in AI. In: Martin, A., et al. (eds.) Proceedings of the AAAI 2021 Spring Symposium on Combining Machine Learning and Knowledge Engineering (AAAI-MAKE 2021), Volume 2846 of CEUR Workshop Proceedings, Stanford University, Palo Alto, CA, USA, 22–24 March 2021. CEUR-WS.org (2021)
13. Smolensky, P.: Connectionist AI, symbolic AI, and the brain. Artif. Intell. Rev. **1**(2), 95–109 (1987). https://doi.org/10.1007/BF00130011
14. Anjomshoae, S., Najjar, A., Calvaresi, D., Främling, K.: Explainable agents and robots: results from a systematic literature review. In: Elkind, E., Veloso, M., Agmon, N., Taylor, M.E. (eds.) 18th International Conference on Autonomous Agents and MultiAgent Systems (AAMAS 2019), Montreal, QC, Canada, 13–17 May 2019, pp. 1078–1088. International Foundation for Autonomous Agents and Multiagent Systems (2019)
15. Calegari, R., Ciatto, G., Dellaluce, J., Omicini, A.: Interpretable narrative explanation for ML predictors with LP: a case study for XAI. In: Bergenti, F., Monica, S. (eds.) WOA 2019–20th Workshop "From Objects to Agents", Volume 2404 of CEUR Workshop Proceedings, Sun SITE Central Europe, RWTH Aachen University, Parma, Italy, 26–28 June 2019, pp. 105–112 (2019)
16. Andrews, R., Diederich, J., Tickle, A.B.: Survey and critique of techniques for extracting rules from trained artificial neural networks. Knowl. Based Syst. **8**(6), 373–389 (1995)

17. Calegari, R., Ciatto, G., Mascardi, V., Omicini, A.: Logic-based technologies for multi-agent systems: a systematic literature review. Auton. Agents Multi-Agent Syst. **35**(1), 1:1–1:67 (2021). https://doi.org/10.1007/s10458-020-09478-3. Collection "Current Trends in Research on Software Agents and Agent-Based Software Development"
18. Hellström, T., Bensch, S.: Understandable robots - what, why, and how. Paladyn J. Behav. Robot. **9**(1), 110–123 (2018)
19. Adadi, A., Berrada, M.: Peeking inside the black-box: a survey on explainable artificial intelligence (XAI). IEEE Access **6**, 52138–52160 (2018)
20. Baarslag, T., Kaisers, M., Gerding, E.H., Jonker, C.M., Gratch, J.: Computers that negotiate on our behalf: major challenges for self-sufficient, self-directed, and interdependent negotiating agents. In: Sukthankar, G., Rodriguez-Aguilar, J.A. (eds.) AAMAS 2017. LNCS (LNAI), vol. 10643, pp. 143–163. Springer, Cham (2017). https://doi.org/10.1007/978-3-319-71679-4_10
21. Jonker, C.M., et al.: An introduction to the pocket negotiator: a general purpose negotiation support system. In: Criado Pacheco, N., Carrascosa, C., Osman, N., Julián Inglada, V. (eds.) EUMAS/AT -2016. LNCS (LNAI), vol. 10207, pp. 13–27. Springer, Cham (2017). https://doi.org/10.1007/978-3-319-59294-7_2
22. Ossowski, S. (ed.): Agreement Technologies. Law, Governance and Technology Series, vol. 3. Springer, Dordrecht (2012). https://doi.org/10.1007/978-94-007-5583-3
23. Baarslag, T., Hendrikx, M.J.C., Hindriks, K.V., Jonker, C.M.: Learning about the opponent in automated bilateral negotiation: a comprehensive survey of opponent modeling techniques. Auton. Agent. Multi-Agent Syst. **30**(5), 849–898 (2015). https://doi.org/10.1007/s10458-015-9309-1
24. Aydoğan, R., Baarslag, T., Hindriks, K.V., Jonker, C.M., Yolum, P.: Heuristics for using CP-nets in utility-based negotiation without knowing utilities. Knowl. Inf. Syst. **45**(2), 357–388 (2014). https://doi.org/10.1007/s10115-014-0798-z
25. Jennings, N., Faratin, P., Lomuscio, A., Parsons, S., Wooldridge, M., Sierra, C.: Automated negotiation: prospects, methods and challenges. Group Decis. Negot. **10**, 199–215 (2001)
26. Aydoğan, R., Marsa-Maestre, I., Klein, M., Jonker, C.M.: A machine learning approach for mechanism selection in complex negotiations. J. Syst. Sci. Syst. Eng. **27**(2), 134–155 (2018). https://doi.org/10.1007/s11518-018-5369-5
27. Ilany, L., Gal, Y.: Algorithm selection in bilateral negotiation. Auton. Agent. Multi-Agent Syst. **30**(4), 697–723 (2015). https://doi.org/10.1007/s10458-015-9302-8
28. Hindriks, K.V., Tykhonov, D.: Opponent modelling in automated multi-issue negotiation using Bayesian learning. In: Padgham, L., Parkes, D.C., Müller, J.P., Parsons, S. (eds.) 7th International Joint Conference on Autonomous Agents and Multiagent Systems (AAMAS 2008), Estoril, Portugal, 12–16 May 2008, vol. 1, pp. 331–338. IFAAMAS (2008)
29. Yu, C., Ren, F., Zhang, M.: An adaptive bilateral negotiation model based on Bayesian learning. In: Ito, T., Zhang, M., Robu, V., Matsuo, T. (eds.) Complex Automated Negotiations: Theories, Models, and Software Competitions. SCI, vol. 435, pp. 75–93. Springer, Heidelberg (2013). https://doi.org/10.1007/978-3-642-30737-9_5
30. Zeng, D., Sycara, K.: Bayesian learning in negotiation. Int. J. Hum Comput Stud. **48**(1), 125–141 (1998)
31. Aydogan, R., Yolum, P.: Ontology-based learning for negotiation. In: 2009 IEEE/WIC/ACM International Conference on Intelligent Agent Technology (IAT 2009), vol. 2, pp. 177–184, January 2009

32. Galitsky, B.A., Kuznetsov, S.O., Samokhin, M.V.: Analyzing conflicts with concept-based learning. In: Dau, F., Mugnier, M.-L., Stumme, G. (eds.) ICCS-ConceptStruct 2005. LNCS (LNAI), vol. 3596, pp. 307–322. Springer, Heidelberg (2005). https://doi.org/10.1007/11524564_21
33. Marsa-Maestre, I., Klein, M., Jonker, C.M., Aydoğan, R.: From problems to protocols: towards a negotiation handbook. Decis. Support Syst. **60**, 39–54 (2014)
34. Oshrat, Y., Lin, R., Kraus, S.: Facing the challenge of human-agent negotiations via effective general opponent modeling. In: 8th International Conference on Autonomous Agents and Multiagent Systems (AAMAS 2009), vol. 1, pp. 377–384. IFAAMAS (2009)
35. Güngör, O., Çakan, U., Aydoğan, R., Özturk, P.: Effect of awareness of other side's gain on negotiation outcome, emotion, argument, and bidding behavior. In: Aydoğan, R., Ito, T., Moustafa, A., Otsuka, T., Zhang, M. (eds.) ACAN 2019. SCI, vol. 958, pp. 3–20. Springer, Singapore (2021). https://doi.org/10.1007/978-981-16-0471-3_1
36. Pasquier, P., Hollands, R., Dignum, F., Rahwan, I., Sonenberg, L.: An empirical study of interest-based negotiation. Auton. Agent. Multi-Agent Syst. **22**, 249–288 (2011). https://doi.org/10.1007/s10458-010-9125-6
37. Kaptein, F., Broekens, J., Hindriks, K., Neerincx, M.: Personalised self-explanation by robots: the role of goals versus beliefs in robot-action explanation for children and adults. In: 2017 26th IEEE International Symposium on Robot and Human Interactive Communication (RO-MAN), pp. 676–682 (2017)
38. Moor, J.H.: The nature, importance, and difficulty of machine ethics. IEEE Intell. Syst. **21**(4), 18–21 (2006)
39. Calvaresi, D., Schumacher, M., Calbimonte, J.-P.: Personal data privacy semantics in multi-agent systems interactions. In: Demazeau, Y., Holvoet, T., Corchado, J.M., Costantini, S. (eds.) PAAMS 2020. LNCS (LNAI), vol. 12092, pp. 55–67. Springer, Cham (2020). https://doi.org/10.1007/978-3-030-49778-1_5

Author Index

Printed in the United States
by Baker & Taylor Publisher Services